Healthcare Systems:
Future Predictions for Global Care

Healthcare Systems:
Future Predictions for Global Care

Edited by

Jeffrey Braithwaite, Russell Mannion,
Yukihiro Matsuyama, Paul G. Shekelle,
Stuart Whittaker, and Samir Al-Adawi

CRC Press
Taylor & Francis Group
Boca Raton London New York

CRC Press is an imprint of the
Taylor & Francis Group, an **informa** business

CRC Press
Taylor & Francis Group
6000 Broken Sound Parkway NW, Suite 300
Boca Raton, FL 33487-2742

First issued in paperback 2021

CRC Press is an imprint of Taylor & Francis Group, an Informa business

ISBN-13: 978-0-367-78126-2 (pbk)
ISBN-13: 978-1-138-05260-4 (hbk)

Library of Congress Cataloging-in-Publication Data

Names: Braithwaite, Jeffrey, 1954- editor.
Title: Health care systems : future predictions for global care / [edited by] Jeffrey Braithwaite [and five others].
Other titles: Health care systems (Braithwaite)
Description: Boca Raton : Taylor & Francis, 2018. | Includes bibliographical references and index.
Identifiers: LCCN 2017055949| ISBN 9781138052604 (hardback : alk. paper) | ISBN 9781315167688 (eBook)
Subjects: | MESH: Delivery of Health Care--trends | Internationality | Quality of Health Care--trends | Global Health--trends
Classification: LCC RA441 | NLM W 84.1 | DDC 362.1--dc23
LC record available at https://lccn.loc.gov/2017055949

Visit the Taylor & Francis Web site at
http://www.taylorandfrancis.com

and the CRC Press Web site at
http://www.crcpress.com

Maps have been adapted from
http://mapchart.net/

Contents

Part I   The Americas
Paul G. Shekelle

Part II   Africa
Stuart Whittaker

Part III   Europe
Russell Mannion

Part IV Eastern Mediterranean
Samir Al-Adawi

Part V South-East Asia and the Western Pacific
Jeffrey Braithwaite and Yukihiro Matsuyama

Preface

It's not the strongest of the species that survives, nor the most intelligent, but the one most responsive to change.

—Charles Darwin

We live in an era of rapid and unprecedented change. Driven by technological innovation and changes in the way we deliver services, the face of healthcare is undergoing a metamorphosis, shifting into a more person-based, technologically enabled, evidence-based, and responsive system.

That is the theory, at least. But are health systems that are changing according to these plans heralding transformative change? And what do some of the best thinkers believe is the profile of their health system over the next 5–15 years? We believe this book represents the best attempt yet to answer those thorny questions.

Very few people could reach into the health systems of 152 countries and territories and orchestrate a book of this magnitude. Jeffrey Braithwaite, as series editor, accompanied by regional editors, Russell Mannion, Yukihiro Matsuyama, Paul G. Shekelle, Stuart Whittaker, and Samir Al-Adawi, and supported by an extremely knowledgeable team at Macquarie University, Sydney, Australia, particularly Dr. Wendy James and Kristiana Ludlow, were just the team to accomplish this.

The omnibus they have created is an invaluable source of predictions about the future scope and shape of health systems across low-, middle-, and high-income countries. It is a treasure trove of important information. People will use it as a practical guide to the future in many ways: it can be read for benefit and learning by region, by theme, and by specific case study exemplars of the kinds of reforms people are enacting in their health systems, extrapolated across the medium-term time horizon. Most books do not do this. The fact that this group has been able to achieve this is an endorsement of the skills, efforts, ingenuity, and expertise of the editors, editorial team, and individual chapter authors.

We commend this book and recommend it as a must-read to many stakeholder groups: students of the system, policy-makers, planners, futurists, and groups representing managers, clinicians, and patients—in fact, all those who have an interest in healthcare and its future success. We enjoyed dipping into it and thinking about its many learning points. We are sure others will too.

Wendy Nicklin
RN, BN, MSc(A), CHE, FACHE, FISQua, ICD.D President, International Society for Quality in Health Care

Clifford F. Hughes
AO, MBBS, DSc, FRACS, FACS, FACC, FIACS (Hon), FAAQHC, FCSANZ, FISQua, AdDipMgt, Immediate Past President, International Society for Quality in Health Care

Acknowledgments

The scope of health systems internationally is vast, and we have, across the pages of this book, endeavored to capture some of its magnitude. Of course, this is a team effort and we are indebted to the many people whose input has been invaluable and instrumental in bringing this book together.

Our thanks go first and foremost to the chapter authors, who willingly gave of their time, skills, and experience to bring the readers each case study. The chapters are inscribed with their wisdom, and bear testament to their dedication to the future of health systems, in their own countries and around the world.

The International Society for Quality in Health Care (ISQua) is to some extent the glue that holds this book together. Many of the authors are fellows, experts, or friends of ISQua, or attendees at its annual conference—and all are in some way connected to ISQua's vast network.

We gratefully acknowledge the editorial team in the Australian Institute of Health Innovation, Sydney, Australia, for their marvelous support and considerable efforts in drawing this book together. Enduring thanks go to Dr. Wendy James, copy editor, for editing every chapter and taking a lead on formatting the book and references, and to Kristiana Ludlow, our tireless coordinator and administrator, who has taken responsibility for editing the individual chapters as well as sourcing the background material. They, in turn, were aided at various times in the development of the book by Dr. Kate Churruca, who assisted with formatting the chapters; Gina Lamprell, who coedited the biographies and produced the chapter data tables; Elise McPherson, who coedited the biographies, conducted background research, and provided periodic administrative support; Jessica Herkes, who provided all-round support at various times and did the figures and maps; Hsuen P. Ting, who researched and produced the introduction data tables; Dr. Louise A. Ellis, who painstakingly checked all the references; Chiara Pomare, who edited the data tables; and Claire Boyling, who joined the team in the final stages of the book, assembled the word pictures, and helped with the proofing.

About the Editors

Series Editor

Jeffrey Braithwaite, BA, MIR (Hon), MBA, DipLR, PhD, FAIM, FCHSM, FFPHRCP (UK), FAcSS (UK), Hon FRACMA, FAHMS, is foundation director, Australian Institute of Health Innovation; director, Centre for Healthcare Resilience and Implementation Science; and professor of health systems research, Faculty of Medicine and Health Sciences, Macquarie University, Sydney, Australia. He has appointments at six other universities internationally, is a board member, and is president-elect of the International Society for Quality in Health Care and a member of the World Health Organization Global Patient Safety Network. His research examines the changing nature of health systems, attracting funding of more than AU$110 million (US$87.5 million). He is interested in healthcare as a resilient system, and applying complexity science to healthcare problems. In addition to this, he is interested in the Anthropocene and the impact of human activity on human and species' health, population, and climate. Professor Braithwaite has contributed more than 900 publications and presented at or chaired international and national conferences, workshops, symposia, and meetings on more than 900 occasions, including more than 90 keynote addresses. His research appears in outlets such as *BMJ*, *Lancet*, *Social Science & Medicine*, *BMJ Quality & Safety*, and the *International Journal for Quality in Health Care*. He has received 39 different national and international awards for his teaching and research. Further details are available at his Wikipedia entry: http://en.wikipedia.org/wiki/Jeffrey_Braithwaite.

Regional Editors

Russell Mannion, BA (Hon), PG Dip Health Econ, PhD, FRSA, FAcSS, has more than 30 years' experience in healthcare research. He is chair in Health Systems, University of Birmingham; visiting professor at the Australian Institute of Health Innovation, Macquarie University; and visiting professor at the Faculty of Medicine, University of Oslo. He was previously director of the Centre for Health and Public Services Management, University of York, and board director of the York Health Economics Consortium. He provides expert advice to various health agencies, including the World Health Organization, Organisation for Economic Co-operation and Development, European Health Management Association, and UK Department of Health. He has authored or edited 10 books and around 200 peer-reviewed publications, many in leading scientific journals including *BMJ*, *Lancet*, and *Milbank Quarterly*. He is associate editor on the editorial board of four international health policy journals and has garnered several international prizes for his research, including the Baxter European Book Award.

Yukihiro Matsuyama, PhD, is research director, the Canon Institute for Global Studies; affiliate professor, Chiba University of Commerce; and honorary professor, the Australian Institute of Health Innovation, Faculty of Medicine and Health Sciences at Macquarie University. His research examines the sustainability of safety net systems in Japan, including healthcare, pension, pandemic crisis, and employment through international comparative analyses. He has served on government committees, including as a member of the Welfare Committee. He has published many books, including *Healthcare Economics in the United States* (1990), which introduced the theoretical concept of Diagnosis Related Group/Prospective Payment System (DRG/PPS) and managed care into Japan for the first time; *AIDS War: Warning to Japan* (1992); *Health Reform in the United States* (1994); *Breakthrough of Japan's Economy under Half-Population* (2002); *Healthcare Reform and Integrated Healthcare Network* (2005), coauthored with Keiko Kono; *Health Reform and Economic Growth* (2010); and *Depth of Healthcare Reform* (2015).

Paul G. Shekelle, MD, PhD, is a staff physician at the West Los Angeles Veterans Affairs Medical Center, and is a professor of medicine at the University of California, Los Angeles School of Medicine. He is widely recognized in the fields of guidelines, quality measurement, and evidence-based medicine. In 1996–1997, he spent a year in England as an Atlantic Fellow in Public Policy. He is a past chair of the Clinical Guidelines Committee of the American College of Physicians.

Stuart Whittaker, BSc, MBChB, FFCH (CM), MMed, MD, founder and former chief executive officer of the Council for Health Service Accreditation of Southern Africa. He pioneered the concepts of a facilitated accreditation program and graded recognition to assist disadvantaged hospitals in southern Africa and other developing countries to comply with professional standards. He has presented at numerous international and national conferences. As a temporary consultant to the World Health Organization, he participated in projects to assess the impact of accreditation on national health systems and choosing Quality Approaches in Health Systems. He was appointed by the minister of health of South Africa in 2013, and reappointed in 2017, to serve on the board of directors for the Office of Health Standards Compliance in South Africa. He is a visiting professor at the School of Public Health and Medicine at the Faculty of Health Sciences, University of Cape Town.

Samir Al-Adawi, BA (Hon), MS, PhD, is a professor of behavioral medicine at the College of Medicine, Sultan Qaboos University, Oman. Previously, he was a Fulbright Senior Scholar at the Department of Physical Medicine and Rehabilitation, Harvard Medical School, Cambridge, Massachusetts, and a research scientist sponsored by Matsumae International Foundation at the Department of Psychosomatic Medicine, Graduate School of Medicine, University of Tokyo, Japan. His doctorate training was at the Institute of Psychiatry, King's College, London, United Kingdom. Dr. Al-Adawi has research interests that focus on non-communicable diseases. His research and publications have specifically focused on the psychosocial determinants of health and ill health. Dr. Al-Adawi is a member of the World Health Organization Expert Consultation Group on Feeding and Eating Disorders, reporting to the International Advisory Group for the Revision of International Classification of Diseases (ICD-10) for Mental and Behavioral Disorders.

About the Contributors

Tine Aagaard is a nurse and holds a PhD in Arctic studies—health and society from Ilisimatusarfik/University of Greenland. She has published on patients' cultural perspectives on healthcare practice in Greenland and on epistemological and methodological questions in the research of users' perspectives. Her current research is dealing with citizens' and healthcare professionals' perspectives on home care in Nuuk, Greenland, and how the connection of different perspectives can contribute to the improvement of healthcare. She also teaches nursing students and is an editor of the journal of the Greenlandic Nurses' Association, *Tikiusaaq*.

William Adu-Krow, MB, ChB, DrPH, is presently the Pan American Health Organization/World Health Organization representative of Guyana. He was previously the World Health Organization country representative for Papua New Guinea (2010–2014) and Solomon Islands (2008–2010). He is a pro bono adjunct professor at the Center for Population and Family Health of the Mailman School of Public Health at Columbia University, New York. He has worked in Ghana, New York, Washington, DC, New Jersey, the Solomon Islands, and Papua New Guinea. His main work has focused on violence against women and the impact of social determinants on global health architecture.

Bruce Agins, MD, MPH, medical director of the NYSDOH AIDS Institute, principal architect of New York's HIV Quality of Care Program, and member of the AIDS Institute executive team. He is the principal investigator of HEALTHQUAL International, the National Quality Center, and oversees the NYLINKS program. Dr. Agins holds academic appointments as full clinical professor in the Division of Epidemiology and Biostatistics in Global Health Sciences at the University of California, San Francisco, and as adjunct professor in the Division of Infectious Diseases and Immunology at New York University School of Medicine. He is a graduate of Haverford College (1975) and Case Western Reserve School of Medicine (1980), and earned his MPH from the Mailman School of Public Health at Columbia University (1994).

Mercedes Aguerrebere, MD, has worked extensively on healthcare delivery and mental health programming in rural Mexico. Dr. Aguerrebere previously served as the mental health coordinator for Partners in Health Mexico, a non-profit partnering with the Ministry of Health to reduce health disparities. Here, she built a model for integrating mental health services in rural primary care clinics. Dr. Aguerrebere is interested in the scaling up of mental healthcare programs, comprehensive primary care, effective universal health coverage, and gender equity in healthcare delivery. She is currently a student in the Harvard Medical School Master of Medical Sciences in Global Health Delivery Program.

Emmanuel Aiyenigba is a physician and certified project management professional, and currently faculty and improvement advisor for the Institute for Healthcare Improvement. He is an ASQ certified quality manager, a certified professional in healthcare quality, a certified professional in patient safety, and a Black Belt Lean Six Sigma practitioner. He is also a fellow of the International Society for Quality in Health Care and a recipient of its prestigious Emerging Leader Award. He is a member of the World Health Organization's Global Learning Laboratory on Quality of Care and Universal Health Coverage, as well as a member of organizations such as the Project Management Institute, the American Society of Professionals in Patient Safety, and the Knowledge Management Institute.

Sara S. H. Al-Adawi is a medical student in Oman, currently completing her final clinical years. She has several publications under her name, and has a particular interest in common public health issues in Oman.

Reem Al-Ajlouni, a health management specialist, currently works with the U.S. Agency for International Development in Jordan. With a great passion to improve the health of others and make her community a better place to live, she has been working in the public health field in Jordan since graduation with both national and international organizations and the United Nations. From work on improving vulnerable groups' access to health, to improving the nutritional and health status of children in camps, to improving health insurance schemes and working on the strengthening of health systems in Jordan, Al-Ajlouni has contributed to the design and implementation of large-scale health system improvement efforts and projects.

Charles Alessi is a general practitioner and senior advisor to Public Health England, and in addition acts in a variety of externally facing positions with government and other agencies. He is also the lead for dementia and risk reduction and recently was appointed as director of antibiotic prescribing for Public Health England as part of the antimicrobial resistance initiative. He has extensive experience working at senior governmental levels nationally, where he is the outgoing chairman of the National Association of Primary Care, and internationally, in both Europe and the Americas. He also has experience in military medicine, until recently fulfilling the role of director of medicine and clinical governance for the British Armed Forces in Germany.

Noora Alkaabi is a community medicine specialist at Hamad General Hospital (HGH) and acting senior medical manager in the Medical Director Office. She is clinical coordinator of the HGH Research Committee and participates on the HGH Medical Ethics Committee. Alkaabi graduated from Jordan University of Science and Technology as a medical doctor in 2005, and interned at Hamad Medical Corporation, 2005–2006. Alkaabi served as co-chairperson of the Community Medicine Residency Programme, chair of the Clinical Competency Committee of Community Medicine Training Program, and vice chair for the End of Life Care Committee. She led the MOPH Public Health Strategy, Healthy Style and Smoking, and was physician lead for the Hospice Centre, Qatar, on its healthy lifestyle project.

Ahmed Al-Mandhari, MD, DTM&H, PhD, MRCGP (Int.), is a senior consultant in family medicine and public health at Sultan Qaboos University Hospital. He was director general of the hospital, 2010–2013. Currently, he is director general of the Quality Assurance Centre at the Ministry of Health. He is a World Health Organization temporary advisor on patient safety and quality. He is the author of *An ABC of Medical Errors Handbook*, as well as chapters in books and many articles on quality in healthcare, patient safety, and health policy.

Yousuf Al Maslamani is a consultant in general surgery and kidney transplantation. He is medical director and director of the Qatar Centre for Organ Transplantation, Hamad General Hospital. He earned his MBBS in 1988 and fellowship in 1993 from the Royal Colleague of Surgeons, Ireland. In 1994, he became a member of the Arab Board of Surgical Specialties, and in 2001 completed his fellowship at the Cleveland Clinic Foundation. He is head of the Kidney Transplant Section, chairman of the Organ Transplant Committee, and one of the chairpersons of the Gulf Cooperation Council Organ Transplantation Congress. He is a member of the Joint Oversight Board, University of Calgary, Qatar, and the International Nursing Advisory Board.

Samia Al Rabhi, MD, currently works as director general of the Quality Assurance Centre at the Ministry of Health, Sultanate of Oman, and works as senior general practitioner in a public, primary healthcare facility. She has clinical experience in primary healthcare programs, communicable and non-communicable disease management programs, elderly care programs, and mother and child health programs. She holds an MSc in quality and safety in healthcare management from Royal College of Surgeons in Ireland. She is working on patient and community engagement and empowerment in healthcare projects and participating in patient safety projects, including the World Health Organization's Patient Safety Friendly Hospital Initiative (PSFHI).

Huda Alsiyabi is a senior consultant in public health and currently works as a director of the Department of Community-Based Initiatives at the Ministry of Health in Oman. She graduated from the Sultan Qaboos University as a medical doctor and then earned her MA and postgraduate diploma in education (health and social care) from the University of Reading, United Kingdom. She is a member of various committees, such as the Technical Committee of His Majesty Award for Volunteerism, National Committee of Non-Communicable Diseases, and GCC Committee of Healthy Cities. She has participated in developing different manuals and guidelines and conducted several projects related to community-based initiatives for health promotion in Oman.

Khaled Al-Surimi, MSc, PhD, is associate professor of health systems and quality management, College of Public Health and Health Informatics, King Saud ben Abduaziz University for Health Sciences, and healthcare management consultant, Saudi Commission for Health Specialities. He is honorary senior research fellow at the Primary Care and Public Health Department, Imperial College London; author of two books and two book chapters; and author and coauthor of more than 50 journal articles and conference papers published in international scientific journals and conferences. His research interests include health services and systems research with a focus on quality improvement, patient safety, patient experience, leadership and change management, human resources for health, public health, and social media.

René Amalberti, MD, PhD. After a residency in psychiatry, Professor Amalberti entered the Air Force in 1977, was appointed to a permanent medical research position in 1982, and became full professor of medicine in 1995. He retired in 2007, dividing his time between the Haute Autorité de Santé (senior advisor, patient safety) and a position as volunteer director of the Foundation for Industrial Safety Culture. He has published more than 150 papers and authored or coauthored 12 books on human error and system safety (most recently, *Navigating Safety*, Springer, 2013; and *Safer Healthcare*, Springer, 2016).

Hugo Arce, MPH, PhD, is a physician. He was president and chief executive officer of the Technical Institute for Accreditation of Healthcare Organizations, 1994–2010, and president of the Argentine Society for Quality in Health Care. He is director of the Department of Public Health, University Institute of Health Sciences, Barceló Foundation; president of the 18th ISQua International Conference, Buenos Aires, 2001 and executive board member in 2002–2003. He is the author of *The Territory of Health Decisions, The Quality in the Health Territory,* and *The Health System: Where It Comes From and Where It Goes,* as well as more than 175 articles and papers on health policy, hospital management, and quality in healthcare.

Anne-Marie Armanteras-de-Saxcé is a hospital director. After several positions as deputy director of Paris Teaching Hospitals, she became executive director of the network of North Paris Teaching Hospitals. She pursued her career first as executive director of Rothschild Ophtalmological Foundation, France, and then as regional director of the federation of non-profit private hospitals (FEHAP). In 2013, she became regional director for hospitals and treatment organization at the Paris Regional Agency for Healthcare. In 2013, she was appointed chief of the National Directorate for Hospitals and Treatment Organization, and in 2017 became a member of the college of directors of the Haute Autorité de Santé.

Jafet Arrieta, MD, MMSc, currently serves as director for the Institute for Healthcare Improvement. She has experience in management and leadership roles within the areas of public health and quality improvement across low-, middle-, and high-resource settings. Dr. Arrieta previously served as director of Operations for Partners in Health Mexico, leading the execution of a health system strengthening strategy to improve access to health care in a highly underserved region of Mexico. She is a third-year student in the Harvard Chan School of Public Health Doctor of Public Health program, and holds a medical degree from Tecnologico de Monterrey School of Medicine, and a Master of Medical Sciences in Global Health Delivery from Harvard Medical School.

Oscar Arteaga, MD, MSc (London), DPH (London), is the dean of the School of Public Health, University of Chile. In addition to his academic career, Dr. Arteaga has extensive experience in healthcare service management at municipal and regional levels. He has also been an active player in health sector reform in Chile, having served at various national commissions, notably the Presidential Advisory Commission for the Assessment and Reform of the Private Healthcare System (Institutos de Salud Previsional). He has also served as a consultant in health systems and care management to the Chilean Ministry of Health, as well as various international agencies, notably the Pan American Health Organization/World Health Organization, Inter-American Development Bank, and World Bank.

Fakhri Athari is a PhD candidate at South Western Sydney (SWS) Clinical School, Faculty of Medicine, University of New South Wales (UNSW), Australia. She is a member of the Simpson Centre for Health Services Research and the Ingham Institute for Applied Medical Research, Liverpool, Australia. Fakhri holds a master of public health degree.

Luis Azpurua, MD, is the clinical research and education director of Clinica Santa Paula, and also professor of clinical and biomedical engineering in the Universidad Simón Bolívar in Caracas, Venezuela. During his 25-year journey throughout the healthcare field, he has served, among others, as a medical director of the Hospital San Juan de Dios of Caracas. He is coauthor of the strategic planning book *De Autoempleados a Empresarios. Como Planificar un Centro de Salud Privado para Potenciar su Desarrollo*. He is passionate about healthcare quality and patient safety culture.

Natasha Azzopardi-Muscat, MD (Melit), MSc public health (Melit), MSc health services management (London), PhD (Maastricht), FFPH (UK), DLSHTM (London), qualified as a doctor (1995), in public health medicine (2003); she obtained a fellowship, Faculty of Public Health, United Kingdom (2006), and a PhD from Maastricht University (2016). She previously held various positions in the Ministry of Health in Malta, including chief medical officer. She is a consultant in public health medicine in Malta and is a senior lecturer at the University of Malta. Her research centers on European Union health policy and health systems in small states, with several publications on these topics. She is president of the European Public Health Association.

Vasha Elizabeth Bachan, MD, an MPH candidate, is presently the principal investigator/program director of the Ministry of Public Health/Centre for Disease Control and Prevention's Cooperative Agreement. She was previously the director of regional health services for the regional and clinical services for the ministry (2014–2015) and served as chronic disease coordinator, Ministry of Health (MoH) (2013–2014); regional health officer for Region 4, MoH (2012–2014); district medical officer (subregion: East Coast Demerara) for Region 4, MoH (2011–2012); and government medical officer working in primary healthcare (2009–2011). Her interest is in public health administration.

Roland Bal is professor of healthcare governance at Erasmus University, the Netherlands. His research interests include science–policy–practice relations and governance infrastructures in healthcare. More recently, he has researched the creation of public accountabilities in healthcare, studying ways in which public service organizations "organize for transparency." He also currently works with the healthcare inspectorate of the Netherlands in research projects on regulation and supervision. Roland has been involved in international comparative research, for example, on hospital quality in Europe, and has a focus on ethnographic, interventionist research methods. He has developed and taught many national and international teaching programs on healthcare (quality) governance.

Marta Ballester Santiago, MSc, is a researcher and project manager in Avedis Donabedian Research Institute. Her main areas of interest cover patient empowerment, self-management, and patient participation. At a European level, she recently coordinated the first Directorate-General Health and Food Safety tender in patient empowerment: Empowering Patients in the Management of Chronic Diseases. Her current research is focused on the promotion of self-care participating in PRO-STEP (a pilot project on promoting self-management for chronic diseases in the European Union) and PISCE (a pilot project on the promotion of self-care systems in the European Union), patient participation mechanisms, and differences of access to treatment across Europe.

Joshua Bardfield, MPH, serves as knowledge management director for HEALTHQUAL International. He has extensive public health communications, research, and technical writing experience, with a focus on reproductive health and HIV/AIDS. Over the last several years, Bardfield's work has focused on implementation science and spreading improvement knowledge, with an emphasis on building a sustainable cross-country peer network for improvement among HEALTHQUAL's 15 target countries. Bardfield has researched and authored a variety of academic publications focused on women's health, HIV, and quality improvement, and earned his master's in public health from Columbia University's Mailman School of Public Health.

Maysa Baroud is a project coordinator in the Refugee Research and Policy in the Arab World Program at the Issam Fares Institute for Public Policy and International Affairs, American University of Beirut. Her current research focuses on the livelihoods of refugee and host communities in Lebanon. Baroud holds a master of public health, concentrating in health management and policy, and a master of science in microbiology and immunology.

Apollo Basenero, MBChB, is currently the chief medical officer, Quality Management (QM) Program in the Ministry of Health and Social Services (MoHSS), Namibia. He is a fellow of the International Society for Quality in Health Care and of the Arthur Ashe Endowment for the Defeat of AIDS. His achievements include being part of the team that developed the MoHSS national QM training curricula for healthcare workers, and consumer involvement in the QM curriculum, QM coaches' curriculum, and QM capacity-building framework for the MoHSS and Infection Prevention and Control guideline. He has participated in various research projects and authored and coauthored a number of abstracts.

Roger Bayingana, MD, MPHM, currently works as a consultant in health systems strengthening, research, quality assurance, and accreditation in Rwanda and the region. He has been a principal clinical investigator on many HIV vaccine research studies and also has worked as head of training and guidelines for the Ministry of Health HIV Research Institute. Dr. Bayingana is an author or coauthor of several scientific papers, mostly on HIV research and programing. He holds a master's degree in research methodologies from the Universite Libre de Bruxelles in Belgium and a medical degree from the National University of Rwanda.

Mustafa Berktaş, MD, is a clinical microbiology professor. He is the managing director of the newly founded Healthcare Quality and Accreditation Institute of Turkey, under Turkish Health Sciences Institute. He lectured on microbiology and fulfilled duties like educational coordinator, director of the ethical committee, chief of the Microbiology Department, deputy chief of the Health Sciences Institute, and vice dean of the Medicine Faculty. He was State Hospital's founding chief physician and medical director of a private hospital. He actively engaged with healthcare quality studies at the Ministry of Health, Turkey. Professor Berktaş has 157 articles published in scientific journals and 141 abstracts presented at conferences, both national and international.

Angelika Beyer, MSc in healthcare management, is a research assistant at the Institute for Community Medicine, Department of Epidemiology of Health Care and Community Health, University Medicine Greifswald, Germany. She has been working in various research projects focusing on the epidemiology of healthcare in pediatrics both in Germany and in the region of Western Pomerania. Her research interests include the improvement of regional healthcare, innovative healthcare concepts (e.g., delegation and substitution of tasks and telemedicine), patient-oriented outcomes, lowering of inequalities in health, and access to healthcare.

Camilla Björk holds a position at QRC Stockholm as senior technical project manager. QRC Stockholm is a regional registry center for National Quality Registries and also forms a strategic cooperation between the Karolinska Institutet and the Stockholm County Council. She holds an MSc from the Royal Institute of Technology in Stockholm. She is involved in procurement testing, and the development of learning eHealth systems for real-time information on quality improvement feedback and safety alert monitoring. She has a background as a statistician, developer, and architect at the Karolinska Institutet, where she has developed systems for epidemiological studies.

Sergi Blancafort Alias holds a PhD in sociology from the Autonomous University of Barcelona and a degree in biology from the University of Barcelona. In 2004, he joined the Josep Laporte Foundation (renamed the Health & Aging Foundation in 2014), where he has been involved in social and educational projects addressed to patients and health professionals, such as the Spanish Patients Forum and the Patients' University. He currently leads the AEQUALIS project, a clinical trial aimed to reduce health inequalities through a group-based intervention that promotes self-management, health literacy, and social support in socioeconomic disadvantaged older adults.

Robert H. Brook, MD, ScD, is a senior advisor and corporate fellow at the RAND Corporation, and professor of medicine and health services at the David Geffen UCLA School of Medicine and UCLA Fielding School of Public Health. An international expert in quality of care, Brook has published more than 560 scholarly articles. He has received numerous awards, including the Gustav O. Lienhard Award, the Health Research & Education Trust Award, the David E. Rogers Award, the Baxter Foundation Prize, the Rosenthal Foundation Award, the Distinguished Health Services Researcher Award, and the Robert J. Glaser Award.

Margaret K. Brown is a project coordinator at the New York State Department of Health AIDS Institute's Office of the Medical Director. Brown graduated cum laude from Haverford College in 2015 with a BA in international studies and French. She first became invested in public health and policy while writing her senior thesis on international sex work regulation. Since starting at the AIDS Institute, her research interests have focused on consumer involvement in healthcare and quality improvement. Out of the office, Brown volunteers for local charity organizations, explores New York City's music scene and occasionally escapes the city to hike and cycle.

Sandra C. Buttigieg, MD, PhD (Aston), FFPH (UK), MSc (public health medicine), MBA, associate professor, and head of the Department of Health Services Management, University of Malta, and consultant in public health medicine, Mater Dei Hospital, Malta. She is an honorary senior research fellow at the University of Birmingham, United Kingdom. She has authored and coauthored numerous articles in peer-reviewed management and health journals and is currently on the Editorial Advisory Board of *Frontiers—Public Health*, the *International Journal of Human Resources Management*, and the *Journal of Health Organization and Management*. Her research centers on performance of health systems and organizations as a function of policy, people, and operating systems.

Tumurbat Byamba is an executive member of the Mongolian Association of Healthcare Quality and a fellow of the International Society for Quality in Health Care. She graduated from the National Medical University of Mongolia with a bachelor of medicine and earned a master's in healthcare service management from the University of Birmingham, United Kingdom. Currently, she is a PhD student in business management. Her experience in policy development, strategic planning, project management, and knowledge management provides support to the activities of the association and to the dissemination of healthcare service quality knowledge among healthcare professionals and patients. Dr. Byamba also has experience in patient safety, external evaluation systems, and risk management issues.

Andrew Carson-Stevens, MBBCh, PhD, is a primary care doctor and founding leader of the Primary Care Patient Safety Research Group at Cardiff University. He has extensive expertise in generating learning from patient safety incident reports. Internationally, he is a member of the Safer Primary Care Expert Group (2011) and the Patient Safety Incident Reporting Expert Advisory Group (2017) at the World Health Organization. He is a visiting professor in the Department of Family Practice, University of British Columbia, and honorary professor at the Australian Institute of Health Innovation, Macquarie University.

Edward Chappy, BA, MPA, FACHE, FISQua, is the project director of the U.S. Agency for International Development–funded Human Resources for Health in 2030 Activity in Jordan. Previously, he was the chief of party of the USAID-funded Jordan Healthcare Accreditation Project, which established the Health Care Accreditation Council, the national healthcare accreditation agency in Jordan. Chappy has worked in several countries as a private hospital CEO and as a consultant implanting quality systems in Eritrea, Indonesia, Pakistan, Turkey, Rwanda, Afghanistan, and Saudi Arabia.

Patsy Yuen-Kwan Chau, MPhil, is research associate at the JC School of Public Health and Primary Care, the Chinese University of Hong Kong. Her background is in statistics and public health. She has been involved in a number of different research projects, such as the impacts of air pollution on mortality and morbidity, the spatial variation of temperature changes on health, the evaluation of hospital accreditation, and the evaluation of regulation of sulfur contents of gasoline.

Americo Cicchetti, director of the Graduate School of Health Economics and Management and professor of healthcare management, Faculty of Economics, Università Cattolica del Sacro Cuore, Rome. He is chief of research for the Health Technology Assessment Unit and Biomedical Engineering, Agostino Gemelli University Hospital, Rome. He was member of the Price and Reimbursement Committee of the Italian National Drug Agency (2009–2015), served as director of Health Technology Assessment International (HTAi) (2005–2008), and is now a member of the executive committee and secretary. He chairs the HTAi's hospital-based health technology assessment interest subgroup and the International Scientific Program Committee Annual Meeting of HTAi (Rome, 2017). He is also president of the Italian Society of Health Technology Assessment.

Silvia Coretti is a postdoctoral research fellow at the Faculty of Economics of the Università Cattolica del Sacro Cuore, Rome, Italy, where she earned her PhD in 2014 and master's in health technology assessment in 2010. In 2013, she earned her master of science at the University of York, United Kingdom. She cooperated with the Pricing and Reimbursement Office of the Italian

Agency of Medicines and with the Health Economics Research Unit at the University of Aberdeen. Her current research interests concern health economics and health technology assessment, with a specific focus on the methodological aspects of outcome measures and patients' preferences.

Jacqueline Cumming is professor of health policy and management and director of the Health Services Research Centre, Faculty of Health, Victoria University of Wellington. She has qualifications in economics, health economics, and public policy; teaches health policy and monitoring and evaluation; and supervises PhD students in health services research. She has led high-profile national evaluations of reforms in health policy and health services. With more than 20 years' experience in health services research, she was president of the Health Services Research Association of Australia and New Zealand (2007–2014). In 2013, she was awarded the association's professional award, awarded biennially for outstanding contributions to the development of health services research in Australia and New Zealand.

Adriana Degiorgi, master in healthcare management, is a member of the executive board, the director of quality and patient safety, and the director of support services at the Ente Ospedaliero Cantonale Hospital Group in Ticino, Switzerland. She is the president of FoQual—the Swiss Quality Forum, and a member of the board of trustees of the Swiss Patient Safety Foundation. She also lectures within the master of advanced studies on economics and healthcare management at the University of Applied Sciences and Arts of Southern Switzerland and at the Center for the Advancement of Healthcare Quality and Patient Safety based at the Università della Svizzera Italiana.

Ellen Tveter Deilkås, MD, PhD, is a consultant in internal medicine with 20 years of clinical experience. She is a senior scientist at Akershus University Hospital, where she conducts research on patient safety culture and improvement. She holds a 20% position as a senior advisor for the Norwegian Directorate of Health, specifically responsible for national measurement of medical injury and patient safety culture. The measurements are part of the government's patient safety program. She led the Patient Safety Committee of the Norwegian Medical Association from 2006 to 2016 and works part-time as a consultant with stroke rehabilitation.

Pedro Delgado, head of Europe and Latin America, Institute for Healthcare Improvement, has been a driving force in the Institute for Healthcare Improvement's global strategy. From work on reducing C-sections in Brazil, to improving early years' education in Chile, to improving patient safety in Portugal and mental health in London, Delgado has led the design and implementation of large-scale improvement efforts globally. He coaches senior leaders and teams and lectures extensively, worldwide, on large-scale change, patient safety, and quality improvement. He holds summa cum laude degrees in psychology and global business, and an MSc in healthcare management and leadership.

Subashnie Devkaran, MScHM, PhD, FACHE, CPHQ, FISQua, BScPT, is a leader in quality, accreditation, and patient experience in the Middle East and Africa region, currently affiliated with Cleveland Clinic Abu Dhabi. She is president of the American College of Healthcare Executives for the Middle East and North Africa Group and an international consultant with Joint Commission International based in Chicago. Dr. Devkaran serves as an associate lecturer with the Royal College of Surgeons in Ireland. She regularly speaks at international conferences and has published several articles on healthcare quality and patient experience. With a passion to reduce the knowledge gap in healthcare quality and patient experience, Dr. Devkaran continues to pursue research in these areas.

Sir Liam Donaldson is an international champion of public health and patient safety. He was the foundation chair of the World Health Organization (WHO) World Alliance for Patient Safety, launched in 2004. He is a past vice chairman of the WHO executive board. He is now the WHO envoy for patient safety. In the United Kingdom, he is a professor of public health at the London School of Hygiene and Tropical Medicine and chancellor of Newcastle University. Prior to this, Sir Liam was the 15th chief medical officer for England from 1998 to 2010.

Liv Dørflinger, MSc, currently runs a program on patient-reported outcomes (PROs) in a partnership with numerous Danish healthcare organizations. She holds a master's degree from the University of Copenhagen, where she graduated in 2013. In 2011, she began working on PROs and has since initiated numerous research and quality improvement projects focused on PROs in the clinical interaction between patients and healthcare professionals. She has been a driving force in setting an agenda and creating awareness of PROs in Danish healthcare.

Thomas E. Dorner is associate professor at the Centre for Public Health, Medical University of Vienna, Austria. He has studied human medicine at the University of Vienna, is trained as a general practitioner, and has graduated as master of public health. He is president of the Austrian Public Health Association and scientific member and consultant of many national boards, like the National Nutrition Commission, the advisory board of gerontology, and the advisory board for a national diabetes strategy. He is a member of the governing board of the European Public Health Association and of the steering committee of the European Network for Prevention and Health Promotion in Primary Care, and was chair of the European Public Health Conference 2016.

Persephone Doupi, MD, PhD, is a senior researcher at the Welfare Department of the National Institute for Health and Welfare, Finland. Currently, her focus is on injury prevention, safety, and public health informatics. She has more than 15 years' experience as a key investigator in European Union–funded and national projects concerned with various aspects of digital health data, cross-border healthcare services, and personalized applications of health information technology. She has served on multiple national, Nordic, and international expert groups on eHealth, quality, and patient safety, and authored or coauthored several scientific articles, books, and book chapters on medical informatics and eHealth topics.

Paul Edwards, M.D, MPH, is the advisor for health systems and services for the Pan American Health Organization (PAHO)/World Health Organization Guyana Office. A national of Belize, he obtained a medicine and surgery degree from San Carlos University, Guatemala, in 1996, and a master's degree in public health, epidemiology, from the National Institute of Public Health, Cuernavaca, Mexico, in 2001. In Belize, his career included medical laboratory technologist, medical director of institutions, national epidemiologist, director of the national AIDS program, deputy director of health services, and central region health manager. He worked in Trinidad and Tobago as the PAHO advisor for HIV and sexually transmitted infection surveillance and thereafter as the advisor for health systems and services, PAHO subregional Office, Barbados.

Carsten Engel, MD, is deputy chief executive at the Danish Institute for Quality and Accreditation in Healthcare. His background is clinical practice as an anesthesiologist, and he has management experience at the departmental and hospital levels. Since 2004, he has been devoted full-time to quality management and improvement in healthcare, taking a leading role in the development and management of the Danish Healthcare Accreditation Programme). Through the Accreditation Council of the International Society for Quality in Health Care, he is engaged in accreditation internationally and serves as an International Society for Quality in Health Care expert.

Jesper Eriksen, MSc, takes special interest in the interaction between the economic and political organization of the healthcare system and the quality of the healthcare efforts. He works from a policy perspective with data-driven quality improvement in cancer treatment, and monitors progress from organizational, clinical, and user-perceived dimensions.

Andrew Evans, chief pharmaceutical officer to the Welsh government since 2016, is a public health specialist who prior to this appointment was principal pharmacist in pharmaceutical public health at the Public Health Wales National Health Service Trust. He previously held positions as a senior pharmacist in the National Health Service in Wales and England, a pharmacy postgraduate education tutor, and a community pharmacist. Andrew is an honorary lecturer at the Cardiff School of Pharmacy and Pharmaceutical Sciences. His research interests are in pharmaceutical aspects of public health.

Paulo André Fernandes, MD, MSc, is director of the National Program for Prevention and Control of Infection and Antimicrobial Resistance in the Directorate-General of Health. Previously, he was a member of the National Program's Direction and Scientific Committee. He is a specialist in intensive care medicine and internal medicine, and is currently the intensive care unit director in Centro Hospitalar Barreiro Montijo, Portugal, where he directs the Quality of Care and Patient Safety Committee and integrates the Pharmaceutical and Therapeutics Committee. From 2009 to 2011, he was the hospital's medication manager. He coordinated, from 1996 to 2013, the local Antimicrobial Therapy Committee and, from 2013 to 2016, the local Committee for Prevention and Control of Infection and Antimicrobial Resistance.

Laura Fernández-Maldonado, BA in information science and master's diploma of advanced studies in sociology, is responsible for the Patients and Citizens Area, Fundació Salut i Envelliment UAB. Her key interests comprise helping patients in their health decision-making, giving them the appropriate knowledge regarding their treatment and conditions, and empowering them. She coordinates the Patients' University Project. She has also contributed to the design of and patients' engagement in a series of advocacy and health educational programs across Spain and Europe, has developed a research link to patients' needs and health literacy, and works with multiple stakeholders in the Spanish health system.

Steven Frost is a registered nurse in the intensive care unit at Liverpool Hospital and a lecturer in the School of Nursing and Midwifery at Western Sydney University. Frost completed his PhD at the Garvan Institute of Medical Research and the School of Public Health and Community Medicine, University of New South Wales, in 2014. His main areas of research are epidemiology, health services research, and risk prediction. Frost has 64 peer-reviewed publications.

Hong Fung, MBBS (HK), MHP (NSW), FRCS (Edin), FHKAM (Surg), FCSHK, FHKAM (community medicine), FHKCCM, FFPHM (UK), FRACMA, FAMS, is professor of practice in health services management at the JC School of Public Health and Primary Care, the Chinese University of Hong Kong. He is codirector of the master of science in health services management, executive director of the Chinese University of Hong Kong Medical Centre, and president of the Hong Kong College of Community Medicine. He is well recognized for his expertise in health leadership, hospital planning, health informatics, and health services management, with prior roles as director of planning and cluster chief executive of New Territories East Cluster in the Hong Kong Hospital Authority.

Ezequiel García-Elorrio, MD, MSc, MBA, PhD, is one of the founders and board members of the Institute for Clinical Effectiveness and Health Policy in Argentina, where he also leads the Department of Health Care Quality and Patient Safety. García-Elorrio has worked extensively in quality of care and patient safety research, education, and implementation projects in Latin America, Sub-Saharan Africa, and Southeast Asia. His main interests are related to patient safety, external evaluation, and improvement methods, focusing on successful implementation in developing countries. He is an active collaborator of the International Society for Quality in Health Care as associate editor of the society's journal from 2009 and a member of the education committee, as well as participating in other activities.

Christine S. Gordon, RN, works at the RM Quality Assurance Unity, Ministry of Health and Social Services, Republic of Namibia.

Victor Grabois, MD, MSc in public health from the State University of Rio de Janeiro, and PhD student in public health at Oswaldo Cruz Foundation. He is the executive coordinator of the Collaborative Center for Quality of Care and Patient Safety (Proqualis) and a representative of the Oswaldo Cruz Foundation in the Implementation Committee of National Patient Safety. He was general manager of federal hospitals in Rio de Janeiro and coordinator of large-scale online courses on health services management. He has published book chapters and scientific articles on hospital management, clinical governance, and patient safety.

Kenneth Grech, MD, MSc (London), MBA, FFPH (UK), DLSHTM (London), consultant in public health with the Department of Health and lecturer with the University of Malta. He holds a master's in public health from the London School of Hygiene and Tropical Medicine and a master's in business administration from the University of Malta, and is a fellow of the Faculty of Public Health Medicine, United Kingdom. He is currently undertaking a PhD on health system performance assessment at the University of Warwick. Previous positions include chief executive officer of St. Luke's Hospital, Malta; director of social welfare standards; and permanent secretary of Malta's Ministry of Health. He is currently the president of the Malta Association of Public Health Medicine.

Catherine Grenier, MD, MPH, MBA, is a public health physician whose 23-year career in medicine has increasingly focused on quality and safety policies and executive management at the national level. She started as a research scientist on quality indicators, working successively for the French School of Public Health, the French National Institute of Health and Medical

Research, and the Unicancer group (French Comprehensive Cancer Centers). In 2011, she joined the Haute Autorité de Santé, the French national authority for health, as director of quality and safety indicators in the Accreditation-Certification Department. In 2016, she then became chief of the department and deputy director of Haute Autorité de Santé.

Girdhar Gyani is widely recognized as the founder of healthcare quality in India. During his tenure as secretary general of the Quality Council of India (2003–2012), Dr. Gyani played a pivotal role in the formulation and operating of the National Accreditation Board for Hospitals and Healthcare Providers, the first of its kind in the country and accredited by the International Society for Quality in Health Care. Currently, Dr. Gyani is working as director general of the Association of Healthcare Providers (India) (AHPI). AHPI represents the vast majority of healthcare providers in India. The mission of AHPI is to build capacity in the Indian health system with a focus on patient safety and affordability.

Ndapewa Hamunime is a senior medical officer of the Namibia Ministry of Health and Social Services and champions the integration of the 3Is (intensified case finding, infection control, and isoniazid prevention therapy) into the national HIV treatment program. The Ministry of Health and Social Services embraces quality improvement as an integral component of the government-led strategy for the implementation of a national framework for tuberculosis prevention, care, and treatment.

Jamie Hayes, BPharm (Hon), ClinDipPharm, PCME, MBA, DipTher, MRPharmS, FFRPS, is a pharmacist; director at the Welsh Medicines Resource Centre and the All Wales Therapeutics and Toxicology Centre; honorary senior lecturer at the School of Medicine, Cardiff University; and clinical author for the Centre for Medicines Optimisation, Keele University. Jamie is an executive coach and experienced medical educator, interested in patient safety, behavioral change, decision-making, and influencing skills. Trained at the Welsh School of Pharmacy, qualifying in 1992, his early career was as a clinical pharmacist in South Wales, with subsequent jobs taking him to New Zealand, North Wales, and England, before finally returning to South Wales. In 2012, he completed an MBA specializing in Lean thinking.

Ken Hillman, AO, MBBS, FRCA, FCICM, FRCP, MD, is professor of intensive care at the University of New South Wales, Australia; director of the Simpson Centre for Health Services Research; and an actively practicing clinician in intensive care. He has authored 170 peer-reviewed publications and 64 chapters in textbooks; coauthored an intensive care textbook; coedited several textbooks; written two books, *Vital Signs: Stories from Intensive Care* and *A Good Life to the End: Taking Control of Our Inevitable Journey through Ageing and Death*; and received more than $20 million in grants. He is internationally recognized as a pioneer in the introduction of the medical emergency team, and in 2005 he helped establish the first international conference on the medical emergency team and has been actively involved since.

Wolfgang Hoffmann is head of the Section Epidemiology of Health Care and Community Health and acting director of the Institute for Community Medicine, University Medicine Greifswald, Germany. In 1995, he qualified for master of public health in epidemiology at the Department of Epidemiology, University of North Carolina, Chapel Hill. In 2002, he was offered a Bundesministerium für Bildung und Forschung (BMBF) (Federal Ministry of Education and Research, Germany) sponsorship—and 2 years later a full professorship—for Epidemiology of Health Care and Community Health at the University Medicine Greifswald, Germany. His research priorities are population-based intervention and prevention, regional concepts of healthcare, central data management, and quality management.

Maria M. Hofmarcher-Holzhacker is an economist and health system expert with a research focus on the economics of health and social care, public finance, health and long-term care supply, efficiency, and comparative health and social care research. She is director of HealthSystemIntelligence and research associate at the Department of Health Economics, Medical University of Vienna. She was coordinator of the European Union FP 7 projects and currently leads work in the area of evaluation of healthcare systems in the context of the European Union project BRIDGE Health. Among her many publications, she is the principal author of *Health System Review, Austria*.

Lise Hounsgaard, PhD in nursing, research leader, and professor at the Institute of Nursing and Health Science, University of Greenland, and the Odense Patient Data Explorative Network (OPEN), Department of Clinical Research, University of Southern Denmark. Her competencies deal with qualitative research methods, fieldwork as a research approach, and global health (Greenland) focusing on patients', relatives', and professionals' perspectives. She often publishes together with her PhD students and outputs include *The Nature of Nursing Practice in Rural and Remote Areas of Greenland, Family Care—Relatives with Parkinson Disease, Relatives' Level of Satisfaction with Advanced Cancer Care in Greenland* and *Telemedicine in Greenland—Citizens' Perspectives on Providing Welfare among Elderly*.

Min-Huei (Marc) Hsu, MD, PhD, is director of the Department of Information Management, Ministry of Health and Welfare, Taiwan. He has been chief information officer at Taipei Medical University and a consultant neurosurgeon at Wanfang Hospital, a 746-bed hospital affiliated with Taipei Medical University. He is the author or coauthor of more than 40 papers and articles in international conferences and scientific journals, focusing on health data, health information technology, eHealth, electronic medical record systems, hospital information management, and patient safety.

Clifford F. Hughes, AO, MBBS, DSc, FRACS, FACS, FACC, FIACS (Hon), FAAQHC, FCSANZ, FISQua, AdDipMgt, is immediate past president of the International Society for Quality in Health Care. He is professor of patient safety and clinical quality at the Australian Institute of Health Innovation at Macquarie University, Sydney, Australia. A former cardiothoracic surgeon and CEO of the Clinical Excellence Commission in New South Wales, Australia, he is a consultant in quality and safety to health services in Australia, New Zealand, the United Kingdom, and the United States. He is passionate about person-based care, better and timely incident management, and the development of clinical leaders. His spare time is devoted to his family and eight grandchildren.

Valentina Iacopino is a postdoctoral researcher in organization at the Faculty of Economics, Università Cattolica del Sacro Cuore, Rome, Italy. Her research interests and publications focus on the adoption process of technological innovations in the healthcare context at institutional, organizational, and professional levels. In her studies, she applies traditional statistical methods as well as social network analysis techniques to explore the adoption and diffusion of technologies in healthcare and to understand interorganizational as well as professional networks' role in the innovation process. Moreover, she deals with policy issues and the governance of technological innovations at both national and local levels.

Usman Iqbal, PharmD, MBA, PhD, FACHI, AFCHSM, is a faculty at College of Public Health, Global Health and Development Department and International Center for Health Information Technology, Taipei Medical University. He is also a fellow of the Australasian College of Health Informatics, Salzburg Global Seminar and Australasian College of Health Service Management. Dr. Usman is on the editorial committee of the *International Journal for Quality in Health Care,* and is an associate editor of *Computer Methods and Programs in Biomedicine.* Facilitating joint collaborative course between TMU & MIT, Harvard, USA. He has been an author of several scientific articles, book chapters and participated in international conferences focused on Health Data and Informatics, Health IT and Patient Safety, and Health Care Quality and Management.

Wendy James works as publications editor for the Australian Institute of Health Innovation at Macquarie University. She has a PhD from the University of New England, and is the author of seven books of fiction, including the best-selling *The Mistake* (Penguin, 2012) and the Ned Kelly Award–winning *Out of the Silence* (Random House, 2005). Her latest novel, *The Golden Child*

(HarperCollins, 2017), was short-listed for the 2017 Ned Kelly Award for crime fiction. Her short stories and assorted non-fiction have been widely published in Australian literary journals, newspapers, and magazines.

Ravichandran Jeganathan is national head of obstetrics and gynaecology services and maternal-fetal medicine, Ministry of Health, Malaysia. He graduated from University Science Malaysia with an MD and master's in obstetrics and gynecology (O&G). He is an associate member of the Royal College of Obstetrics and Gynaecologists. He is the chair of the National Specialist Registry (O&G), Malaysia, as well as the Confidential Enquiry into Maternal Deaths, Malaysia. He is a life member of the Malaysian Society for Quality in Health and president-elect of the Obstetrical and Gynaecological Society of Malaysia. He serves as associate professor of O&G with Monash Malaysia Medical School and has been a temporary consultant with the World Health Organization and United Nations Population Fund.

Ravindran Jegasothy is dean of the Faculty of Medicine, MAHSA University, Malaysia. He graduated MBBS from the University of Malaya and is a fellow of the Royal College of Obstetricians and Gynaecologists, London. He has been awarded the MMed (obstetrics and gynecology) from the National University of Singapore, as well as fellowships from the Academy of Medicine of Malaysia and the Indian Academy of Medical Sciences. He worked in the Ministry of Health, Malaysia, where he was closely associated with the efforts in the reduction of maternal mortality. He is on the International Accreditation Panel of the International Society for Quality in Health Care.

Ruth Kalda, DrMedSci, is professor of family medicine at the University of Tartu. She is head of the Institute of Family Medicine and Public Health. She is also a part-time family doctor and belongs to the board of the Estonian Society of Family Doctors. She is a member of the European Academy of the Teachers in General Practice/Family Medicine. She is the author and coauthor of more than 100 scientific and research articles, among them 45 papers published in international peer-reviewed scientific journals. The main topics of her research interest are the quality of primary healthcare, the continuity of care, and care integration.

İbrahim H. Kayral, PhD, has a BSc degree in economics, MBA, and doctorate in business administration. He is working for the Healthcare Quality and Accreditation Institute of Turkey, coordinating International Society for Quality in Health Care International Accreditation Programs for the institute and establishing accreditation programs. He has different business experiences. After business development consultation for small and medium-sized enterprises, Dr. Kayral served for the Ministry of Health, Turkey, in developing standard sets, training programs, pay-for-performance systems, and strategic plans. He also represented the Ministry of Health in International Society for Quality in Health Care International Accreditation Programs. He is an honorary advisor in the Confederation of International Accreditation Commission. Dr. Kayral has authored numerous published books, articles in scientific journals, and presentations in national and international conferences in healthcare.

Rawya Khodor is a project coordinator in the Refugee Research and Policy in the Arab World program at the Issam Fares Institute for Public Policy and International Affairs, Beirut, Lebanon. Her prior work experience includes working with Lebanese ministries (Ministry of Social Affairs, and Ministry of Public Health), UN agencies (UNDP, and UNRWA), and international non-governmental organizations on several social and health projects. She holds a bachelor of science in nutrition and dietetics, and a master of public health. She had also completed a program on Global Health Delivery at Harvard. Her research and practice interests are directed towards the planning, design, and implementation of health interventions/programs, community supports, and health policy-making.

Janne Lehmann Knudsen, MD, specialist in public health medicine and community medicine, PhD, MPM, is currently head of Pharmacovigilance & Medical Devices at the Danish Medicines Agency. She has a long track record of being a leader in healthcare, working with healthcare regulation, research, and quality. For 8 years, she held a leading position in the Danish Cancer Society, putting up programs for patient-centered care, including patient-reported outcome measures. She has been an associate professor and worked internationally for many years. She is also a member of the board of the International Society for Quality in Health Care. She has published several scientific papers and edited textbooks.

Madelon Kroneman, PhD, is a senior researcher at the Netherlands Institute of Health Services Research. She graduated from Wageningen University and obtained her PhD from Utrecht University in sociology. She specialized in international comparative health services research and has broad knowledge of the Dutch healthcare system and other European healthcare systems. She is coauthor of the "Netherlands Health System Review" from the World Health Organization's European Observatory on Health Systems and Policies and coordinator of the Dutch page of the Health Services and Policy Monitor (www.hspm.org).

Grace Labadarios graduated from the University of Stellenbosch in 1992 and was a general practitioner in the United Kingdom until her return to South Africa in 2011. She earned her diploma from the Royal College of Obstetricians and Gynaecologists in 1996, and a certificate from the Joint Committee on Postgraduate Training for General Practice in 1999, and became a member of the Royal College of General Practitioners in March 2010. She joined the Council for Health Service Accreditation of Southern Africa (COHSASA) in 2011 as the GP Accreditation Programme coordinator and assumed responsibility for all standards development activities from 2014 on. In 2015, she joined the Office of Health Standards Compliance as the director of Health Standards Design and Development. She is currently completing her MMed in public health at the University of Cape Town (UCT).

Gavin Lavery is clinical director of the HSC Safety Forum, a lead organization for quality improvement and patient safety across health and social care, and is also a consultant in critical care medicine at Belfast HSC Trust. Dr. Lavery is a graduate of the Advanced Training Programme at Intermountain Healthcare and is a member of the Health Foundation's College of Assessors. Previous roles include lead clinician for the Northern Ireland Critical Care Network, president of the Intensive Care Society of Ireland, and an examiner for several colleges and faculties. Dr. Lavery has published more than 60 peer-reviewed papers and book chapters.

Wui-Chiang Lee is the director of the Department of Medical Affairs and Planning, Taipei Veterans General Hospital, Taiwan. He was the director general of the Department of Medical Affairs, Ministry of Health and Welfare (2013–2014), and chief executive officer of the Joint Commission Taiwan (2011–2013). Dr. Lee has actively participated in many national and international quality improvement and patient safety programs for the past 10 years. He has been the president of the Asian Society for Quality in Health Care and also on the board of the International Society for Quality in Health Care since 2013. Dr. Lee earned his PhD and MHS from Johns Hopkins University, Baltimore, Maryland, and MD from National Yang-Ming University, Taiwan.

Margus Lember, MD, PhD, is professor and head of the Department of Internal Medicine and dean of the Faculty of Medicine at the University of Tartu. He was a president of the Estonian Society of Family Doctors and Society of Internal Medicine, and an advisor to the Ministry of Social Affairs in the 1990s, when the Estonian primary care reform was carried out. He is the chairman of the National Practice Guidelines Advisory Board, is a member of the editorial board of several medical journals, and has been a consultant to the World Bank in healthcare reforms in several countries.

Yu-Chuan (Jack) Li is a pioneer of medical informatics research in Asia, dean of the College of Medical Science and Technology, and a professor at the Graduate Institute of Biomedical Informatics. He obtained his MD from Taipei Medical University and PhD from the University of Utah. He was designated as one of the Ten Outstanding Young Persons of the Year in 2001, became an elected fellow of the American College of Medical Informatics (2010) and the Australian College of Health Informatics (2010), and was president of the

Asia Pacific Association for Medical Informatics (2006–2009). Author of 130 scientific papers and three textbooks, he is the editor-in-chief of *Computer Methods and Programs in Biomedicine* and the *International Journal for Quality in Health Care*.

Kristiana Ludlow, BPsych (Hon), MRes, is a PhD candidate and research assistant at the Centre for Healthcare Resilience and Implementation Science, Australian Institute of Health Innovation, Macquarie University, Sydney, Australia. She has a background in psychology, graduating in 2015 with first-class honors, and completed her master of research in health innovation in 2017. Her research interests include residential aged care, patient-centered care, shared decision-making, priority-setting, and using complexity science to improve health systems sustainability.

Marcella Marletta is currently general manager of the General Directorate of Medical Devices and Pharmaceutical Service of the Italian Ministry of Health. She graduated in medicine and specialized in ophthalmology. In 1990, she became a medical director of the Ministry of Health. In 1995, she was appointed deputy minister counselor of the General Directorate of Public Hygiene and director of Division II, Hospital General Management, Ministry of Health. From 1997 to 2004, she led the Medical Devices Office of the Ministry of Health as director and became a delegate to the community activities of MDEG, MSOG, and CTEG in Brussels.

Susanne Mayer, PhD, is assistant professor at the Department of Health Economics, Centre for Public Health, Medical University of Vienna, Austria. She holds a PhD in economics and master's degrees in economics and socio-economics from the Vienna University of Economics and Business, Austria. She has also studied at McGill University, Canada, and worked as visiting researcher at the Department of Health Services Research, Maastricht University, the Netherlands. Specific research areas include the interplay of providers in the outpatient sector, pharmaceutical consumption by socioeconomic status, and unit costs of health services. She lectures extensively on health economics and health services research.

Laetitia May-Michelangeli, MD, is head of patient safety and quality indicators at Haute Autorité de Santé (HAS), the French accreditation agency. She started her career in 1999 as the infection, health and safety manager of a public hospital, and then moved in 2006 to the Office of Quality and Safety at the French Ministry of Health. She joined the HAS in 2013 to lead the Patient Safety Mission. In 2016, she expanded her mission to the HAS Office of Quality and Safety Indicators. In 2016, she coauthored the chapter on risk management in the French reference manual of public health, *Traité de Santé Publique*.

Lizo Mazwai, MBChB, FRCSEd, associate fellow of the College of Surgeons of South Africa, emeritus professor of surgery and former dean of the Medical

School, Walter Sisulu University, Mthatha, South Africa. Mazwai is former president and current honorary fellow of the Colleges of Medicine of South Africa. He has served as chairman of the South African Medical Research Council, president of the South African Medical Association, member of the Health Professions Council of South Africa and National Health Ethics Research Committee, and board chairman of the Office of Health Standards Compliance. He has received seven honorary academic fellowships and authored 30 conference presentations, 16 journal articles, and a chapter in the *Textbook of Tropical Surgery*.

Charlotte McArdle, MSc, BSc, PGCert, RGN, is chief nursing officer, Department of Health, Northern Ireland. As head of the nursing and midwifery professions, she is responsible for the professional leadership, performance, and development of the professions in Northern Ireland, including allied health professions. Previously, Charlotte was executive director of nursing and primary care, South Eastern Health and Social Care Trust. She has a keen interest and track record in healthcare management, professional leadership, and the field of quality improvement. Charlotte played a key leadership role in the development of the Attributes Framework for Leadership in Quality Improvement and Safety. She is a visiting professor at Ulster University Faculty of Health Science.

Cathy McCusker, MBA, BSc, RN, is a senior professional officer for professional development at the Northern Ireland Practice and Education Council for Nursing and Midwifery. Cathy was the project manager for the development of the Northern Ireland Quality 2020 Attributes Framework and continues to support its implementation across health and social care. She also leads a wide range of projects supporting the development of nurses and midwives in Northern Ireland, including frameworks for specialist and advanced practitioners, preceptorship, competence assessment tools, Northern Ireland Practice and Education Council online portfolio, and Northern Ireland's Nursing and Midwifery Careers website.

Elise McPherson, BA, BSc (Hons), is a Master of Research candidate and research assistant at the Centre for Healthcare Resilience and Implementation Science, Australian Institute of Health Innovation, Macquarie University, Sydney, Australia. She has a background in English literature and neuroscience, with honours in auditory neuroscience. Her research interests include molecular mechanisms associated with noise-induced hearing loss, the implementation of genomic medicine into clinical practice, and health systems sustainability.

Walter Mendes, MD, MSc in health policy and planning from the State University of Rio de Janeiro, and PhD in public health from Oswaldo Cruz Foundation (Fiocruz). He is a faculty member of the graduate program in public health at Fiocruz and coordinator of the International Course on Quality in Health and Patient Safety. He is a consultant on health services management, a surveyor of health services in accreditation, and an author of books and articles on healthcare quality evaluation, patient safety, and home care. He is a representative of Fiocruz on the Implementation Committee of the National Patient Safety Program and Partner of Proqualis.

Ali Mohammad Mosadeghrad, PhD, is associate professor of health policy, management and economics at Tehran University of Medical Sciences, Iran. He was the general director of Accreditation of Healthcare Organizations, Ministry of Health. He is an author, speaker, and professional management consultant and trainer. Mosadeghrad has developed and reviewed hundreds of strategic plans for public and private healthcare organizations. He has facilitated many functional and cross-functional teams to improve the quality and safety of healthcare services. His research interests include global health, strategic management, quality management, and health reform organizational change. Mosadeghrad has written extensively on many aspects of healthcare organization and management, covering a full spectrum of subjects in strategy formulation, implementation, and evaluation.

Bafana Msibi is a qualified registered nurse with more than 16 years' experience implementing and managing healthcare projects, and leading compliance inspection work and standards development. He holds a BCur in health science education and health science management and master of public health. He is currently executive manager of compliance inspections for the Office of Health Standards Compliance, South Africa. He contributed to the development of the National Core Standards, reviewing the current assessment tools and norms and standards regulations. Previously, he was director of compliance inspection and deputy director of women's health, National Department of Health. In addition, Msibi was the district coordinator of maternal child and women's health, Free State Department of Health, Thabo Mofutsanyana District.

Julie Taleni Neidel is a registered nurse by profession who works as a quality coaching coordinator in the National Quality Management Programme, Namibia Ministry of Health and Social Services, supported by HEALTHQUAL International. Her work entails building capacity of healthcare workers and consumers in quality improvement by facilitating several requisite training workshops, providing coaching and mentoring to quality improvement teams. She is a fellow and member of the International Society for Quality in Health Care. Her abstract, titled "Consumer Involvement in the Quality of HIV Care, the Namibian Experience," was presented as an oral presentation at the 34th Annual ISQua Conference.

Wendy Nicklin, RN, BN, MSc(A), CHE, FACHE, FISQua, ICD.D, is president of the International Society for Quality in Health Care. She is adjunct professor at both Queen's University and the University of Ottawa, and a member of the advisory committees at Queen's University and the University of Toronto for graduate programs in safety and quality. Former president and CEO of Accreditation Canada and having held senior healthcare leadership and governance positions in Canada, with extensive knowledge and experience in quality and safety, she is currently a healthcare consultant. Wendy is strongly committed to contributing to the improvement of the quality of healthcare globally. In addition to the International Society for Quality in Health Care board, Wendy is a board member of a regional healthcare organization in her home city of Ottawa, Ontario, and a board member of the Health Insurance Reciprocal of Canada.

Richard Norris is director of the Scottish Health Council, which promotes patient focus and public involvement in the National Health Service. The council produces guidance, standards, and tool kits for participation, and provides practical support and quality assurance through its 14 local offices across Scotland. It also gathers public views on a range of topics and is a part of Healthcare Improvement Scotland. Previously, Norris was director of policy at the Scottish Association for Mental Health, and prior to this, in the 1990s, Norris was chief executive of the Centre for Scottish Public Policy, an independent center for policy development, based in Edinburgh.

John Øvretveit is director of research and professor of healthcare improvement implementation and evaluation at the Medical Management Centre, Karolinska Institute, Stockholm, and previously professor of health policy and management at Bergen University Medical School, Norway, and the Nordic School of Public Health, Gothenburg, Sweden. He is a board member for Joint Commission Resources/International and chair of the standards committee, a board member of the Global Implementation Initiative, and chair of the Global Implementation Society. Some of his 300 peer-reviewed scientific papers and books have been translated into nine languages. He was conferred the 2014 Avedis Donabedian international quality award for his work on quality economics.

José-Artur Paiva, MD, PhD, is currently the medical director and member of the administration board of Centro Hospitalar São João, Porto, Portugal. Before that, he was director of the Emergency and Intensive Care Management Unit of the same hospital for 9 years. He is a specialist in intensive care medicine and internal medicine and certified in the management of healthcare systems by the Portuguese Medical Council. He presided over the National Commission for the Reorganisation of

Emergency Care and Departments in Portugal in 2012. From 2013 to 2016, he was the director of the National Programme for Prevention and Control of Infection and Antimicrobial Resistance, Directorate-General of Health.

Ana Luiza Pavão, MD, PhD in epidemiology from the State University of Rio de Janeiro, is a researcher at Oswaldo Cruz Foundation and runs patient safety projects. She was a teacher-tutor in the International Course on Quality in Health and Patient Safety, mentoring patient safety plans for several hospitals' teams. She is author of articles on healthcare quality and patient safety, associate editor for the *Electronic Journal of Communication, Information and Innovation in Health*, partner of the Collaborative Center for Quality of Care and Patient Safety (Proqualis), and member of the International Health Literacy Association.

Lilisbeth Perestelo-Pérez is a leading investigator in the development, evaluation, and implementation of decision aids for shared decision-making in Spain with a PhD in clinical and health psychology. She completed postdoctoral work at the KER Unit (Mayo Clinic) and at Health Decision Sciences Center (Massachusetts General Hospital), focusing on SDM and health services research. She works as a clinical psychologist–health services researcher at the evaluation unit of the Canary Islands Health Service and is collaborating with other health technology assessment agencies (HTA) in Spain and Europe. She is a member of the Health Services Research on Chronic Patients Network and cofounder of the Latin American Network for shared decision-making.

Nataša Perić is researcher and PhD candidate, Department of Health Economics, Centre of Public Health, Medical University of Vienna, Austria. Nataša holds a master's degree in international business administration from the University of Vienna, where she specialized in healthcare and international management. Currently, she is working on the European project BRIDGE Health and focuses on the prioritization of indicators for structured monitoring of health system performance in the European Union. Her research interests include evaluation of healthcare systems, health system performance assessment, and social return on investment analysis in public health.

Holger Pfaff is director of the Institute for Medical Sociology, Health Services Research, and Rehabilitation Science, and director of the Centre for Health Services Research Cologne—joint institutions of the Faculty of Medicine and Faculty of Human Sciences, University of Cologne. Pfaff studied social and administrative sciences at the Universities of Erlangen-Nürnberg, Konstanz, and Michigan (Ann Arbor). He is now professor for quality improvement and evaluation in rehabilitation, University of Cologne. Since 2012, he is an elected member of the Review Board for Medicine of the German Research

Foundation. Appointed by the German Federal Ministry of Health in 2016, he is chairman of the Board of Experts of the German Innovation Funds.

Kaja Põlluste, MD, PhD, is senior researcher of internal medicine at the University of Tartu. She has been lecturer in healthcare organization and quality management, and expert and advisor of healthcare quality at the Ministry of Social Affairs of Estonia. Her research is related to the quality of care and the quality of life of people with chronic conditions. She is author and coauthor of more than 100 scientific papers and has published 25 articles in international peer-reviewed scientific journals. She is also a coauthor of *Quality Policy of Estonian Health Care.*

Margareth Crisóstomo Portela, BSEE from the Federal University of Bahia, MSc in biomedical engineering from the Federal University of Rio de Janeiro, and PhD in health policy and administration from the University of North Carolina at Chapel Hill. She is a senior researcher and a faculty member of the Graduate Program in Public Health at Oswaldo Cruz Foundation. She has run healthcare quality improvement research projects, is the general coordinator of the Collaborative Center for Quality of Care and Patient Safety (Proqualis), and is author of articles on health technology assessment, health services research, and healthcare quality improvement.

Martin Powell, BA, PhD, is professor of health and social policy at the Health Services Management Centre, University of Birmingham, United Kingdom. His main research interests are in policy evaluation of the British National Health Service, and he is the author of some 80 articles and 10 books on British social policy. He co-wrote the chapter on England in *Healthcare Reform, Quality and Safety: Perspectives, Participants, Partnerships and Prospects in 30 Countries* (Ashgate Publishing Ltd., 2015). His most recent book is *Dismantling the National Health Service* (edited with Mark Exworthy and Russell Mannion, Policy Press, 2016).

Shivani Ranchod is a healthcare actuary at a consultancy and a part-time academic at the University of Cape Town. She has wide-ranging technical, strategic, and policy-related experience across the South African healthcare system. Over the last 16 years, she has worked and consulted for, among others, government, regulators, funders, managed care organizations, and healthcare providers. She is passionate about strengthening the South African healthcare system, and at present her energy is focused on work in the public sector. As part of this role, Shivani has been appointed to the Government Technical Advisory Centre.

Claudine Richardson-Sheppard is a registered nurse and full-time lecturer at the School of Nursing, Faculty of Medical Sciences, University of the West Indies. She teaches quality management for healthcare, organizational behavior, and leadership and management. She is a fellow of the International Society for Quality in Health Care. She holds a BA in business management and a master's in health administration from the University of Trinidad and Tobago, where she is also currently a PhD student with an interest in quality management systems. Previously, she has held positions as chief executive officer, North-West Regional Health Authority (2009–2011); hospital administrator, Port of Spain General Hospital (2008–2009); and general manager of quality, North-West Regional Health Authority (2006–2008).

Ånen Ringard, BA, MSc (political science), PhD (health policy), is senior consultant at Rud Pedersen Public Affairs Company in Norway. Before joining Rud Pedersen, he was a senior adviser at the Association of the Pharmaceutical Industry in Norway, and a scientist at the Health Services Research Unit, Akershus University Hospital. He has been employed by the Ministry of Health and Care Services, the Directorate of Health, and the Norwegian Knowledge Centre for the Health Services, where he was lead author of the Norwegian Health System in Transition report (commissioned by the World Health Organization/European Observatory of Health Systems and Policies). His research focuses on patient empowerment, hospital choice, priority setting, quality and safety, and health system analysis.

Viviana Rodríguez, MD, PhD, graduated from the Medical University of La Plata and is a specialist in internal medicine and infectious diseases. She has an MA in clinical effectiveness from the University of Buenos Aires. She is chief of the department of infectious diseases at the German Hospital of Buenos Aires, where she is coordinating the Patient Safety Committee. Currently, she is the coordinator of the Department of Health Care Quality and Patient Safety, Institute for Clinical Effectiveness and Health Policy. She has been fellow of the International Society for Quality in Health Care, and she is a certified professional in patient safety by the National Patient Safety Foundation (United States).

Jorge J. Rodríguez Sánchez, MD (1972), holds a specialty in psychiatry (1978), a master's degree in the organization of public health and social medicine (1982), and a doctorate (PhD) in health sciences (1990). Dr. Rodríguez has had extensive experience in the academic field and in international technical cooperation. From 2006 to 2014, he was unit chief in the Mental Health and Substance Use Unit at the Pan American Health Organization/World Health Organization in Washington, DC. He is currently a member of the Pan American Health Organization Group of Experts on Mental Health, senior advisor on mental health of the Group of Health International Advisors (United States), and visiting professor at the School of Medicine, Loma Linda University, California.

Antoni Salvà Casanovas earned his PhD in medicine specializing in geriatrics and gerontology from the Universitat Autònoma de Barcelona, is general director of Fundació Salut i Envelliment UAB (Health and Aging Foundation UAB), and is president of the advisor committee of the Pla Director Sociosanitari (Social and Health Care Steering Plan), Department of Health. His research areas include the study of different syndromes and geriatric problems, such as malnutrition, falls, or cognitive impairment; the organization of services for older people, particularly social services, and policies and models of care for the elderly and people with cognitive impairment; and the promotion of active and healthy aging. He has participated in health promotion projects and the design of programs to prevent disability.

Tomás J. Sanabria, MD, is consultant clinical and interventional cardiologist and director of the Cardiac Cathertization Lab, Centro Médico de Caracas, and founder and director of the MANIAPURE Foundation and Telesalud International. He has been visiting professor at the University of Massachusetts, University Paul Sabathier, and Miller School of Medicine. Sanabria is a fellow of the American College of Physicians and the American College of Cardiology, and a member of the French Cardiological Society, with awards from the American College of Physicians (2001), Venezuela Sin Limites (2011), and Schwab Foundation (2012). He is a lecturer at the Wharton and Harvard Kennedy School, World Economic Forum, and its social branch Schwab Foundation meetings (2013–2018). He is also author of more than 60 articles, presentations, and book chapters on cardiology, telemedicine, and social entrepreneurship.

Karin Schnarr is an assistant professor of policy at the Lazaridis School of Business and Economics at Wilfrid Laurier University in Waterloo, Canada. She holds a PhD in strategic management and an MBA from the Ivey School of Business at Western University, an MA (English literature) from Queen's University, and a BA (Hon English literature) and a BSc from the University of Waterloo. Prior to academia, Dr. Schnarr served in senior roles in the government of Ontario and the private sector. Her research focuses on the intersection of healthcare and management, as well as corporate governance.

Nagah Abdelaziz Selim, MBBCh, MSc, MD, is consultant physician of public health and preventive medicine; associate program director of community medicine residency training, Primary Healthcare Corporation; and associate professor of public health and preventive medicine, Cairo University. She is chair of the Institutional Review Board and Research Committee at Primary Health Care Corporation, Qatar, and a member on numerous boards, including the Arab Board Exam Committee and Graduate Medical Education Committee at Hamad Medical Corporation, the Egyptian Medical Syndicate, and the Curriculum Development Team at the Faculty of Medicine, Cairo University. Dr. Nagah's responsibilities include curriculum development and evaluation and thesis mentorship. She has published numerous works in peer-reviewed journals, particularly in the area of public health.

Syed Shahabuddin, MBBS, FCPS (GS), FCPS (CDS), FACS (US), is a consultant cardiothoracic surgeon and assistant professor at the Aga Khan University Hospital, and is involved in the teaching of medical students, nurses, and residents and the training of residents as a residency program coordinator. He is a fellow of the College of Physicians and Surgeons Pakistan and a fellow of the American College of Surgeons. He has interests in outcome research, databases, and quality. He is a member of the Quality Improvement Committee and is involved in monitoring clinical quality indicators to comply with standards of care. He has produced a number of publications and abstract presentations in national and international meetings.

Paulinus Lingani Ncube Sikosana, MPH, MBA, CIRM, FRSPH, FAAN, a physician, former permanent secretary for health in Zimbabwe, and recently coordinator and team leader for health systems and social determinants for health in the World Health Organization country office in Papua New Guinea until his retirement in September 2017. He has more than 30 years' experience in public health, health systems development, and health sector reforms. During his career, he has worked in various capacities in Botswana, Ethiopia, Kenya, Lesotho, Malawi, Mozambique, South Africa, and Zambia. He is the author of the book *Challenges in Reforming the Health Sector in Africa: Reforming Health Systems under Economic Siege—The Zimbabwean Experience* (2010).

Anne W. Snowdon, BScN, MSc, PhD, FAAN, is professor of Strategy and Entrepreneurship, academic chair of the World Health Innovation Network, and scientific director/CEO of SCAN Health, a global network of partners from industry, academia, health systems, and government focused on advancing supply chain infrastructure in health systems to strengthen quality, safety, and system performance. Dr. Snowdon leads over 15 innovation research initiatives across Canada, collaborating with government, health professionals, private industry, foundations, and families. She is a member of the Ontario Health Innovation Council, is vice chair of the Board of the Directors for Alberta Innovates, a board member of the Ontario Centres of Excellence, and is an expert advisor to the Canadian Space Agency, Space Health and Innovation.

Paulo Sousa, MPH, PhD, is currently professor at the National School of Public Health, Universidade Nova de Lisboa, where he coordinates the master's in public health and the International Course in Quality Improvement and Patient Safety (a joint program with the National School of Public Health in Brazil). In recent years, he has been involved in, and coordinated, several initiatives and research projects in the areas of quality improvement, patient safety, and health outcomes evaluation. He is the author of many chapters in books and articles published in peer-reviewed national and international journals. He is a member of the editorial board of the *Portuguese Journal of Cardiology* and of the *International Journal for Quality in Health Care.*

Anthony Staines, PhD, currently runs patient safety improvement projects for the Hospital Federation Vaud (Switzerland) and advises a number of hospitals on patient safety strategies. He holds an MBA from the European Institute of Business Management (France) and an MPA from the Swiss Graduate School of Public Administration (IDHEAP) (Switzerland), as well as a PhD in management from the Institute for Education and Research in Healthcare and Social Service Organizations at the University of Lyon (France), where he lectures on quality improvement and patient safety. He is coauthor, with John Øvretveit, of a book on improving value in healthcare and serves as deputy editor for the *International Journal for Quality in Health Care.*

David R. Steel, OBE, MA, DPhil, FRCP Edin, is honorary senior research fellow at the University of Aberdeen and a senior associate of the Nuffield Trust. He worked for 25 years in National Health Service management and was chief executive of NHS Quality Improvement Scotland from its creation in 2003 until 2009. He is author of the Scottish health systems review published in 2012 as part of the European Observatory's *Health Systems in Transition* series. In 2008, he was awarded an Order of the British Empire for services to healthcare.

Jacqui Stewart, chief executive of the Council for Health Service Accreditation of Southern Africa (COHSASA), has worked in the public and private healthcare sectors in South Africa and England. Having qualified as a registered nurse in Cape Town she went on to specialize in cardiothoracic nursing in England. She then held various managerial roles in the NHS, including quality improvement and service planning. Stewart joined COHSASA in 2005 as operations manager and was appointed CEO in April 2016. She holds a masters in professional studies (health) from Middlesex University, London. Stewart serves on the Accreditation Council and Board of the International Society for Quality in Health Care (ISQua) and is an ISQua Expert.

Mark Swaim, MD, PhD, is a hepatologist, gastroenterologist, internist, and biotechnology consultant. He earned both advanced degrees from the National Institutes of Health Medical Scientist Training Program at Duke University with honors, where he was also elected to Alpha Omega Alpha. He has held faculty positions at Duke University Medical Center and University of Texas MD Anderson Cancer Center, in addition to running a private clinical trials institute. He was elected to fellowship in the American College of Physicians. He has known coauthor Vasiliy V. Vlassov, whom he regards as a mentor, for more than 20 years.

Ganesh Tatkan is an economist attached to the Planning and Project Implementation Unit of the Ministry of Public Health and is currently pursuing postgraduate studies in Health Economics and Pharmacoeconomics. He was previously employed at the Department of Public Service of the Ministry of the Presidency as a student's affairs officer (2007–2014) and worked with the private sector firm IMEX International as an auditor (2006–2007). His interests are in health economics and farming. He is also a member of the advisory board on the elimination of microfilaria in Guyana.

Andrew Thompson, BSc, PhD, is a social scientist. He is professor of public policy and citizenship in the Department of Politics and International Relations at the University of Edinburgh. His main research interests are in two distinct areas: (1) citizenship and public policy (especially health services), in relation to quality improvement, and participatory and deliberative democracy, and (2) European public administration. He is a member of the Participatory and Deliberative Democracy Group of the UK Political Studies Association and the European Consortium for Political Research. He was a regional editor for the *International Journal for Quality in Health Care*.

Daniel Tietz is manager for consumer affairs at the New York State Department of Health AIDS Institute. Tietz has been working for the AIDS Institute since 1991. He is a valued member of the senior management team responsible for formulating policy, establishing goals and priorities, and furnishing guidance and technical assistance related to consumer involvement in quality management. Tietz holds a bachelor's degree from the SUNY Empire State College in Business, Management and Economics.

Tsolmongerel Tsilaajav is a health economist and former director of policy planning, Ministry of Health, Mongolia. Tsilaajav holds a master's degree from the University of Manchester, United Kingdom. For the last 15 years, she has been extensively engaged in the reforms of healthcare financing and social health insurance in Mongolia. She led the process to develop long-term policy for social health insurance in Mongolia and introduced a consensus-building approach among different stakeholders to resolve difficult reform agendas. Tsilaajav undertook several policy studies, including the distribution of catastrophic and impoverishing health payments and hospital services costing in Mongolia, Philippines, and Lao People's Democratic Republic. She currently works as a freelance consultant in healthcare financing and health insurance systems.

Mary E. Vaiana, PhD, is a senior communications analyst at the RAND Corporation. She helps to shape communications strategies for RAND Health, develops dissemination plans for key research projects, and creates materials to support funding and outreach efforts. She works across all RAND Health's research areas, helping to design communication products that are accessible to a wide range of stakeholders, including public decision makers, healthcare providers, and the general public. She is also skilled at developing derivative materials, such as policy briefs and fact sheets that highlight key findings in non-technical language appropriate for multiple audiences.

Enrique Valdespino is a medical doctor interested in healthcare systems, medical sciences, and global health who has worked as a primary healthcare physician delivering care for vulnerable populations. Dr. Valdespino has experience serving and collaborating directly with the Ministry of Health as a clinician and researcher. He was a regional supervisor for Partners in Health supporting and instructing physicians in charge of community medical units. More recently, Dr. Valdespino pursued a master of medical science at Harvard Medical School, where he conducted research in health system strengthening and the professional workforce through medical education to deliver high-value healthcare.

John Van Aerde, MD, PhD, is a neonatologist who has fulfilled several leadership roles, including program integration in several Canadian provinces. He holds appointments at the Universities of Alberta and British Columbia, at Royal Roads University, and at the Physician Leadership Institute, helping physicians learn those skills not covered during clinical training. He is the immediate past president of the Canadian Society of Physician Leaders and the editor in chief of the *Canadian Journal of Physician Leadership*. From forest regeneration and from living in a self-sustainable house, he discovers models and applications for the Canadian healthcare system as a complex adaptive system.

Neeltje van den Berg is a health services researcher and geographer. Since 2005, she has been working at the Institute for Community Medicine, Department of Epidemiology of Health Care and Community Health at University Medicine Greifswald, Germany. Since 2011, she has been the deputy head of the department. She is the coordinator of the research area Innovative Care Concepts and Regional Health Care and heads the Integrated Telemedicine Centre of the University Medicine Greifswald. Her research focuses on regional healthcare; geographical analyses; the development, implementation, and evaluation of innovative care concepts; and telemedical functionalities.

Vasiliy V. Vlassov, MD, DMedSc, is professor of public health at the National Research University Higher School of Economics, Moscow, and president of the Society for Evidence-Based Medicine (Russia). He serves as a member of the World Health Organization Euro Advisory Committee on Health Research, and as an expert to the Russian Academy of Sciences. He is the author of a number of books, including *Effectiveness of Diagnostic*

Tests, Reaction of the Organism to External Stimuli, Health Care under Deficit of Resources, and *Textbook of Epidemiology* (all in Russian), as well as more than 200 articles on epidemiology, public health, and evidence-based medicine.

Cordula Wagner, PhD, physiotherapist, sociologist, is executive director of the Netherlands Institute of Health Services Research in Utrecht and works as a professor of patient safety at VU Medical Center in Amsterdam. She is also head of the patient safety research center, Safety 4 Patients, a collaboration of EMGO+, the VU Medical Center, and the Netherlands Institute of Health Services Research. For the last 20 years, Professor Wagner has been involved in many projects, including European Union research focusing on the implementation of quality systems and the evaluation of national quality programs, for example, improvement activities such as guidelines; team training and breakthrough projects; the relation between quality systems, care process, and clinical outcomes; and risk management and patient safety.

Eliza Lai-Yi Wong, RN, MPH, PhD, is professor at the JC School of Public Health and Primary Care, the Chinese University of Hong Kong. Her research areas include the patient experience of using healthcare, patient-reported outcome measures (EQ-5D), the acceptability of human papilloma-virus self-sampling, and service delivery from a health system approach. She is the codirector of the diploma of health services management and master of science in gerontology. She is an accredited mediator and is running applied mediation skill workshops for healthcare workers of hospital authority to enhance their communication with patients.

Nasser Yassin is director of research at the Issam Fares Institute for Public Policy and International Affairs and professor of policy and planning, Health Management and Policy Department, American University of Beirut (AUB), Lebanon. He holds a PhD from University College London, an MSc from the London School of Economics, and an MSc and BSc from the AUB. He co-chairs the AUB4Refugees Initiative aiming to bring together and build synergy among faculty and departments in AUB responding to the Syrian refugee crisis and is currently leading a research project on understanding the informal adaptive mechanisms among refugees and their host communities in the Middle East. He is author of more than 30 internationally published articles and reports.

Eng-Kiong Yeoh, FRCP (Edin), FHKCP, FRCP (Lond), FRCP (Glasg), FRACP, FHKAM, FHKCCM, FFPH (UK), FRACMA, is head of the Division of Health System, Policy and Management and director at the JC School of Public Health and Primary Care, the Chinese University of Hong Kong. He was the first chief executive of the Hong Kong Hospital Authority, responsible for the management and transformation of the public hospital system. Prior to joining Chinese University of Hong Kong, he was secretary for health, welfare and food, government of the Hong Kong Special Administrative Region. His research is in health systems, services, and policy and in applying systems thinking in studying how complex components of health systems interact and interrelate to improve health.

Hao Zheng, PhD, MD, MBA, is currently leading the Clinical Indicator Program at the Australian Council on Healthcare Standards. She has been working in the area of safety and quality improvement in healthcare since her experience at the Patient Safety Programme (Service Delivery and Safety), World Health Organization headquarters in Geneva, leading and coordinating international quality and safety projects. She has also been one of the initiators and core members of the Patient Safety Collaborative of China—the first nationwide patient safety initiative, launched in 2014—and guided the development and updates of National Patient Safety Goals and Standards.

Contributors

Tine Aagaard
Institute of Nursing and Health
  Science, Ilisimatusarfik/
  University of Greenland
Nuuk, Greenland

William Adu-Krow
Pan American Health
  Organization/World Health
  Organization (PAHO/WHO)
Georgetown, Guyana

Bruce Agins
HEALTHQUAL International;
  University of California San
  Francisco, Division of Global
  Health Sciences
San Francisco, California

Mercedes Aguerrebere
Department of Global Health
  and Social Medicine, Harvard
  Medical School
Boston, Massachusetts

Emmanuel Aiyenigba
Institute for Healthcare
  Improvement (IHI)
Ilorin, Nigeria

Sara S. H. Al-Adawi
Sultan Qaboos University
Muscat, Oman

Samir Al-Adawi
College of Medicine & Health
  Sciences, Sultan Qaboos University
Muscat, Oman

Reem Al-Ajlouni
Independent
Amman, Jordan

Ahmed Al-Mandhari
Family Medicine and Public Health
  Department, Sultan Qaboos
  University Hospital
Muscat, Oman

Yousuf Al Maslamani
Hamad Medical Corporation
Doha, Qatar

Samia Al Rabhi
Directorate General of Quality
  Assurance Center, Ministry
  of Health
Muscat, Oman

Khaled Al-Surimi
King Saud ben Abdualizi
  University for Health Sciences
Riyadh, Saudi Arabia
Primary Care & Public Health,
  School of Public Health,
  Faculty of Medicine,
  Imperial College London
London, United Kingdom

Charles Alessi
Odette School of Business,
  University of Windsor
Windsor, Ontario, Canada

Noora Alkaabi
Hamad Medical Corporation
Doha, Qatar

Huda Alsiyabi
Department of Community
  Based Initiatives,
  Ministry of Health
Muscat, Oman

René Amalberti
Haute Autorité de Santé (HAS)
Saint-Denis, France

Hugo Arce
University Institute of Health
  Sciences, Barceló Foundation
Buenos Aires, Argentina

Anne-Marie Armanteras-de-Saxcé
Haute Autorité de Santé (HAS)
Saint-Denis, France

Jafet Arrieta
Institute for Healthcare
  Improvement, Harvard T.H.
  Chan School of Public Health
Boston, Massachusetts

Oscar Arteaga
School of Public Health, University
  of Chile
Santiago, Chile

Fakhri Athari
Simpson Centre–Ingham Institute
  Health Research, University of
  New South Wales
Sydney, Australia

Luis Azpurua
Grupo Medico Santa Paula
Caracas, Venezuela

Natasha Azzopardi-Muscat
Department of Health Services
  Management, Faculty of Health
  Sciences, University of Malta
Msida, Malta

Vasha Elizabeth Bachan
Ministry of Public Health
Georgetown, Guyana

Roland Bal
Erasmus School of Health
  Policy and Management,
  Erasmus University
Rotterdam, The Netherlands

Marta Ballester Santiago
Avedis Donabedian Research
  Institute, Universitat Autònoma
  de Barcelona
Red de Investigación en Servicios de
  Salud en Enfermedades Crónicas
  (REDISSEC)
Barcelona, Spain

Joshua Bardfield
HEALTHQUAL International;
  University of California San
  Francisco, Division of Global
  Health Sciences
San Francisco, California

Maysa Baroud
Issam Fares Institute for Public
  Policy and International
  Affairs, American University
  of Beirut
Beirut, Lebanon

Apollo Basenero
Ministry of Health and Social
  Services Namibia
Windhoek, Namibia

Roger Bayingana
MedVet Health Solutions
Kigali, Rwanda

Mustafa Berktaş
Institute of Quality and
  Accreditation in Healthcare of
  Turkey
Ankara, Turkey

Angelika Beyer
University Medicine Greifswald
Greifswald, Germany

Camilla Björk
Department of Healthcare
  Development, Public Health Care
  Services Committee, Stockholm
  County Council
Stockholm, Sweden

Sergi Blancafort Alias
Fundació Salut i Envelliment
Barcelona, Spain

Jeffrey Braithwaite
Australian Institute of Health
  Innovation, Macquarie
  University
Sydney, Australia

Robert H. Brook
The RAND Corporation
Santa Monica, California

Margaret K. Brown
New York Department of Health
  AIDS Institute
New York, New York

Sandra C. Buttigieg
Department of Health Services
  Management, Faculty of Health
  Sciences, University of Malta
Msida, Malta

Tumurbat Byamba
Mongolian Association of
  Healthcare Quality
Ulaanbaatar, Mongolia

Andrew Carson-Stevens
Cardiff University
Cardiff, Wales

Edward Chappy
Human Resources for Health 2030
  (HRH2030)
Amman, Jordan

Patsy Yuen-Kwan Chau
JC School of Public Health and
  Primary Care, The Chinese
  University of Hong Kong
Hong Kong, People's Republic
  of China

Americo Cicchetti
Università Cattolica del Sacro Cuore,
  ALTEMS
Rome, Italy

Silvia Coretti
Università Cattolica del Sacro Cuore,
  ALTEMS
Rome, Italy

Jacqueline Cumming
Victoria University of Wellington
Wellington, New Zealand

Adriana Degiorgi
Ente Ospedaliero Cantonale
Bellinzona, Switzerland

Ellen Tveter Deilkås
Health Services Research Centre,
  Akershus University Hospital
Lørenskog, Norway

Pedro Delgado
Institute for Healthcare
  Improvement
Cambridge, Massachusetts

Subashnie Devkaran
Cleveland Clinic Abu Dhabi
Abu Dhabi, The United Arab Emirates

Sir Liam Donaldson
London School of Hygiene and
  Tropical Medicine
London, United Kingdom

Liv Dørflinger
Danish Cancer Society
Copenhagen, Denmark

Thomas E. Dorner
Institute of Social and Preventive
  Medicine, Centre for Public
  Health
Vienna, Austria

Persephone Doupi
National Institute for Health &
  Welfare – THL
Helsinki, Finland

Paul Edwards
Pan American Health
  Organization/World Health
  Organization (PAHO/WHO)
Georgetown, Guyana

Carsten Engel
IKAS – Danish Institute for Quality
  and Accreditation in Healthcare
Aarhus, Denmark

Jesper Eriksen
Danish Cancer Society
Copenhagen, Denmark

Andrew Evans
Welsh Government
Cardiff, Wales

Paulo André Fernandes
Direção-Geral da Saúde
Lisbon, Portugal

Laura Fernández-Maldonado
Fundació Salut i Envelliment – UAB
Barcelona, Spain

Steven Frost
Intensive Care, Liverpool Hospital
Western Sydney University
Sydney, Australia

Hong Fung
JC School of Public Health and
  Primary Care, The Chinese
  University of Hong Kong
Hong Kong, People's Republic
  of China

Ezequiel García-Elorrio
Institute for Clinical Effectiveness
  and Health Policy
Buenos Aires, Argentina

Christine S. Gordon
Division of Quality Assurance,
  Ministry of Health and Social
  Services (MoHSS)
Windhoek, Namibia

Victor Grabois
Fundação Oswaldo Cruz
Rio de Janeiro, Brazil

Kenneth Grech
Department of Health Services
  Management, Faculty of
  Health Sciences, University
  of Malta
Msida, Malta

Catherine Grenier
Haute Autorité de Santé (HAS)
Saint-Denis, France

Girdhar Gyani
Association of Healthcare Providers
  (India)
New Delhi, India

Ndapewa Hamunime
Ministry of Health and Social
  Services Namibia
Windhoek, Namibia

Jamie Hayes
Welsh Medicines Resource Centre
Penarth, Wales

Ken Hillman
Intensive Care Unit, Liverpool
  Hospital
Sydney, Australia

Wolfgang Hoffmann
University Medicine Greifswald
Greifswald, Germany

Maria M. Hofmarcher-Holzhacker
HS&I HealthSystemsIntelligence
Medical University of Vienna
Vienna, Austria

Lise Hounsgaard
Institute of Nursing and Health
  Science, Ilisimatusarfik/
  University of Greenland
Nuuk, Greenland

Min-Huei (Marc) Hsu
Department of Information
  Management, Ministry of Health
  and Welfare
Taipei, Taiwan

Clifford F. Hughes
International Society for Quality in
  Health Care (ISQua)
Dublin, Ireland
Australian Institute of Health
  Innovation, Macquarie University
Sydney, Australia

Valentina Iacopino
Università Cattolica del Sacro Cuore,
  ALTEMS
Rome, Italy

Usman Iqbal
College of Public Health, Global
  Health and Development
  Department, Taipei Medical
  University
Taipei, Taiwan (ROC)
Health Informatics Department,
  COMSATS Institute of
  Information Technology
Islamabad, Pakistan

Wendy James
Australian Institute of Health
  Innovation, Macquarie University
Sydney, Australia

Ravichandran Jeganathan
Ministry of Health Malaysia
Johor Bahru, Malaysia

Ravindran Jegasothy
MAHSA University
Kuala Lumpur, Malaysia

Ruth Kalda
University of Tartu, Institute of
  Family Medicine and Public
  Health
Tartu, Estonia

İbrahim II. Kayral
Institute of Quality and Accreditation
  in Healthcare of Turkey
Ankara, Turkey

Rawya Khodor
Issam Fares Institute for Public
  Policy and International Affairs,
  American University of Beirut
Beirut, Lebanon

Janne Lehmann Knudsen
Danish Medicines Agency
Copenhagen, Denmark

Madelon Kroneman
NIVEL
Utrecht, The Netherlands

Grace Labadarios
Office of Health Standards
  Compliance
Pretoria, South Africa

Gavin Lavery
HSC Safety Forum, Public Health
  Agency
Lisburn, Northern Ireland

Wui-Chiang Lee
Taipei Veterans General Hospital
and National Yang-Ming
University
Taipei, Taiwan

Margus Lember
University of Tartu, Institute of
Clinical Medicine
Tartu, Estonia

Yu-Chuan (Jack) Li
College of Medical Science and
Technology, Taipei Medical
University
Taipei, Taiwan

Kristiana Ludlow
Australian Institute of Health
Innovation, Macquarie University
Sydney, Australia

Russell Mannion
University of Birmingham
Birmingham, England

Marcella Marletta
Italian Ministry of Health
Rome, Italy

Yukihiro Matsuyama
The Canon Institute of Global
Studies
Tokyo, Japan

Susanne Mayer
Center for Public Health,
Department of Health
Economics, Medical University
of Vienna
Vienna, Austria

Laetitia May-Michelangeli
Haute Autorité de Santé (HAS)
Saint-Denis, France

Lizo Mazwai
Walter Sisulu University
Mthatha Eastern Cape Province,
South Africa

Charlotte McArdle
Department of Health
Belfast, Northern Ireland

Cathy McCusker
Northern Ireland Practice and
Education Council for Nursing
and Midwifery (NIPEC)
Belfast, Northern Ireland

Elise McPherson
Australian Institute of Health
Innovation, Macquarie University
Sydney, Australia

Walter Mendes
Fundação Oswaldo Cruz
Rio de Janeiro, Brazil

Ali Mohammad Mosadeghrad
Tehran University of Medical Sciences
Tehran, Iran

Bafana Msibi
Office of Health Standards
Compliance
Pretoria, South Africa

Julie Taleni Neidel
Ministry of Health and Social
Services Namibia
Windhoek, Namibia

Wendy Nicklin
International Society for Quality in
Health Care (ISQua)
Dublin, Ireland

Richard Norris
Scottish Health Council, Healthcare
Improvement Scotland
Glasgow, Scotland

John Øvretveit
Karolinska Institutet
Stockholm, Sweden

José-Artur Paiva
Department of Emergency and
    Intensive Care Medicine,
    Centro Hospitalar São João,
    Faculty of Medicine, University
    of Porto
Porto, Portugal

Ana Luiza Pavão
Oswaldo Cruz Foundation (Fiocruz)
Rio de Janeiro, Brazil

Lilisbeth Perestelo-Pérez
Evaluation Unit. Canary Islands
    Health Service
Red de Investigación en Servicios de
    Salud en Enfermedades Crónicas
    (REDISSEC)
Santa Cruz de Tenerife, Spain

Nataša Perić
Department of Health Economics,
    Centre of Public Health, Medical
    University of Vienna
Vienna, Austria

Holger Pfaff
University of Cologne
Cologne, Germany

Kaja Põlluste
University of Tartu, Institute of
    Clinical Medicine
Tartu, Estonia

Margareth Crisóstomo Portela
Fundação Oswaldo Cruz
Rio de Janeiro, Brazil

Martin Powell
University of Birmingham
Birmingham, England

Shivani Ranchod
University of Cape Town
Cape Town, South Africa

Claudine Richardson-Sheppard
University of the West Indies
St. Augustine, Trinidad and Tobago

Ånen Ringard
Rud Pedersen Public Affairs Company
Oslo, Norway

Viviana Rodríguez
IECS
Buenos Aires, Argentina

Jorge J. Rodríguez Sánchez
PAHO External Consultant for
    Regional Office
Washington, DC

Antoni Salvà Casanovas
Fundació Salut i Envelliment—UAB
    (Universitat Autònoma de
    Barcelona)
Barcelona, Spain

Tomás J. Sanabria
Fundacion Proyecto Maniapure
Caracas, Venezuela

Karin Schnarr
Lazaridis School of Business &
    Economics, Wilfrid Laurier
    University
Waterloo, Ontario, Canada

Nagah Abdelaziz Selim
Cairo University
Giza, Egypt

Syed Shahabuddin
Aga Khan University Hospital
Karachi, Pakistan

Paul G. Shekelle
VA Greater Los Angeles Healthcare
    System
Los Angeles, California

Paulinus Lingani Ncube Sikosana
World Health Organization, PNG
    Country Office
Port Moresby, Papua New Guinea

Anne W. Snowdon
World Health Innovation Network,
    Odette School of Business,
    University of Windsor
Windsor, Ontario, Canada

Paulo Sousa
National School of Public Health,
    Universidade Nova de Lisboa
Lisbon, Portugal

Anthony Staines
IFROSS, Lyon France; and FHV,
    Switzerland
Prilly, Switzerland

David R. Steel
University of Aberdeen
Edinburgh, Scotland

Jacqui Stewart
The Council for Health Service
    Accreditation of Southern
    Africa NPC
Cape Town, South Africa

Mark Swaim
N.C. Hepatobiliary Institute
Durham, North Carolina

Ganesh Tatkan
Ministry of Public Health
Georgetown, Guyana

Andrew Thompson
University of Edinburgh
Edinburgh, Scotland

Daniel Tietz
New York Department of Health
    AIDS Institute
New York, New York

Tsolmongerel Tsilaajav
Mongolian Association of
    Healthcare Quality
Ulaanbaatar, Mongolia

Mary E. Vaiana
The RAND Corporation
Santa Monica, California

Enrique Valdespino
Harvard Medical School
Cambridge, Massachusetts

John Van Aerde
Canadian Society of Physician Leaders
Ottawa, Ontario, Canada

Neeltje van den Berg
University Medicine Greifswald
Greifswald, Germany

Vasiliy V. Vlassov
National Research University
    Higher School of Economics
Moscow, Russia

Cordula Wagner
NIVEL and VU University Medical
    Center
Utrecht, The Netherlands

Stuart Whittaker
School of Public Health and
    Medicine, Faculty of Health
    Sciences, University of Cape Town
Cape Town, South Africa

Eliza Lai-Yi Wong
JC School of Public Health and
   Primary Care, The Chinese
   University of Hong Kong
Shatin, Hong Kong

Nasser Yassin
Issam Fares Institute for Public
   Policy and International Affairs
   and the Health Management and
   Policy Department, American
   University of Beirut
Beirut, Lebanon

Eng-Kiong Yeoh
JC School of Public Health and
   Primary Care, The Chinese
   University of Hong Kong
Shatin, Hong Kong

Hao Zheng
Australian Council on Healthcare
   Standards (ACHS)
Sydney, Australia

Introduction

Jeffrey Braithwaite, Russell Mannion, Yukihiro
Matsuyama, Paul G. Shekelle, Stuart Whittaker, Samir
Al-Adawi, Kristiana Ludlow, and Wendy James

CONTENTS

Background

This is the third book in a series on international health systems reform. The first book, *Healthcare Reform, Quality and Safety: Perspectives, Participants, Partnerships and Prospects in 30 Countries*, documented the reform efforts of multiple countries at national levels. It examined a question pertinent to all health systems wanting to improve the care they provide: What is the relationship between quality and safety initiatives at the meso- and micro-levels of systems, and the macro-level reforms that key stakeholders are striving to achieve in the long-term, aiming at change and improvement across the whole of their system? The cases examined in this book revealed that strategic initiatives are frequently not related. National-level reforms often go on, disconnected from the middle- and lower-level changes envisaged under the quality and safety umbrella. The take-home message was that national initiatives require intersectoral effort, and that evaluation of reform activities and improvement measures is all too often neglected. Improvement, in summary, is too often assumed and not sufficiently measured or evaluated (Braithwaite et al., 2015, 2016).

The second compendium, *Health Systems Improvement Across the Globe: Success Stories from 60 Countries*, extended the range of health systems involved, doubled the number of countries, and shifted the focus. It asked a deceptively simple question of each team of authors: where have you been successful in implementing change in your health system? Authors were invited to narrate, among the potential myriad successful exemplars, a story of improvement that might benefit and inspire change efforts in sectors and countries seeking to make similar changes. The book concluded that there were four recurring success factors that permeated the cases. The first we called the acorn-to-oak-tree principle, whereby a small-scale or pilot initiative, once demonstrated, could, depending on the circumstances, be rolled out as a system-wide reform. The second factor, as information technology (IT) systems and data generation become more prevalent in health systems, was that the transformation of data into information and then intelligence is crucial for making evidence-based decisions. The third factor related to the notion of success requiring many hands: successful change initiatives are never the product of individual heroes or even one group, but rather, meaningful change is founded on the contributions of multiple stakeholders. The final success factor, and the most important of all, is the patient as preeminent player principle. This principle propounds that without the patient as the starting point and the locus of improvement initiatives, reform initiatives can become meaningless (Braithwaite et al., 2017a,b).

This third volume is the culmination of the first two. It asks 148 contributing authors a question rarely mobilized in the health systems improvement literature: how might you induce sustainable change in your health system over time? To answer this, the authors were asked to identify an aspect of their healthcare system that shows promise of improving in the next 5–15 years, and from which other countries can learn. The compendium covers 152 countries and territories, across the World Health Organization (WHO) regions, in 57 chapters, producing a vision of the future not previously articulated. Graphically, we show the coverage of the book, with the shaded areas as the included countries and regions (Figure I.1).

Some of Our Views on What the Future Holds for Healthcare

It was Neils Bohr, the famous Danish physicist of the twentieth century, who quipped, "Prediction is very difficult, especially about the future." He was discussing quantum mechanics, and the joke, of course, is that prediction is always about the future and is almost always wrong. We cannot foreshadow what will happen next, nor peer into a crystal ball to see across the years. Things are just too complex, discontinuous, and, well, *un*predictable. However, in the major scale of systems change, there are trends, tendencies,

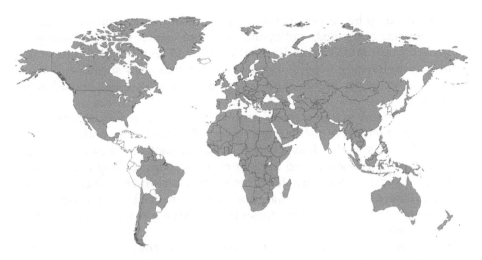

FIGURE I.1
Coverage of the health systems in the book (shaded = included). (Adapted from https://map-chart.net/.)

inclinations, movements, and broad-based directions, and in the minor scale of systems change, there are specific case studies of reform and improvement that are pointers to the future. We will get to the case studies of change when we lay out the 57 chapters, but as far as broad themes go, we see five pronounced trends as we read the health services literature:

1. The question of the sustainability of health systems
2. The genomics revolution
3. Technologically based solutions, aides, improvements, or substitutes for the way we currently do things
4. Shifts in population demands, health, and services
5. Alternate modes of care, including home- and community-based care

While we do not claim that this is a comprehensive list, these trends are frequently represented in research and commentary. So, while we could add or subtract from this list, let us work with it and discuss some of these issues, by way of presaging the case studies to follow.

How Sustainable Are Health Systems?

A sustainable health system is one that adapts and endures across time, constantly adjusting to changing pressures (Coiera and Hovenga, 2007; Braithwaite et al., 2016). The pressures that every health system faces include stretched resources and workforces, new demands from patients, and the need to be responsive to various stakeholder groups, including governments, clinicians, policy-makers, and patients. Contributing to improved population and individual health and well-being, while simultaneously delivering a triple bottom line of financial, social, and environmental value for money, and return on investment, is a considerable challenge.

To cope, health systems must optimize their workforces; acquire new, relevant, and effective technology; reduce financial risk; and maximize the effectiveness of the delivery system. Redundant practices and models of care must be disestablished in favor of innovations leading to better services to patients. All too often, however, innovations and new models of care fail to deliver or lead to only marginal improvements, but with substantial cost increases. In combination, these forces create a state of affairs that is unsustainable.

In one sense, all health systems that exist in every country are sustainable: no country fails to provide a system of sorts to provide care to its citizens. However, this is a weak conception of the sustainability we must produce in every country if we are to be true to the WHO's noble aim of healthcare for all and the provision of effective, affordable, and responsive care systems to populations needing treatment and services.

Perhaps a more realistic framing of the question of sustainability is how do we provide, on a long-term basis, a solid fiscal and economic foundation for all health systems to deliver care to patients? And how do we provide effective models of care so that services are available to all population groups, whether in urban, rural, or remote settings, whether rich or poor, whether in a majority or minority group, and regardless of gender, political preference, race, religion, and socioeconomic status? In short, how do we fund health systems, with growing cost pressures and demands, and still deliver good care to all those who need it, now and in the future?

The Genomics Revolution

What about the role of genetics? In the early 2000s, researchers were generating the first reference human reference genome. It is estimated that these efforts cost over US$1 billion (Lander et al., 2001; Venter et al., 2001). By the mid-2010s, thousands of people had had their genomes sequenced despite the challenge of a human having approximately 3 billion base pairs, and the cost of sequencing fell to US$1000 (Mattick et al., 2014). Since then, costs have fallen even further, and very rapidly.

The genomics revolution, as exciting as it is, is not just expressed in terms of the science or the amazing reduction in costs to sequence a genome. There is also much excitement around new systems of care exploiting DNA sequencing via genetic counseling, clinical diagnosis, therapeutic decision-making, precision medicine, and gene-based disease treatments and interventions (Feero et al., 2010; Guttmacher and Collins, 2002). According to the futurists, a whole new population of patients will benefit.

The portents of this have animated and mobilized many researchers, commentators, and patient groups. Everyone is looking forward to the day when we eradicate some diseases, treat other conditions much earlier in the disease process, and are able to provide more appropriate, targeted medicine.

What will we need to deal with this revolution? At least three things will be required. One is the capacity to pay, to ensure that many people benefit, and not just the wealthy. The second is to alter—even transform—the delivery system so that clinicians and patients understand the benefits and risks of early diagnosis and new treatment regimes, and participate in the new medical era. The third is to deal with the ethical issues that surround the genomics medicine revolution: how to control both individual and "big" data, and how to manage the political and economic challenges that ensue as the transition unfolds.

Technologically Based Solutions, Aides, Improvements, or Substitutes for the Way We Currently Do Things

Other forms of technology are changing the way we provide healthcare to patients. While there is much nuanced literature on this topic, technology essentially falls into two categories: digital technology, often labeled information and communications technology (ICT), and diagnostic and treatment technology, which often centers on pathology tests, imaging, new drugs and methods of drug delivery, and less invasive surgical and care procedures.

Both categories are determining factors in producing new healthcare services to patients. Digital eHealth technology is centrally concerned with the documentation and management of patients' treatment and experience, and becomes a tool for effective communication between clinicians. It also makes data available to various parties, especially the patient. It heralds improvements to the way we intervene with patients, including providing data to health providers about their patients so they can provide better and more focused care (Coiera, 2015).

Clinical technology, the second category, relates to the use of technology in the diagnosis, prevention, monitoring, and treatment of disease or injury. As medical knowledge advances by leaps and bounds, so the range and capability of technologies for medical diagnosis and treatment grows at speed. It is not only in hospital settings that we are seeing an increase in the adoption of technology; it is also on the rise in ambulatory care, including telemedicine, and home-based care. The extent to which clinical technologies are successfully and effectively implemented in health services will play a vital role in shaping and improving health services for patients over future generations.

Shifts in Population Demands, Health, and Services

These last two trends are largely supply-side questions. What about demand? The world's population continues to rise beyond the current 7.6 billion. It is projected to peak at 11.2 billion in 2100 (United Nations, 2015). The population is also aging (Amalberti et al., 2016; International Society for Quality in Health Care, 2016), and within countries, demographics are shifting. Many countries are becoming more multicultural. The movement of people internationally, despite political rhetoric and laws that inhibit migration, continues to accelerate. People are traveling and changing citizenship more frequently than ever before.

While the world has never been wealthier, there are still inequities, both within countries and between nations: around 10% of the global population still eke out a subsistence existence, living in poverty. Inequality may be shrinking in some countries, but in others it is rapidly rising. Despite ongoing efforts to create equity in many places, the capacity to pay for future expensive healthcare is likely to continue to be unevenly distributed both within countries and across the world's regions.

Changes in demographics and wealth patterns such as these, and demands for health services, pose significant challenges to health systems. The magnitude of these challenges has never been greater.

Alternate Modes of Care, Including Home- and Community-Based Care

So, there are middle of supply-side and demand-side challenges, and hospital care is expensive. A health system with many hospitals and with eager specialists in plentiful supply can end up overservicing the population. People get knee arthroscopies, some of which will be unnecessary (Siemieniuk et al., 2017). Too many women will have caesarean sections instead of natural births (Betrán et al., 2016). People will be overtested and overscanned, for example, for normal, everyday back pain (Emery et al., 2013). In addition to exposing patients to additional risks (the more treatments patients receive, the greater the possibility of things going wrong), acute care is very costly, and if there is excessive supply, the additional benefits to patients will be marginal.

Underservicing can also be a problem. Hospitals tend to cluster in major cities. Given that healthcare is largely provided through hospital-based services, access to healthcare in regional or remote settings is often severely impeded. Decentralizing the provision of services, and integrating primary care into community-based services, could potentially help to deliver appropriate medical care to populations previously underserviced.

In addition, aging populations and the associated increase in comorbid and chronic conditions create mounting pressures on healthcare services to move beyond acute settings and into home- and community-based care. This type of shift in service provision reflects a reprioritization of the patient journey, with an emphasis on the need for ongoing medical care to support long-term conditions, and preserving quality of life, rather than curative treatments (Amalberti et al., 2016).

The ideal is to have a balanced health system that does not privilege only the expensive and mainly acute care provided in hospitals, but has more equitable distribution of services, providing care in the setting that is most optimal. Thus, home care, primary care, and family medicine (collectively, community-based care) become an important emphasis for those designing future health systems.

This trend is already underway. Many countries are de-emphasizing the focus on hospital-based care and putting more resources into non-hospital-based services. This is often cheaper, more focused on the needs of patients—providing care where it is most needed—and less risky. This means that some developed countries are reducing their acute care services, while others in a more developmental stage of healthcare provision are giving primacy to community-based services before they allocate significant resources to creating an extensive hospital system.

In the past, hospitals were the focus of attention and resources for a variety of reasons. In the minds of many patients, hospitals are synonymous with healthcare: alternative modes of care are rarely considered. Politicians like to fund hospitals in big cities because it appeals to voters (if not taxpayers). Specialist doctors enjoy the access to technology and advanced techniques.

There is no doubt that acute care still has an important role to play; however, the real test of a good health system is that it provides the right care, to the right patients, at the right time, for the right cost, in the right setting.

The Participant Countries and Regions in Context

Having discussed future trends that are theorized to be changing healthcare, in Tables I.1 through I.4, we present profile data on the countries and regions in the book. This helps further contextualize the stories in the chapters that follow.

These data frame the countries and regions represented in the book. It highlights the diversity of the countries and regions, and provides a backdrop to the stories that are presented in the subsequent chapters.

TABLE I.1

The World by Region and Income Group: Population and Income Data

		Population				Gross National Income per Capita, Atlas Method[a] (US$) 2016	Gross National Income per Capita[a] (PPP int. $) 2016[b]	Gross Domestic Product per Capita[a] (US$) 2016	Gross Domestic Product per Capita[a] (PPP int. $) 2016
		Total (Thousands) 2013	Population under 15 (%) 2013	Population over 60 (%) 2013	Median Age (Years) 2013				
WHO region	Africa	927,371	42	5	19		3,682		
	The Americas	966,495	24	14	32		28,962		
	Eastern Mediterranean	612,580	33	7	24		10,968		
	Europe	906,996	18	21	39		27,369		
	South-East Asia	1,855,068	29	9	27		5,987		
	The Western Pacific	1,857,588	19	15	37		14,238		
World Bank income group	Low income	848,668	39	6	21	612	1,646	615	1,683
	Lower middle income	2,554,925	31	8	25	2,079	6,767	2,075	6,800
	Upper middle income	2,449,819	21	13	34	8,202	16,553	7,937	16,746
	High income	1,272,686	17	22	40	41,046	46,965	40,678	46,704
Global	World	7,126,098	26	12	30	10,299	16,101	10,151	16,143

Source: All data are from World Health Organization (http://apps.who.int/iris/bitstream/10665/170250/1/9789240694439_eng.pdf) unless stated otherwise. Data accessed in September 2016 and again in September 2017, with latest available data presented in all cases. Table template adapted from World Health Statistics, 2015.

[a] Data for World Bank income group and Global are from World Bank (https://data.worldbank.org/).

[b] Data for WHO region are from 2013.

TABLE I.2

The World by Region and Income Group: Life Expectancy and Mortality Data

		Life Expectancy at Birth[a] (Years) 2015			Healthy Life Expectancy at Birth (Years) 2015			Healthy Life Expectancy at Age 60 (Years) 2015			Neonatal Mortality Rate (per 1000 Live Births) 2015	Infant Mortality Rate (Probability of Dying by Age 1 per 1000 Live Births) 2015	Under-Five Mortality Rate (Probability of Dying by Age 5 per 1000 Live Births) 2015	Adult Mortality Rate (Probability of Dying between 15 and 60 Years of Age per 1000 Population) 2015			Maternal Mortality Ratio (per 100,000 Live Births) 2015
		Both Sexes 2015	Male 2015	Female 2015	Both Sexes 2015	Male 2015	Female 2015	Both Sexes 2015	Male 2015	Female 2015				Both Sexes 2015	Male 2015	Female 2015	
WHO region	Africa	60	58	62	52	51	54	12	12	13	28.0	55.4	81.3	300	318	269	542
	The Americas	77	74	80	67	65	69	17	16	18	7.7	12.5	14.7	124	158	87	52
	Eastern Mediterranean	69	67	70	60	59	61	14	13	14	26.6	40.5	52.0	155	174	130	166
	Europe	77	73	80	68	66	71	16	16	19	6.0	9.8	11.3	124	168	76	16
	South-East Asia	69	67	71	61	60	62	14	13	14							
	The Western Pacific	77	75	79	69	67	70	17	16	18	6.7	11.3	13.5	93	110	73	41
World Bank income group	Low income	62	60	64							26.9	53.1	76.1				
	Lower middle income	68	66	70							25.8	40.0	52.8				
	Upper middle income	75	73	78							8.9	15.2	19.1				
	High income	81	78	83							3.7	5.8	6.8				
Global	World	72	70	74	63	62	65	16	15	17	19.2	31.7	42.5	149	176	117	216

Source: All data are from World Health Organization (http://apps.who.int/iris/bitstream/10665/170250/1/9789240694439_eng.pdf) unless stated otherwise. Data accessed in September 2016 and again in September 2017, with latest available data presented in all cases. Table template adapted from World Health Statistics, 2015.

a Data for World Bank income group and Global are from World Bank (https://data.worldbank.org/).

TABLE I.3

The World by Region and Income Group: Health Expenditure

	Total Expenditure on Health as % of Gross Domestic Product 2014	General Government Expenditure on Health as % of Total Government Expenditure 2014	Private Expenditure on Health as % of Total Expenditure on Health 2014	General Government Expenditure on Health as % of Total Expenditure on Health 2014	External Resources for Health as % of Total Expenditure on Health 2014	Social Security Expenditure on Health as % of General Government Expenditure on Health 2014	Out-of-Pocket Expenditure as % of Private Expenditure on Health 2014	Out-of-Pocket Expenditure as % of Total Expenditure on Health 2014	Private Prepaid Plans as % of Private Expenditure on Health 2014
WHO region Africa	5.5	10.0	52.2	47.8	9.6	10.5	60.0	31.3	28.5
The Americas	14.2	18.1	50.2	49.8	0.0	74.8	27.3	13.7	59.7
Eastern Mediterranean	4.8	7.9	43.2	56.8	1.2	8.8	81.3	35.1	9.6
Europe	9.5	15.5	24.6	75.4	0.0	50.7	69.4	17.1	22.4
World Health Organization South-East Asia	4.0	6.1	62.9	37.1	1.6	9.0	85.7	53.9	3.7
The Western Pacific	7.1	14.3	33.6	66.4	0.1	67.3	74.3	25.0	12.7
World Bank income group Low income	5.9	10.5	58.8	41.2	28.3	5.2	65.5	38.5	3.4
Lower middle income	4.5	6.7	63.8	36.2	3.3	15.0	87.5	55.9	3.6
Upper middle income	6.1	10.2	43.8	56.2	0.2	48.9	68.4	30.0	20.6
High income	12.0	17.6	38.1	61.9	0.0	65.1	37.6	14.3	50.2
Global World	9.9	15.5	39.9	60.1	0.2	61.7	45.5	18.2	42.5

Health Expenditure Ratios

Source: All data are from World Health Organization (http://apps.who.int/iris/bitstream/10665/170250/1/9789240694439_eng.pdf) unless stated otherwise. Data accessed in September 2017, with latest available data presented in all cases. Table template adapted from World Health Statistics, 2015.

TABLE I.4

The World by Region and Income Group: Per Capita Health Expenditures

		Per Capita Health Expenditures			
		Per Capita Total Expenditure on Health at Average Exchange Rate (US$) 2014	Per Capita Total Expenditure on Health at Average Exchange Rate (PPP int. $) 2014	Per Capita Government Expenditure on Health at Average Exchange Rate (US$) 2014	Per Capita Government Expenditure on Health at Average Exchange Rate (PPP int. $) 2014
WHO region	Africa	108	228	52	111
	The Americas	3730	3959	1858	1983
	Eastern Mediterranean	243	598	138	318
	Europe	2426	2580	1828	1884
World Health Organization	South-East Asia	79	263	29	94
	The Western Pacific	746	1007	495	635
World Bank income group	Low income	37	92	15	38
	Lower middle income	89	268	32	95
	Upper middle income	487	869	274	495
	High income	4539	4608	2811	2825
Global	World	1058	1273	636	739

Source: All data are from World Health Organization (http://apps.who.int/iris/bitstream/10665/170250/1/9789240694439_eng.pdf) unless stated otherwise. Data accessed in September 2017, with latest available data presented in all cases. Table template adapted from World Health Statistics, 2015.

Conclusion

In what follows, we present a book of case studies, covering 152 countries and territories, in 57 chapters, which together provide a comprehensive view of global health system improvements across a 5 to 15-year horizon. We have divided the book into five sections: the Americas, Africa, Europe, Eastern Mediterranean, and South-East Asia and the Western Pacific. Many of the stories contained in the chapters map to the five trends discussed above. Some fall outside these trends. But all provide insights into how health systems are orienting themselves to a new future.

Part I

The Americas

Paul G. Shekelle

The continents of North and South America and their proximate islands make up the World Health Organization region of the Americas—an area that exceeds 16 million square miles and extends 8700 miles from north to south. The geography encompasses every extreme: from arctic tundra in the north, through deserts and plains, vast mountain ranges, and large river basins, to tropical rainforests in the south. The combined population of the 35 countries and 24 dependent territories that make up the Americas is more than 1 billion. The population is as demographically, culturally, and economically diverse as the landscape and climate, and includes those in the wealthiest nations of the world, such as the United States and Canada, along with those in some of the poorest, such as Haiti.

Authors from nine countries from the region—Argentina, Brazil, Canada, Chile, Guyana, Mexico, Trinidad and Tobago, the United States, and Venezuela—have contributed to this book, and the overall message is that there is a broad range of possible future innovations and reforms. While the discussion is expansive and each chapter has its own particular focus, shifts in global trends in demographics, economics, and technology mean that there are a number of shared concerns. These include improved training and education (Trinidad and Tobago and Brazil), reforms to health system financing (Chile, Mexico, and Argentina), primary and mental healthcare (Venezuela and Guyana), and the importance of developing patient-oriented systems (Canada and the United States of America).

Concerning potential improvements in training and education, Walter Mendes, Ana Luiza Pavão, Victor Grabois, and Margareth Crisóstomo Portela describe a Brazilian distance education training program designed to improve patient safety and increase research, with an emphasis on collaboration. Claudine Richardson-Sheppard describes ways in which nurse education in Trinidad and Tobago could be improved with the introduction of competency-based training, particularly when it comes to leadership roles.

With regard to how care is paid for, Oscar Arteaga describes Chile's struggle to create a more fully integrated health insurance system, while Hugo Arce, Ezequiel García-Elorrio, and Viviana Rodríguez home in on the goal of achieving universal healthcare coverage through the targeting of priority populations in Argentina. Jafet Arrieta, Enrique Valdespino, and Mercedes Aguerrebere describe how Mexico might use conditional cash transfers and universal healthcare coverage to tackle the growing burden of non-communicable diseases.

In Venezuela, Pedro Delgado, Luis Azpurua, and Tomás J. Sanabria present an innovation using training and technology to improve primary care in rural clinics. William Adu-Krow, Vasha Elizabeth Bachan, Jorge J. Rodríguez Sánchez, Ganesh Tatkan, and Paul Edwards describe a shift from institutional care to community care when it comes to treating patients with psychological disorders in Guyana.

In Robert H. Brook and Mary E. Vaiana's radical vision of the future, the U.S. health system has been completely reoriented and adopts a patient-focused approach, with all social and health services integrated. Anne W. Snowdon, Charles Alessi, John Van Aerde, and Karin Schnarr also look forward to a more personalized health system in Canada, where all medical information is accessible and transparent, and health decisions can be made by the informed consumer.

Thus, in the Americas, a wide range of potential future innovations is envisioned, at different stages of development, and influencing care from specialized populations to the entire country. These aggregate to a set of remarkable visions, some of which are more speculative, and others more grounded in the present.

1

Argentina

Achieving Universal Coverage

Hugo Arce, Ezequiel García-Elorrio, and Viviana Rodríguez

CONTENTS

Argentinian Data

- Population: 43,847,430
- GDP per capita, PPP: $19,934.4
- Life expectancy at birth (both sexes): 76.3 years
- Expenditure on health as proportion of GDP: 4.8%
- Estimated inequity, Gini coefficient: 42.7%

Source: All data are from the World Health Organization and World Bank. Latest available data used as at October 2017.
GDP = Gross Domestic Product
PPP = Purchasing Power Parity

Background

Argentina, like most countries, is in the process of epidemiological and demographic transition. Argentina's population is largely urbanized (about 80%), and around 32.2% of the population is impoverished. Urbanized areas are served by both ambulatory and hospital services—including public and private, for profit and non-profit (Table 1.1).

During the last 2 years, there has been a marked increase in epidemic outbreaks of dengue, Zika, chikungunya, and even yellow fever due to the increase in population and displacement of their common source, the *Aedes aegypti* mosquito, whose natural reservoir is the jungle of the Mato Grosso. Preventive strategies are multipronged and include education as well as the elimination of the insects and their habitats. Low-income populations in the north and center of the country, where *Triatoma infestans* (vinchuca) transmit Chagas disease, are particularly vulnerable. Malaria cases, however, are uncommon in the north of the country, as a result of an intense campaign deployed in the 1940s and 1950s.

Chronic non-communicable diseases (NCDs), such as obesity, diabetes, hyperlipidemias, hypertension, and chronic obstructive pulmonary diseases (COPDs), are simultaneously becoming increasingly common in urban populations. This has forced a shift in urban primary care services' traditional emphasis on acute and infectious disease control to the management of chronic conditions.

High maternal mortality rates persist in Argentina, due to the prevalence of unchecked chronic conditions, adolescent pregnancies, and illegal abortions, and despite the fact that more than 97% of births occur in hospitals. In 1990, as a signatory to the United Nations (UN) Millennium Development Goals, Argentina committed to reducing its maternal mortality rate by 75%, aiming to reach 1.3 deaths per 10,000 live births by the year 2000. However,

TABLE 1.1

Healthcare Organizations

Type of Organization	Acute Care[a]		Long-Term[b]		Total	
	n	%	n	%	n	%
Hospital	3,323	100.0	1,795	100.0	5,118	21.4
Public	1,312	39.5	209	11.7	1,521	6.4
Social security	30	0.9	2	0.1	32	0.1
Private	1,980	59.6	1,582	88.1	3,562	14.9
Mixed	1	0.0	2	0.1	3	0.0
Ambulatory	18,814	100.0	n/a	—	18,814	78.6
Public	8,802	46.8	n/a	—	8,802	36.8
Social security	303	1.6	n/a	—	303	1.3
Private	9,688	51.5	n/a	—	9,688	40.5
Mixed	21	0.1	n/a	—	21	0.0
Total	22,137	—	16,085	—	23,932	100.0

Source: Ministerio de Salud [Argentina]. (n.d.). Sistema Integrado de Información Sanitaria Argentina. Retrieved from https://www.msal.gob.ar.

Note: n/a = not applicable.

[a] All types of organizations with and without admission services, with the exception of centers for geriatric care.

[b] Might include some mental health and addiction centers.

the rate is currently 3.2%, with abortion a leading cause of maternal death, especially among teenagers. An estimated 500,000 illegal abortions are carried out every year.

Healthcare Organizations

Argentina has 3300 acute care hospitals, of which 1300 are public and 2000 are private. There are approximately 134,000 beds: 73,500 (55%) public beds and 60,500 (45%) private beds. However, over the past 25 years, the total number of beds per 1000 inhabitants has decreased: from 4.5 in 1995 to 3.1 in 2016. These figures do not take into account the numbers of geriatric patients currently hospitalized, which increases the number to 4.9 per 1000 (Arce, 2010). Several studies indicate that despite superior capacity, public hospitals account for less than 40% of the current system's resources. There are also 1800 geriatric institutions with about 80,000 beds where the public patients account for less than 12% of the total. There has been a notable increase in the number of outpatient clinics over the last three decades: in total, there are 18,800 establishments with 8,800 public centers primarily concerned with general primary care, and 10,000 private diagnostic and specialist treatment centers (Table 1.2).

TABLE 1.2

Types of Hospitals Beds by Type of Organization, 2016

Type of Organization	Acute Care[a]		Long-Term[b]		Total	
Available Beds	n	%	n	%	n	%
Public	73,731	54.9	16,333	20.3	90,064	41.9
Average number of beds/ organization	56.2	—	78.1	—	59.2	—
Private	60,599	45.1	64,211	79.7	124,810	58.1
Average number of beds/ organization	30.1	—	40.5	—	34.7	—
Total	134,330	100.0	80,544	100.0	214,874	100.0
Beds per 1000 inhabitants	3.1	—	1.8	—	4.9	—

Source: Ministerio de Salud [Argentina]. (n.d.). Sistema Integrado de Información Sanitaria Argentina. Retrieved from https://www.msal.gob.ar.

Note: All jurisdictions, municipal, provincial, and national. Includes non-profit and profit organizations, and social security. Estimated population as of March 1, 2016: 43,590,368.

[a] All types of Organizations with and without admission services with the exception of centers for geriatric care.

[b] Might include some mental health and addiction centers.

Human Resources

In terms of human resources, the country has a large and increasing number of doctors and nurses, partially due to immigration, particularly from other South American countries (Table 1.3). Despite the relatively high doctor-to-population ratio of 1:253, numbers are concentrated in urban

TABLE 1.3

Physicians and Nurses in 2001 and 2013

Profession	Training	2001		2013	
Doctor	Specialist[a]	108,000	n/d	166,187	74,622
	Non-specialist		n/d		91,565
Nurses	Licensed[b]	86,000	29,000	177,974	91,901
	Non-licensed[c]		57,000		86,073

Source: Organización Panamericana de la Salud, 2001, Recursos Humanos en Salud en Argentina—2001. Retrieved from http://publicaciones.ops.org.ar/publicaciones/coleccionOPS/pub/pub53.pdf; Williams, G., Intersectorialidad en las Políticas de Recursos Humanos en Salud. Experiencia argentina y perspectivas. Regulación en el ejercicio profesional. Reunión Regional de Recursos Humanos para la Salud. Buenos Aires: Ministerio de Salud de la Nación, 2015.

Note: n/d = no data.

[a] Different levels of specialization, including general internal medicine.

[b] Includes nurses with tertiary degree and university degree.

[c] Includes nurses with auxiliary degree and non-licensed.

areas and there are significant shortages in geographically isolated communities. Forty-five percent of doctors have specialist training; however, relatively few are specialists in general or family medicine. On the other hand, among the 55% of doctors without specialist training, a large number practice general medicine and others practice as specialists without formal qualifications or registration. While the number of university-trained nurses remains very low (11%), the proportion of nursing staff without formal training has decreased from almost one-third in the 1990s to less than 10%.

Twenty-First-Century Health Initiatives

In order to strengthen primary care in Argentina, the Ministry of Health (MoH), with the support of international banks, implemented a number of community health programs in the first years of the new century. The *Programa de Médicos Comunitarios* (Community Physicians Program) was designed to enhance the community orientation of general practitioners. The *Remediar* Program involved the provision of a reserve of basic medications recommended by WHO that were provided free of charge to patients in outpatient facilities. The *Nacer* Plan introduced a mandatory protocol for maternal and child health (covering pregnancy, childbirth, puerperium, and children aged up to 5 years old). This program has a mixed payment model that includes fee-for-service and capitation. The *Nacer + Sumar* Plan extended this coverage to 18-year-olds. All programs were implemented by government agencies (Ministerio de Salud [Argentina], 2016).

In addition to these government programs, Argentinian communities have embraced a number of non-governmental initiatives aimed at improving the quality of medical care and ensuring patient safety (Arce et al., 2017).

While there are little published data, it appears that private services have been markedly more sensitive to expectations of improvement in the quality of patient care and safety, both in external evaluation modalities (accreditation, International Organization for Standardization [ISO], specialized quality standards, and quality indicators) and in improvements in administrative and care management (management consulting, procedure standards, clinical practice guidelines, and registry of hospital infections). As with other Latin American countries that have implemented accreditation programs, there have been no incentives to encourage the improvement of services in Argentina. The exception to this is Chile, where a hospital ranking system has been established to encourage quality improvement (Iturriaga and Valdivia, 2017). However, there have been some gradual advances on this front: while there is little in the way of overt government support, there is clearly increased interest from all involved parties in improving the quality and safety of hospital care.

Proposed Health System Reforms

Subsequent to the appointment of a new government administration, at the end of 2015, the implementation of universal health coverage, or *Cobertura Universal de Salud* (CUS), had been proposed but not yet implemented. As in many other countries, a significant number of Argentinians work in the informal economy. While this sector does not contribute to social security taxes and is not registered in the current tax regime, its workforce will still be able to access all the benefits of the public sector, including healthcare. Under the new model, all citizens, regardless of income or work affiliation, will be entitled to

- Public-sector healthcare
- Public-sector rehabilitation services
- Pharmaceuticals

It should be noted that as almost all public hospital and outpatient services in Argentina are administered by provinces and municipalities, these entities will have to conform to national requirements.

For those who belong to the formal economy, healthcare coverage is already provided via the social security sector. This sector is divided into more than 300 entities named *Obras Sociales*. The majority were originally controlled by labor unions that administer their respective health plans; membership accounts for approximately 40% of the population (Bermúdez, 2016). Although these entities are centralized in a single umbrella organization, the body in charge of regulating the social security health plans does not intervene in the distribution of funds. In addition, there are 24 social security health plans for provincial state employees, who constitute 15.6% of the population, as well as separate coverage for retirees and pensioners—called PAMI—which protects another 10.5%. There is also private insurance, including prepaid medicine enterprises, which covers 4.2% of the population.

Based on these data, the first government estimate was that CUS would be required by 29.8% of the population (more than 12.5 million inhabitants). However, these figures may vary depending on how much of the population is living below the poverty line; this number currently exceeds 32%. Taking into account the unregistered economically active population (EAP), the estimated number of households eligible for CUS would be around 11.3 million inhabitants (Universidad Católica Argentina, 2016).

Another measure established by the new government is a project to create a technology assessment agency (AgNET) in collaboration with stakeholders from the national health system and specialized agencies. Such an agency would evaluate new technology and determine reimbursement rates based on criteria relating to effectiveness, equity, and cost efficiency. The evaluations would apply to services covered by the mandatory minimum healthcare benefits provided by the state via the *Programa Médico Obligatorio* (PMO), and would help ease existing tensions between insurers and providers.

Achieving Universal Coverage

The strategy developed to gradually achieve universal coverage involved the implementation of programs targeting priority segments of the population and paying for those services most in demand as needed. This system requires that users are served by public services, that computer records are generated, and that service providers present their invoices to the state to receive the reimbursement for the care provided. This process began in 2004 and was completed in 2015 (Table 1.4).

Coverage provided by the plans in March 2017, according to the percentage of the target population incorporated, was as follows:

- Children from 0 to 5 years: 65.4%
- Children from 6 to 9 years: 40.4%
- Teens up to 18 years: 35.2%
- Adult women: 26.1%
- Adult men: 14.5%

TABLE 1.4

Progressive Improvements in Coverage of Priority Populations

Plan Year	Nacer—2004	Nacer—2007	Sumar—2012	Sumar—2015
Population/ objective	Maternal and child care—Northern region	Maternal and child care—Entire country	Children, adolescents, and adult women	Adult men
Potential reach	700,000	2,000,000	10,000,000	15,000,000

Conclusion

Due to the fragmented nature of the Argentine health system, the gradual targeting of segments of the population appears to be the most appropriate way to achieve universal health coverage and make healthcare accessible to the entire population (Rubinstein, 2017).

2

Brazil

Patient Safety: Distance-Learning Contribution

**Walter Mendes, Ana Luiza Pavão, Victor Grabois,
and Margareth Crisóstomo Portela**

CONTENTS

Brazilian Data

- Population: 207,652,865
- GDP per capita, PPP: $15,127.8
- Life expectancy at birth (both sexes): 75.0 years
- Expenditure on health as proportion of GDP: 8.3%
- Estimated inequity, Gini coefficient: 48.4%

Source: All data are from the World Health Organization and World Bank. Latest available data
       used as at October 2017.
GDP = Gross Domestic Product
PPP = Purchasing Power Parity

Background

In 2013, the Brazilian Ministry of Health established the *Programa Nacional de Segurança do Paciente* (National Patient Safety Program) (PNSP), a program that aims to contribute to healthcare quality in all public and private health facilities in the country in accordance with the resolution approved during the 57th World Health Assembly of the World Health Organization (WHO) (Ministério da Saúde [Brazil], 2013). The PNSP is organized along four axes:

1. Encouraging safe care practice
2. Engaging citizens to ensure their own safety
3. Incorporating the theme in education
4. Increasing safety research

The Implementation Committee of the National Patient Safety Program (CIPNSP), which consists of representatives from government, the community, organizations, and universities, was also established to promote and support patient safety initiatives in different healthcare areas (Ministério da Saúde [Brazil], 2013). Three education and training areas were identified and targeted for investment in the medium- and long-term: the training of pharmacists and nursing professionals, and a patient safety specialization course for health professionals.

There are about 104,000 healthcare units under the PNSP's remit, including more than 6,000 hospitals, of which only 4% are accredited. The implementation of essential protocols—the main objective when encouraging safe care practice—can therefore be a challenge, requiring the efforts of both the PNSP and the CIPNSP. Safe care practice, as per WHO's framework, has been incorporated into many areas, including education and training, with some success. This is despite Brazil's difficult economic situation and the ensuing health system challenges, including scarce resources, insecure facilities, and numerous personnel problems (Ministério da Saúde [Brazil] et al., 2013). Such problems directly affect healthcare processes, hamper improvement efforts, and compromise desired outcomes (Donabedian, 1978). The underfunding and unprofessional management of the politically compromised Brazilian health system also poses major challenges for patient safety actions.

Success Story

The story told here is that of the patient safety specialization course for health professionals, mentioned above, which was contracted to the Fundação Oswaldo Cruz (Oswaldo Cruz Foundation) (Fiocruz), a Brazilian federal organization dedicated to science, technology, innovation, and education in the health field, at the end of 2013. Fiocruz, a member of the CIPNSP, had previously developed (in collaboration with the Universidade Nova de Lisboa, Portugal) a textbook on the subject, as well as an online course in patient safety linked to the WHO. The new course, the International Quality in Health and Patient Safety Specialization Course, was also conceived in partnership, utilizing a distance-learning format. In its initial delivery, the course required students to design patient safety plans to be implemented in their healthcare facilities across the country, as their *trabalho de conclusão de curso* (mandatory course conclusion work) (TCC).

This initiative represents Brazil's most comprehensive and largest-scale training course for health professionals in the field of quality and safety. The Brazilian Ministry of Health sponsored the pilot course, helping to define the selection criteria for hospitals, and mediating initial contact with higher management. The course objective was to develop competencies and to provide scientific knowledge in quality and patient safety issues. Incidentally, it also encouraged the strengthening of patient safety centers within hospitals, and reinforced the partnership between the Brazilian and Portuguese institutions involved. One thousand places were offered, covering all regions of Brazil, as well as an additional 50 places delivered in Portugal and Portuguese-speaking African countries. The course was conducted in a blended learning mode, which combines different learning technologies and methodologies, mixing online and classroom training (MacDonald, 2008). Training involves four interdependent mediums:

1. Teaching material—a two-volume book totaling 655 pages
2. The virtual learning environment—specifically tailored for the course
3. The academic-pedagogical monitoring system (already established via Fiocruz)

4. Learning through mentoring and guidance

Fifty tutors or teachers were selected for the course, with one tutor for every 20 students. Student–tutor relationships were established within the virtual learning environment, but other means of communication were used, such as telephone, fax, mail, and Internet tools, such as e-mail, WhatsApp, and Skype. Seven teaching advisors were appointed to mentor the tutors, ensuring that they were successfully running the course and helping students achieve their objectives. Team training, focusing on care teams rather than isolated students, was also an important teaching strategy, and guided the selection of students. In Brazil, groups made up of at least four professionals were nominated by the selected hospitals. In Portugal and Africa, numbers engaged in training were smaller, and this arrangement was not possible. At the end of the course, students and tutors were able to evaluate it through an online questionnaire and a seminar.

The Brazilian context is one of serious economic and political crisis, and improving the quality of healthcare and patient safety in the public health system is challenging, to say the least. Nevertheless, establishing an initial critical mass of professionals dedicated to these improvements will help to keep issues of quality and safety on the healthcare agenda, and increase the demand for continuing training and capacity building. Success factors include

- The formation of a critical mass of professionals able to plan and implement initiatives geared to the quality of care
- The establishment of a collaborative network of graduates
- The participation of students in the Committee for the Implementation of Patient Safety in each state

Impact

One of the most important elements of graduate training was the utilization of both teams and individuals (the "student–team" concept). This concept guided the entire course, from student selection to final paper (TCC), in which, as has already been mentioned, students and teams were asked to develop a patient safety plan to be implemented in the hospital in which they were based. This meant the course was able to play a direct part in assisting in the consolidation of patient safety centers within hospitals. According to course evaluation data obtained from 42 of the 50 tutors, and 565 of the 1000 students,

- 91.5% of respondents reported that the course changed their perspective on patient safety.

- 82.1% said they had observed some change in their workplace from the knowledge acquired in the course.
- 98.6% said they would recommend the course to someone else.

These results point to the probability of improved quality of care in these hospitals; however, such conjectures need to be investigated fully in future studies.

The next challenge will be to establish, consolidate, and maintain collaborative networks of graduates in order to further invest in the education, training, and capacity building of health professionals in the areas of quality in health and patient safety.

The course developers originally planned to establish a permanent education program through Proqualis (https://proqualis.net/), a portal established by Fiocruz to disseminate knowledge about patient safety and quality of care; however, the Ministry of Health instead opted to promote graduates' continuing education through the establishment of a community of practice. The initiative was expected to provide a conduit for the experiences of Brazilian public health system professionals—encouraging a permanent dialogue between those involved in the course, from students to authors. Unfortunately, the community of practice has never been implemented; Proqualis currently performs a similar function, but without a specific program.

If the benefits of the International Quality in Health and Patient Safety Specialization Course are to be consolidated in the medium- and long-term, it is essential that both continuing graduate education and the development of a collaborative network remain a priority. Improvements to the structure of services are also crucial to advances in healthcare quality and safety.

Implementation and Transferability

The first International Quality in Health and Patient Safety Specialization Course, conducted in 2014 and 2015, prioritized the participation of hospitals linked to the public network and selected according to size: only hospitals with a capacity of 200 or more beds were selected.

A modified version of the course, targeting professionals involved in the Brazilian urgent and emergency network, including hospitals, emergency care facilities (UPAs), and the Mobile Emergency Care Service (SAMU), which was launched in October 2016. It was first launched in October 2016, and is expected to conclude in October 2017. The course TCC will focus on the implementation of the basic protocols of patient safety recommended by the Ministry of Health: hand hygiene, safe surgery, patient identification, safety protocols for prescriptions, the use and administration of medications, fall prevention, and pressure sore prevention protocol.

Moreover, the Brazilian Agência Nacional de Vigilância Sanitária (National Health Surveillance Agency) (ANVISA) has recently requested an updated version of the course, which will involve a new edition of the book and online content.

There is potential to expand the scope of the course to other levels of care, mainly in the context of primary care, and focusing on outpatient services. Another course, targeting safe childbirth, and conducted in distance mode, has been scheduled for the second half of 2017. There are expectations that patients will become more involved in their own care, with initiatives underway to improve quality and promote safer care environments. Health education is also a priority, with patients being encouraged to learn more about their health (health literacy), thereby increasing their autonomy.

Another important aspect to highlight is the potential of the course to expand its reach into other Portuguese-speaking countries that still lack initiatives aimed at improving the quality of care and patient safety.

Prospects

Further initiatives are needed to strengthen and sustain the success of the International Quality in Health and Patient Safety Specialization Course and to ensure continuing improvement efforts. Such initiatives should expand continuing education; strengthen collaborative networks and processes; extend the scope of the course to other levels of care, such as primary care; and target patient populations. Studies are required to evaluate the effects of these training initiatives on final health outcomes (Donabedian, 1978).

While conditions across Brazil vary, and the differences between Brazil and Portugal and their respective health systems are considerable, the Fiocruz–Universidade Nova de Lisboa partnership created a collaborative learning experience marked by strong exchange among students and tutors during classes and in forums. It was very enriching, and is considered one of the strengths of the course according to students' surveys. Undoubtedly, the high value placed on diversity and the exchange of experiences from multiple backgrounds should be preserved in further steps.

Brazil's ongoing political and economic crisis has led to great uncertainty in the public sphere, particularly when it comes to social welfare, and there has been considerable damage to the public health system. In the face of these seemingly insurmountable structural problems, efforts to improve healthcare quality and patient safety are even more critical.

Conclusion

The early success of the International Quality in Health and Patient Safety Course demonstrates the efficacy of distance learning in the training and qualification of health professionals. The course has been designed to be accessible to different audiences, and can be effectively utilized by professionals in other Portuguese-speaking countries seeking to improve the quality of healthcare and patient safety.

3

Canada

The Future of Health Systems: Personalization

Anne W. Snowdon, Charles Alessi, John Van Aerde, and Karin Schnarr

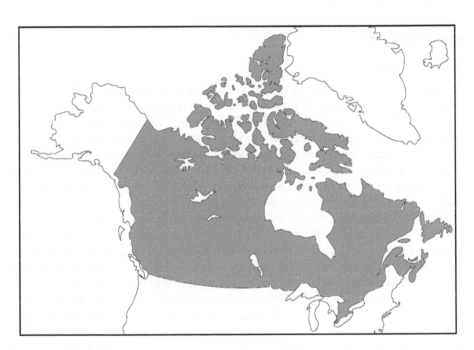

CONTENTS

Canadian Data

- Population: 36,286,425
- GDP per capita, PPP: $44,025.2
- Life expectancy at birth (both sexes): 82.2 years
- Expenditure on health as proportion of GDP: 10.5%
- Estimated inequity, Gini coefficient: 33.7%

Source: All data are from the World Health Organization and World Bank. Latest available data
used as at October 2017.
GDP = Gross Domestic Product
PPP = Purchasing Power Parity

Background

In Canada, equity and access to care for all citizens, regardless of their social, cultural, political, or economic circumstances, is a significant source of national pride. Canadian citizens, who are the "shareholders" of the publicly funded system, expect value, quality, and strong performance as the return on their investment. Yet, despite the expectations for value, current rankings for health system performance place the Canadian health system 10th out of 11 Organisation for Economic Co-operation and Development (OECD) countries studied, while demands for care continue to grow as the population ages and the prevalence of chronic illness steadily increases (Davis et al., 2014; Elmslie, 2012).

A sustainable, high-performing health system that delivers value to the population is more likely to be achieved through personalization, which involves designing a system that is person-centric, connected, and accountable, and that tailors health services to the unique needs, values, and circumstances of every citizen. Personalization in health systems requires a transformational shift from the current provider-centric system organized around teams of caregivers focused on disease management, toward a system that delivers personalized care focused on supporting person-centric self-management of health and wellness. Personalized systems are open and transparent, enabling the person seeking care to maintain control over his or her health decisions and hold the health system accountable for the quality and value of care focused on health, wellness, and quality of life that is meaningful and valued. The key drivers of personalization are described in the following section.

Key Drivers of the Future Personalized Health System

A key driver of personalization across all business sectors is the rapid development of communication and information technologies that have "connected" citizens to the world around them, virtually in real-time. Since the mid-1990s, populations around the world have had access to health information via the World Wide Web, which hosts a number of online sites, tools, products, and health services. The online application (app) sector is the largest-growing business sector in the world, with gross revenue expected to exceed US$100 billion in 2020 (App Annie, 2016), and more than 100,000 online health management apps are now available. This explosion of information technologies and access to information in real-time is driving a consumer revolution across many business sectors, including healthcare.

Until the early 1990s, health provider teams were the sole source of health expertise, evidence, and overall health. Health providers were widely viewed as the experts to whom citizens deferred for prescription of best available treatments and health services to manage disease. Health providers were, and perhaps still are, socialized to be expert decision makers, while citizens were expected to follow the prescribed care. However, it has become increasingly clear that best evidence for disease management is effective for only a fraction of the population (Topol, 2017). Empowered citizens are negotiating and determining the care options that will offer the best fit for their unique circumstances and values. Consumers in pursuit of personalized health goals are shifting from the traditional role of patient as passive recipient of care, characterized by "blind trust" in provider expertise, toward the empowered consumer role, characterized by "earned trust" and strategic partnerships with providers.

If health systems are to remain relevant, they must transition to a person-centric consumer strategy, whereby consumers manage their own health, control their own health information, and hold health providers accountable as partners in care. An additional shift toward personalization has been facilitated by extraordinary advances in genetic and genomic-based therapies, which has further contributed to heightened consumer expectations of personalized approaches to managing illness and disease.

Personalized Health System

The five key features of a personalized health system that Canada is moving toward, and current manifestations in the Canadian health system structure, are outlined below.

Democratization of Health Information

In the everyday lives of Canadians, access to information is readily available using online tools on smartphones or tablets. Individual ownership of health data is critical to achieving the personalization people strive for, and rely on, in their everyday lives. People inherently strive for health goals that are meaningful to them, and hold decisions imposed on them by health providers in varying degrees of regard (Snowdon et al., 2014). Hence, it is not surprising that searching for health information is the third most common online activity for adults (Fox, 2011). Online tools enable people to customize and personalize their choice of web-based products and services. Technologies such as data mining, artificial intelligence, and rule-based matching will enable citizens in future health systems to further engage and use their personal health information for making decisions, sharing information with provider teams, and connecting to the services that support self-management efforts. Opening access to health records will further democratize the health system for citizens and support them in making informed decisions. Although progress is slow, Canada's health systems are beginning to democratize information by giving citizens access to their complete health record. Currently, the Canadian provinces of Alberta and Nova Scotia are establishing online access to the full health record of every citizen and lead the country in democratizing health information, offering a critical first step toward the future health system.

Accountability to Citizens

As health system information is democratized and made available to all citizens, the accountability of health systems will naturally shift toward provider teams being held accountable by the citizens receiving healthcare services. Access to health system information creates transparency of health records, which empowers citizens to inform provider team decisions and assessments, and holds providers accountable for meeting expectations. If health systems are to be successfully personalized, it is essential to customize services that are accountable for achieving targeted outcomes (McClellan et al., 2013). Accountability for outcomes must reach across the entire continuum of care to ensure that every provider has a stake in achieving the personalized goals and aspirations of the individuals, families, and communities for whom they provide care.

The province of Alberta has an integrated health information system that enables Albertans to access all health records, including laboratory results, diagnostic information, progress notes, medications, and procedural records (personal communication, 2017). The chief information officer for Alberta Health Services identifies the fundamental goal of this open access to information as being a key accountability of the Alberta health system to its citizens (personal communication, 2017).

Personalized Approaches to Care

Personalization of healthcare requires a change in mindset from primarily diagnosing and treating illness to determining which health solutions will enable and empower people to achieve their health and wellness goals. In a personalized health system, the quality of health outcomes is measured in terms of accuracy and relevance to the individual's values, rather than by more typical measures of wait times, access, quality, and safety or errors related to adverse events or disease management. As health teams adopt technologies to connect to their patients in a more personalized manner, the practice structures and facilities will adapt and shift toward a more "connected" and personalized healthcare system. Relationships with provider teams will shift from the current patriarchal, "top-down," decision-making provider approach to a collaborative model that empowers individuals to design and manage the care needed to meet their personal health, wellness, and quality of life goals and expectations. Healthcare practitioners will need to recognize that engagement with patients means much more than their compliance with provider-prescribed therapies (Chase, 2013).

Value-Based Resource Allocation

A key enabler for system personalization will be a shift in the business model of healthcare from transactional funding approaches (i.e., fee-for-service) to value-based, accountable care models. These are beginning to emerge in a number of countries (McClellan et al., 2013). Accountable care funding models reimburse health systems to achieve priority population health outcomes, personalized to the unique needs and values of defined populations (McClellan et al., 2013). Such an approach demonstrates a very important shift from the current "one size fits all," which focuses on standardizing care for patients using evidence-based clinical pathways for disease management. In a personalized system, provider teams help citizens select the combination of strategies needed to fit with their unique situation. When people define their own goals, informed by provider expertise, they are far more likely to adopt positive health behaviors to achieve their goals, embracing a "nothing about me without me" approach to personalized care. Financial models incentivize health system players to help people achieve their personal health goals, rather than reimbursing them for simply conducting provider visits or transactions that have no defined target or outcome.

Visibility via Tracking and Traceability

One of the greatest challenges facing global healthcare systems today is patient safety. Medical error has now become the third leading cause of death in North America, right behind heart disease and cancer (Makary and Daniel, 2016). Not only are adverse events devastating for patients and their families, but also they are very costly to the patient, the organization, and the healthcare system. While there is growing awareness worldwide of the challenge of patient safety, with little or no access to better health information it is impossible for citizens to understand the care they have received, track their own progress, or report outcomes that may inform and anticipate risk that provider teams can more effectively address (Snowdon and Alessi, 2016). In a personalized health system, risk of error is proactively identified to enable interventions that mitigate the risk of error or adverse events in order to prevent them.

Open access to health information, coupled with automated supply chain infrastructure, offers citizens complete transparency, facilitating tracking of every product (e.g., implanted devices and medication) used in every procedure (Snowdon and Alessi, 2016). Full transparency of health information is further extended when using supply chain processes consisting of unique device identifiers (e.g., barcodes) on products, further enabling citizens to systematically report outcomes or adverse events that can be linked to product failure, thus holding the system accountable and preventing similar errors (Snowdon and Alessi, 2016).

Conclusion

The emergence of the empowered consumer, enabled by automated access to health information to self-manage his or her health, with the support of provider teams, focused on personalized health goals, is the hallmark of the personalized health system in Canada. This cultural shift in consumer expectations provides the fuel for health system leaders to transform healthcare systems, reorienting the focus to engaging citizens to work toward what they value most: personal health, wellness, and quality of life.

The growing prevalence of information technology and social networking platforms offers a golden opportunity for health systems to personalize care models and approaches. Digital tools empower people to adopt health behaviors that are meaningful and consistent with their personal needs and values. Embedding and scaling healthcare tools and services into the day-to-day lives of citizens mobilizes and empowers them to achieve health, wellness, and quality of life using techniques personalized to their unique values and life circumstances.

4

Chile

The Struggle for an Integrated Health Insurance System

Oscar Arteaga

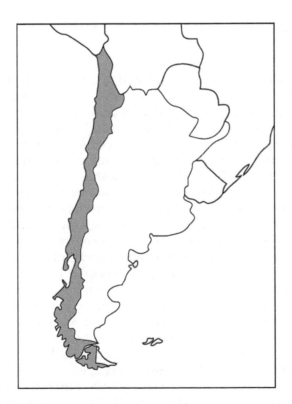

CONTENTS

Chilean Data

- Population: 17,909,754
- GDP per capita, PPP: $23,960.3
- Life expectancy at birth (both sexes): 80.5 years
- Expenditure on health as proportion of GDP: 7.8%
- Estimated inequity, Gini coefficient: 50.5%

Source: All data are from the World Health Organization and World Bank. Latest available data
used as at October 2017.
GDP = Gross Domestic Product
PPP = Purchasing Power Parity

Background

Social insurance legislation was established in Chile as early as 1924. This initial system, which followed the German model, made Chile the first developing nation to provide a social security program for workers (Roemer, 1991). In 1981, during the military dictatorship (1973–1990), a private insurance market was created, with private companies, or *Institutos de Salud Previsional* (ISAPREs), administering the compulsory healthcare contributions of those who voluntarily moved from the public health insurer, the National Health Fund (FONASA). Private companies, or *Administradoras de Fondos de Pensiones* (AFPs), were also created to administer compulsory pension contributions. These reforms, which were deemed to be ideologically driven (Reichard, 1996), had the objective of breaking down the government monopoly of social insurance by encouraging the development of private agents within the insurance sector (Ministry of National Planning and Economic Policy [Chile], 2000).

During its first 9 years, the ISAPRE system was not accountable to any regulatory authority, nor were there any guidelines as to which health plans the ISAPREs were obliged to cover. This absence of restrictions created exceptional conditions that favored the increase in private-sector involvement in health (Ministry of National Planning and Economic Policy [Chile], 2000).

It was not until 1990 that the state regulatory agency for the ISAPRE market, the Superintendence of ISAPREs, was created. Over the last two decades, there have been a number of attempts to change the legal framework that regulates the functioning of ISAPREs at the policy level. In the last 10 years, three high-level presidential commissions have proposed reforms to the ISAPRE sector. Recommendations from the commissions have ranged from making marginal changes to the ISAPRE system to implementing comprehensive reforms to the entire public health system (Presidential Commission, 2010, 2014).

There are a number of factors behind this failure to achieve meaningful reform, including

- Capture of the policy-making decisions by stakeholders with a vested interest in maintaining the status quo
- Lack of technical consensus on the visions on the future of the system and/or ways to transition
- Competition from other government sectors that are also in need of reform (e.g., education, taxes, labor, the constitution)

The current system remains highly inequitable, with ISAPREs catering to the small minority with a high income, and FONASA covering the remaining 80% or so of the Chilean population.

Challenges

Although ISAPREs receive compulsory contributions, since their creation they have been legally permitted to establish discrete health plans according to individuals' risks, and charge additional individual premiums. This runs counter to traditional social security arrangements whereby individuals' risks are spread over the entire population, thus severing the link between premium and individual risk (Barr, 2012). While ISAPREs may deny coverage to people due to income and health status, FONASA cannot reject anyone seeking health insurance coverage—thus effectively providing reinsurance for ISAPREs. This has led to the Chilean health insurance system being criticized for its lack of equity (Hsiao, 1995; López, 1997; OECD, 2016b; Reichard, 1996; WHO, 2000).

FONASA covers a higher concentration of health risk–prone population, such as older individuals and women of reproductive age. FONASA also covers a greater share of people who report poor health—16.9% of beneficiaries as opposed to the ISAPREs' 6.9% (Presidential Commission, 2014). FONASA is also more likely to cover those with disabilities (7%) than ISAPREs (2.3%) (Presidential Commission, 2014). Thus, the public insurance system overwhelmingly covers the most economically disadvantaged population, while ISAPREs cover only those with minimal risks, or apply higher premiums to people with greater needs (OECD, 2016b), as Table 4.1 illustrates.

Over the last three decades, Chile has experienced significant economic growth, associated with social, political, and cultural changes. Chile's entry into the OECD in 2010 has led to a shift in orientation, with global standards becoming increasingly significant. In recent years, an

TABLE 4.1

Chilean Health Insurance System, Selected Indicators, 2012

Indicators	FONASA	ISAPREs	CHILE
Total population			17,444,799
Population distribution per type of insurance	76.5%	17.5%	100%[a]
Population aged <15 years	21%	22%	22%
Population aged 15–59 years	63%	71%	65%
Population aged ≥60 years	17%	8%	14%
Dependency index 100* (Inactive/Active)	60	35	49
Aging index 100* (≥60/<15)	82	35	63
Sex ratio 100* (male/female)	90	116	98

Source: Adapted from Presidential Commission. (2014). Informe Final: Comisión asesora presidencial para el estudio y propuesta de un nuevo régimen jurídico para el sistema de salud privado. Retrieved from https://www.researchgate.net/publication/303255601_ INFORME_FINAL_COMISION_ASESORA_PRESIDENCIAL_PARA_EL_ESTUDIO_ Y_PROPUESTA_DE_UN_NUEVO_REGIMEN_JURIDICO_PARA_EL_SISTEMA_DE_ SALUD_PRIVADO_2014

[a] The population not covered by FONASA or ISAPREs (6%) is covered by other specific arrangements (e.g., Armed Forces).

increasingly empowered and politically active population has applied considerable pressure when it comes to matters of equity, demanding guaranteed rights to quality public education (Bachelet, 2013; United Nations Development Programme, 2015) and protesting unequal distribution of income, for instance (United Nations Development Programme, 2017a). Moreover, during the past three presidential elections most of the candidates running for office addressed equity issues in their proposed government programs. This is a relevant indicator of change, since equity was traditionally a topic considered only in political proposals from more progressive political sectors.

Private companies for administering compulsory pension contributions (AFPs) were also created in 1981. The creation of these companies required a new paradigm based on privately managed individual accounts, also called capitalization, which replaced the previously existing public pay-as-you-go pension system. In this scheme, employers do not make contributions. The system has never fulfilled its original promise of delivering better pensions, and 40 years on is undergoing serious reevaluation due to public pressure (Financial Times, 2016). In 2008, a minimum means-tested government-funded benefit was guaranteed for financially disadvantaged older patients (age 65+) who are unable to satisfy minimum contribution requirements, along with a pension top-up that increases pension benefits for those who did contribute to the AFP system (Behrman et al., 2011; Kritzer, 2008). The current government has recently passed legislation that will force employers to contribute 5% of workers' wages to the system, with 3% directed to individual accounts and 2% to a solidarity fund. This additional contribution will not be administered by the private AFPs but by an autonomous public entity.

Although the proposed pension system reform may appear modest, it is an important step toward a more equitable system. Further reforms to the health insurance system are likely to be a priority for future governments. Table 4.2 summarizes some of the most significant problems of the ISAPRE system.

TABLE 4.2

Relevant Problems of the ISAPRE System, Chile, 2014

Problem	Description
Dual health insurance system	The existing ISAPRE and FONASA scheme involves opting out of the public system for those who choose an ISAPRE. This means that any person who chooses ISAPRE coverage is removed from the population pool, taking out a private contract with a company that competes against the other ISAPREs. A segmented system is thereby created, with ISAPREs legally permitted to discriminate according to the income and health status of beneficiaries
Captivity	As a consequence of age, sex, or poor health, some beneficiaries cannot choose the institution they wish to affiliate with. This means that a beneficiary of a given ISAPRE cannot move to other ISAPREs
Weak cost containment culture	The prevailing method used by ISAPREs to pay providers is fee-for-service, which gives providers a strong incentive to deliver more and lucrative items (WHO, 2007). In turn, ISAPREs translate the higher costs into higher health plan premiums
Market concentration and vertical integration	81% of ISAPRE membership is held by the four largest insurers. Some of the ISAPREs are vertically integrated, with private hospital and healthcare networks belonging to the same companies. The ISAPRE market creates barriers to the entry of new companies, which reduces competition
Insurers can arbitrarily increase prices of health plans	ISAPREs can unilaterally increase the prices of the health plans they offer every year. Since 2007, beneficiaries have been able to fight these decisions in the courts. Both the appeal and supreme courts have been favorable to beneficiaries' demands

Source: Author's own elaboration on information from Presidential Commission. (2014). Informe Final: Comisión asesora presidencial para el estudio y propuesta de un nuevo régimen jurídico para el sistema de salud privado. Retrieved from https://www. researchgate.net/publication/303255601_INFORME_FINAL_COMISION_ ASESORA_PRESIDENCIAL_PARA_EL_ESTUDIO_Y_PROPUESTA_DE_UN_ NUEVO_REGIMEN_JURIDICO_PARA_EL_SISTEMA_DE_SALUD_PRIVADO_2014.

Future Perspectives

One of the most pressing questions when it comes to the struggle for an integrated health insurance system in Chile is whether ISAPREs are part of a social security arrangement for health. If they are to remain a part of the social security system, ISAPREs can no longer discriminate according to income or health status. If, however, they wish to continue setting premiums according to individual risks, they can no longer expect funding from compulsory health contributions and will have to become private voluntary insurance companies.

While there is widespread consensus that the Chilean health insurance system needs reform, there is less agreement on just how things can be improved. Nevertheless, there is a growing sense that a social security arrangement for health that provides the same system for the entire population will provide the most effective solution.

Even representatives of the ISAPRE Association have incorporated social security concepts into their discourse (Institutos de Salud Previsional, 2016). Following the last presidential commission, which included experts with diverse political views, a unanimous conclusion was that any alternative solution to the current problems of the ISAPRE system should be oriented to building a social security system for the entire population, with fair financing of a universal fund a requirement of any proposed design (Presidential Commission, 2014). Figure 4.1 summarizes the Commission's minimal requirements.

While major changes to the Chilean health insurance arrangement are required, it is unlikely that health financing will follow the tax-funded Beveridge Model, even though the Chilean public healthcare provision

---

- Universality of the system. Everyone has the same right to health.
- Financing based on solidarity of the system.
- Comprehensive and integral healthcare activities including primary care as the basis for the development of the health system.
- Open affiliation to insurers and no discrimination according to individual risks.
- Insurance coverage for the entire life cycle.
- Community risk assessment for premium pricing instead of individual assessment.
- Providers payment mechanisms to ensure efficiency and sanitary efficacy.

FIGURE 4.1

Minimal agreed-upon requirements for a future health insurance system in Chile. (From Presidential Commission, Informe Final: Comisión asesora presidencial para el estudio y propuesta de un nuevo régimen jurídico para el sistema de salud privado, 2014. Retrieved from https://www.researchgate.net/publication/303255601_INFORME_FINAL_COMISION_ASESORA_PRESIDENCIAL_PARA_EL_ESTUDIO_Y_PROPUESTA_DE_UN_NUEVO_REGIMEN_JURIDICO_PARA_EL_SISTEMA_DE_SALUD_PRIVADO_2014.)

system has been modeled on the British National Health Service. On the contrary, the health insurance will be funded on the basis of payroll contributions. The specific design, however, is still controversial—as is illustrated by the following quote from the Presidential Commission: "The views of majority for the future is one of a single national fund with a national insurance arrangement and private insurers being complementary insurance; the views of minority propose a multi-fund system, which is also a social security arrangement" (Presidential Commission, 2014: 27).

Conclusion

Over its long history, the Chilean health system has provided a model for other countries. For many decades, it provided a model of an integrated system for developing countries (Roemer, 1991), and more recently, the privatization policies adopted during the 1980s have also been followed by other Latin American countries (Weiland, 2007). Current efforts to reinstate an integrated system will certainly be of interest globally.

5

Guyana

Paradigm Shift: From Institutional Care to Community-Based Mental Health Services

William Adu-Krow, Vasha Elizabeth Bachan, Jorge J. Rodríguez Sánchez, Ganesh Tatkan, and Paul Edwards

CONTENTS

Guyanese Data

- Population: 773,303
- GDP per capita, PPP: $7,818.9
- Life expectancy at birth (both sexes): 66.2 years
- Expenditure on health as proportion of GDP: 5.3%
- Estimated inequity, Gini coefficient: 44.5

Source: All data are from the World Health Organization and World Bank. Latest available data
used as at October 2017.
GDP = Gross Domestic Product
PPP = Purchasing Power Parity

Background

"Without mental health there can be no true physical health" (Chisholm,
1954). The relationship between physical and mental health (MH) is multifaceted. Psychological disorders increase the risk of contracting communicable and non-communicable diseases, and contribute to unintentional
and intentional injuries. Furthermore, many pathologies increase the risks
for psychological disorders, and this comorbidity not only complicates help
seeking and treatment, but also influences prognosis.

Psychological and psychoactive substance-related disorders are highly
prevalent throughout the world and are major contributors to morbidity, disability, and premature mortality. However, the resources allocated by countries to tackle this burden are generally insufficient, inequitably distributed,
or inefficiently used. This has led to a treatment gap that, in many countries, is
over 70%, as shown in Figure 5.1. The stigma attached to psychological disorders—which may lead to social exclusion and discrimination—compounds
this situation.

Community-based mental health (CMH) models that integrate MH components into primary healthcare (PHC) are the most effective way of closing
the treatment gap and ensuring that those in need receive MH care (Caldas
de Almeida et al., 2015). CMH also helps to ensure that available resources
are used efficiently.

Mental Health: Burden versus Response

Psychological disorders are a growing health problem in Guyana and the
Americas. In 2010, psychological and neurological disorders accounted for 13.9%
of the total burden of disease in Latin America and the Caribbean (LA&C),

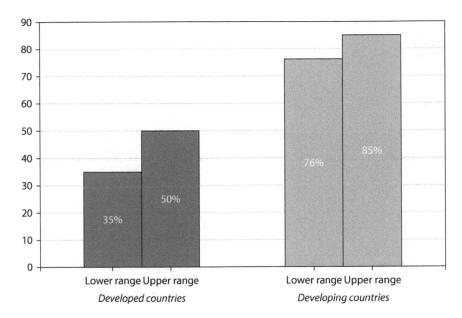

FIGURE 5.1
Treatment gap: people with psychological disorders receiving no treatment in LA&C, 2015. (From WHO [World Health Organization], *Investing in Mental Health: Evidence for Action*, WHO, Geneva, 2013).

as measured by disability-adjusted life years (DALYs), with depression being among the first three leading causes of DALYs region-wide (WHO, 2008a). A WHO projection shows that without any significant intervention, depression will be among the three leading causes of years lost due to ill health by 2030.

Based on a recent review of the most relevant epidemiological studies of psychological disorders conducted in LA&C, the estimated average prevalence rates in the adult population are 1.0% for non-affective psychoses, 4.9% for major depression, and 5.7% for alcohol abuse or dependence. The studies also revealed that large numbers of patients had not received any medical treatment, whether specialized or general: more than a third of those with non-affective psychosis, more than half of those with depression, and approximately three-quarters of those with substance abuse problems. In practical terms, only a minority of people who need MH care actually receive it.

The Mental Health System in Guyana

The bulk of Guyana's MH services are provided by two hospitals. The National Psychiatric Hospital (NPH) in Canje, Region 6, is a 150-bed facility that provides care for patients with chronic psychological conditions. The

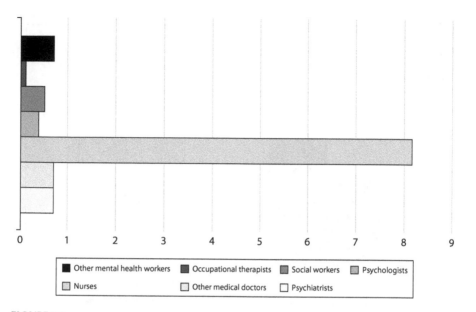

FIGURE 5.2
Guyana MH workforce, 2014 (rate per 100,000 population). (From WHO [World Health Organization], *Mental Health Atlas 2014*, WHO, Geneva, 2015).

Georgetown Public Hospital Corporation (GPHC), a national referral hospital, provides mainly outpatient care, but it also has six beds, split between men and women, and is generally only used for acute cases. Figure 5.2 shows the MH human resources in the country and their ratios per 100,000 of population for 2014. A small number of patients are seen at mobile clinics at Suddie in Region 2 and Parika in Region 3.

MH is increasingly recognized by governments as an area of emerging priority (DeSilva et al., 2014). Guyana has a decentralized healthcare system, where central ministries and regional administrations are given separate budgets. The MH budget for Guyana is split into three different budgetary agencies—the central Ministry of Public Health (MoPH), the GPHC Mental Health Unit, and the NPH in Region 6—with other regions given very little or nothing for MH.

Response of the Health System

Despite the magnitude of the burden of psychological disorders, the responsiveness of health services is still limited in LA&C. This is due to financial constraints and the structural limitations outlined above. As in many other low- and middle-income countries, where on average between 0.5% and 1.9% of the health

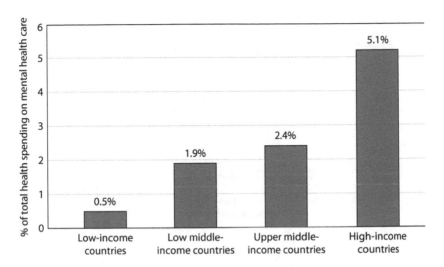

FIGURE 5.3

MH spending as a proportion of total health spending in 2013. (From WHO [World Health Organization], *Investing in Mental Health: Evidence for Action*, WHO, Geneva, 2013).

budget is spent on MH—sometimes less than US$2 per capita—levels of public expenditure on MH are very low in LA&C. This is significantly below spending levels in high-income countries, where 5.1% of the health budget is spent on MH—which may be upwards of US$50 per capita (WHO, 2013) (Figure 5.3).

The average proportion of the health budget spent on MH in those evaluated countries in the region is less than 2.0%, with 67% of these funds going to psychiatric hospitals. In 2011, 0.11% of the total Guyanese health budget was allocated to MH. The situation improved slightly in 2017, when this rose to 1.16%, 75% of which was allocated to psychiatric hospitals (Ministry of Public Health [Guyana], 2017).

The median number of MH workers—namely, psychiatrists, psychologists, and occupational therapists—is commensurately low: less than one per 100,000 of population. MH-trained PHC staff are essential, as they can strengthen the diagnostic and treatment capacity of patients with common psychological disorders.

Justification for Changing to a Community-Based Mental Health Model

An analysis of the structure and features of MH disorders by expenditure (Figure 5.4) shows the following:

- The annual prevalence of all psychological disorders is around 15–20% of the adult population.

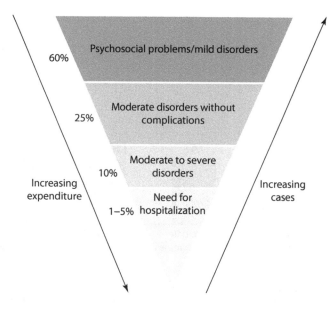

FIGURE 5.4

Distribution of prevalence of MH disorders in adults. (Adapted from Rodríguez, J., Community-based mental health model: Conceptual framework and experiences, presented at the Mental Health Situation in Guyana, Georgetown, Guyana, October 17, 2016.)

- 60% of that prevalence corresponds to mild disorders (disorders related to stress, mild depression, anxiety, somatic symptoms, etc.) or borderline conditions, like non-pathological psychosocial problems. These conditions do not usually require specialized care and should be handled in PHC or receive counseling, orientation, and psychosocial support.
- Moderate to severe psychological disorders constitute 25% of the remaining disorders, which usually need ambulatory care (outpatient services) by MH professionals, and may require prescription of psychotropic drugs.
- Only about <5% of cases have serious conditions that justify hospital admission.

As mentioned, the MH needs of the population require a better response at the PHC level, together with decentralized outpatient services.

CMH Model

Implementing a CMH service model is the most viable way to close the treatment gap and ensure that people get the MH care they need. The model

- Emphasizes primary care, including clinical care, with a focus on the prevention of both physical and psychological disorders
- Promotes comprehensive and community-based psychosocial rehabilitation

The paradigm shift will address existing health system limitations and barriers, including

- MH's current low priority in the public health agenda
- Heavily centralized health services
- Practical problems relating to the integration of MH into PHC
- Limited resources (financial, human, and infrastructural)
- MH-related social stigma and discrimination

The paradigm shift will include the coordination of health services, including the addition of psychotropic medications to the essential drug list, designing clear mechanisms for referral and counterreferral between PHC and MH services, and arranging patients' social reintegration after acute-phase hospitalization.

Additional input from other areas of the health and social systems will be required. These include

- Active collaboration, participation, and commitment from stakeholders, such as faith-based organizations (FBOs), non-governmental organizations (NGOs), and the community, to assist in the integration at the local level.
- Data collection and analyses with appropriate tools to assist in ongoing monitoring and evaluation for future planning.
- Academic collaboration, educational outreach, interprofessional education, problem-based learning, coaching, point-of-care reminders, and self-regulated learning. These should be applied to MH training programs to establish a successful MH workforce to deliver quality services (Lyon et al., 2011).

The general adult MH services will include outpatient and ambulatory clinics, community MH services, acute inpatient care, and long-term community-based residential care (Thornicroft et al., 2016).

The Reorganized MH System

The new Guyanese model will incorporate a MH unit made up of a central coordinating unit and a MH institute (Figure 5.5) and will include the following reforms:

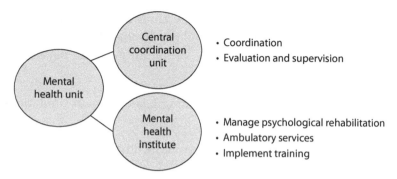

FIGURE 5.5
Structure of the reorganized MH system.

- General MH care services will be provided at PHC and general hospitals, eliminating the need for large centralized psychiatric institutions, and will be financed from the regional budget.
- Community residential facilities will be created to meet the needs of people with severe psychological disorders who show marked disability and/or difficulties with social reintegration for residential therapeutic services (RTS) for middle- to long-term care.
- MH services will be integrated and coordinated with many important sectors of government, including education, housing, employment, justice and social welfare, and local government, and with civil society, FBOs, the community, users, and relatives. There must also be coordination of care between the formal and informal caregivers of healthcare.
- Monthly stipends will be paid to stabilized and de-institutionalized patients to assist with living expenses and support their socialization back into society.

The reform will include the closing of the NPH and the review and reclassification of current inpatients in three dimensions, namely, clinical, disability, and social, to evaluate their needs and potential social reintegration. Those who have serious psychological disorders or who are not yet ready for discharge will be relocated to three new RTS facilities (halfway houses) that will provide comprehensive psychosocial services, including rehabilitation, for up to eight residents per facility, with care provided by psychiatric nurse practitioners. The integrated approach to the treatment of persons with psychological disorders will include psychosocial and pharmacological interventions.

Community care will be provided by MH nurses who will go into the community (patients' homes), and who will be supported by patients' families and communities. The components of the CMH program are shown in Figure 5.6.

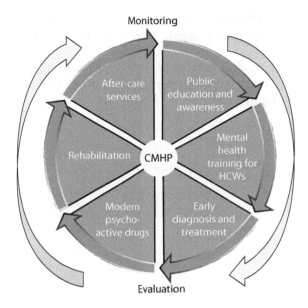

FIGURE 5.6
Components of the CMH program within PHC. HCWs, healthcare workers.

As in most other lower-middle income countries (LMICs), most of Guyana's MH services are currently provided by non-specialized staff (WHO, 2015b). Specialist MH training for PHC staff is therefore essential to the success of CMH. Adequate supervision, ongoing on-the-job training, effective and efficient referral and counterreferral network settings, regular monitoring and evaluation of patients, and the reliable supply of psychotropic medications will help to improve MH services (Hanlon et al., 2010).

Conclusion

Analysis of the existing situation will provide an accurate picture of what is required for the transition to CMH (Caldas de Almeida et al., 2015). However, it is clear that the coordination of all government sector and community healthcare services, as well as the coordination of formal and informal health networks, will be vital. Early diagnoses, ensuring appropriate and adequate treatment of patients, and the establishing of an active, accessible, and equitable CMH program will reduce the current MH burden and the need for inpatient beds.

A modern MH service must have a pragmatic balance of community and hospital care. Guyana has the conditions to move forward in an effective process of MH services reform, and the steps outlined above can catapult Guyana into a future that includes effective and sustainable MH.

With this paradigm shift, 85% of MH resources will therefore be used for 85% of MH patients instead of 85% being used for 15% of the MH population, as highlighted in Figure 5.4.

6

Mexico

Leveraging Conditional Cash Transfers and Universal Health Coverage to Tackle Non-Communicable Diseases

Jafet Arrieta, Enrique Valdespino, and Mercedes Aguerrebere

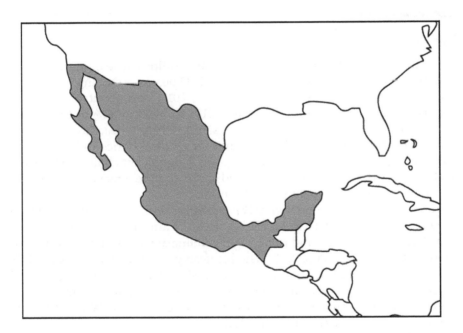

CONTENTS

Mexican Data

- Population: 127,540,423
- GDP per capita, PPP: $17,861.6
- Life expectancy at birth (both sexes): 76.7 years
- Expenditure on health as proportion of GDP: 6.3%
- Estimated inequity, Gini coefficient: 48.2%

Source: All data are from the World Health Organization and World Bank. Latest available data used as at October 2017.
GDP = Gross Domestic Product
PPP = Purchasing Power Parity

Background

Non-communicable diseases (NCDs) are the leading cause of mortality and disability worldwide, with nearly 80% of NCD deaths occurring in low- and middle-income countries (Beaglehole et al., 2011), and about 40% of them occurring before the age of 70 (WHO, 2011). In Mexico, 82% of the burden of disease is due to NCDs, and diabetes and ischemic heart disease represent the top two causes of death (Institute for Health Metrics and Evaluation, 2015). This growing burden of NCDs poses a significant challenge to health systems and demands innovative and more comprehensive and effective approaches to NCD prevention and control.

In 2003, Mexico launched *Seguro Popular* (SP) as the centerpiece of its System of Social Protection in Health (SSPH), with the aim of providing universal health coverage (UHC)* and reducing the financial risk for the Mexican population (Knaul et al., 2006, 2012). The Mexican government accompanied the implementation of SP with a conditional cash transfer (CCT)† program, initially known as *Oportunidades*, and currently *Prospera*. This program created a social support network for patients aimed at reducing social and financial barriers to healthcare (Knaul et al., 2012).

By 2012, the Mexican government announced that it had achieved UHC, with more than 50 million people, who were previously uninsured, newly enrolled in SP, and another 5.6 million of the most materially disadvantaged population now enrolled in *Oportunidades* (Knaul et al., 2012).

---

* "Universal health coverage is defined as ensuring that all people have access to needed promotive, preventive, curative and rehabilitative health services, of sufficient quality to be effective, while also ensuring that people do not suffer financial hardship when paying for these services" (WHO, 2017b).
† CCT programs provide cash transfers to female heads of household on the condition that their members engage in certain behaviors related to maternal and child care, health promotion and prevention activities, and health education.

However, despite the success of the health system in achieving UHC and reducing the health-related financial burden for the Mexican population, significant differences in access to healthcare between individual states, between rural and urban areas, and among people of varying socioeconomic status persist (Lozano et al., 2007; Nigenda et al., 2015). Also, despite the high prevalence of NCDs, approximately 50% of people living with NCDs do not receive any healthcare intervention, and the quality of care is variable among those with access to treatment (Lozano et al., 2007).

As the NCD epidemic continues to worsen—the prevalence of diabetes, for instance, reached 15% in 2012 (OECD, 2016d)—UHC presents a unique opportunity for both developing and developed countries to respond to the challenges the epidemic poses. However, given that the greatest health and economic burden of NCDs falls on the poor (Stevens et al., 2008), UHC will need to be accompanied by interventions aimed at reducing social and financial barriers to health service access. This chapter describes the potential of combining UHC with a CCT program to effectively respond to the NCD epidemic, and outlines some key lessons to guide their implementation based on the experience of Mexico.

Impact of Universal Health Coverage and Conditional Cash Transfers on Health Outcomes

The combination of UHC with a CCT program was chosen as an exemplar based on its proven effectiveness with long-term outcome evaluations, the capacity of the innovation to be integrated within the health system, the potential for this innovation to further address existing inequities in access and quality of healthcare, and the feasibility of the implementation, transferability, and scalability of this innovation to other countries (DeFulio and Silverman, 2012; Dolor and Schulman, 2013; Gertler et al., 2012; Ranganathan and Lagarde, 2012).

With the launch of SP, public investment in Mexico's healthcare system increased from 2.4% to 3.2% of the GDP between 2003 and 2013 (World Bank, 2015b). This increase in public investment allowed the government to develop the physical and human resources infrastructure needed to deliver healthcare services to the newly insured population (Knaul et al., 2012). Taken together, these interventions have helped to reduce supply-side barriers to healthcare services (Knaul et al., 2012; Nigenda et al., 2015; Nigenda, 2013).

Similarly, CCT programs in Mexico have proven effective in decreasing demand-side barriers to preventative and healthcare services, which has translated into more accessible reproductive and child health services, better health outcomes for children and older individuals, and increased preventative visits to screen for diabetes and hypertension (Fernald et al., 2008, 2009;

Gertler et al., 2012). Additionally, CCTs have enabled people living in hard-to-reach rural communities to enroll in UHC by ensuring that this sector of the population has more contact with the healthcare system (Biosca and Brown, 2015). Furthermore, evidence from Mexico and other countries suggests that CCTs have a positive impact on the utilization of health services, health-related behaviors, and health outcomes in low- and middle-income countries, by reducing or eliminating financial barriers to healthcare access (Bärnighausen and Bloom, 2009; DeFulio and Silverman, 2012; Dolor and Schulman, 2013; Gertler et al., 2012; Lagarde et al., 2009; Ranganathan and Lagarde, 2012).

Following these and other health system interventions between 2003 and 2010, the under-five mortality rate decreased from 28.5 to 16.8 per 1000 live births, and the mortality rate associated with communicable diseases and maternal, perinatal, and nutritional issues decreased from 15.4% to 10.8% (Knaul et al., 2012). Similarly, by 2013, 22.8 million people had received a screening for diabetes and hypertension, and those who screened positive had been linked to care (World Bank, 2015b). In that same period, the proportion of out-of-pocket (OOP) expenditure decreased from 58.2% to 35.7% for SP recipients (Knaul et al., 2012).

This evidence suggests that interventions to strengthen the supply side, such as UHC, should be implemented in combination with demand-side incentives, such as CCTs, thus boosting both health and economic benefits to the population.

Implementation and Transferability

Even though the combination of UHC and CCTs has proven effective in improving access to preventative and healthcare services, operating these programs effectively demands strong administrative coordination, robust financing mechanisms, and strong regulatory, accountability, and governance systems to ensure that resources are managed and distributed adequately (Baird et al., 2011; Nigenda et al., 2015; Secretaría de Gobernación, 2016).

In Mexico, SP and the CCT program have always operated autonomously, run by the Ministry of Health (MoH) and Ministry of Social Development, respectively; this has led to inefficiency, duplicity, and fragmentation. Furthermore, these challenges have been amplified in poorer states, where regulatory and administrative government systems are already weaker, leading to varying degrees of access and quality of healthcare services between individual states. Since SSPH was implemented through the decentralized structure of the MoH, current recentralizing efforts at the federal level to enhance accountability and efficiency have been met with resistance by state-level actors (Nigenda et al., 2015).

Such factors should be considered when trying to replicate this innovation in low- and middle-income countries with limited organizational structure and weak regulatory and governance systems.

Prospects

Efforts to strengthen the supply side through UHC approaches, along with the expansion of cash transfer programs within the primary care delivery platform, have the potential to unlock the capacity of these two innovations to address the rising epidemic of NCDs in Mexico and globally. Furthermore, by reaching the poorest segment of the population, this combination is strategically positioned to make sure that access to care and health outcomes is equitably distributed among the population.

Conclusion

The magnitude and rapid spread of the NCD epidemic poses a significant challenge to the health systems of low- and middle-income countries. However, this epidemic also offers an opportunity for countries to adopt innovative, comprehensive, and effective approaches to preventing and controlling NCDs. As CCTs bring the population closer to the health system, and UHC enhances the health system's capacity to respond to their needs; coupling these two programs more effectively will be crucial to changing the course of the epidemic, and to avoiding the future health, economic, and social impact of NCDs and their complications. If lessons from the Mexican experience are embraced, this combination of UHC and CCT programs, when well organized, coordinated, and regulated, has the potential to create responsive and adaptive health systems that will achieve better health outcomes for all.

7

Trinidad and Tobago

Nurse Training: A Competency-Based Approach

Claudine Richardson-Sheppard

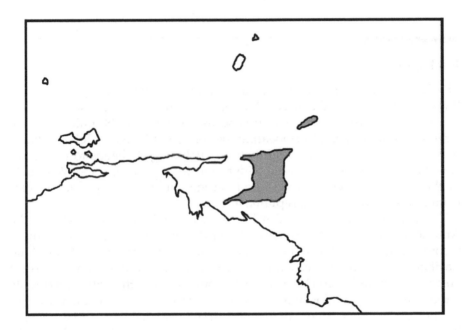

CONTENTS

Trinidadian and Tobagonian Data

- Population: 1,364,962
- GDP per capita, PPP: $31,907.8
- Life expectancy at birth (both sexes): 71.2 years
- Expenditure on health as proportion of GDP: 5.93%
- Estimated inequity, Gini coefficient: 40.3%

Source: All data are from the World Health Organization and World Bank. Latest available data
     used as at October 2017.
GDP = Gross Domestic Product
PPP = Purchasing Power Parity

Background

The Republic of Trinidad and Tobago is a twin-island country situated off
the northern edge of the South American mainland, lying just 11 km (6.8
miles) off the coast of northeastern Venezuela and 130 km (81 miles) south of
Grenada. Bordering the Caribbean to the north, Trinidad and Tobago shares
maritime boundaries with other nations, including Barbados to the north-
east, Grenada to the north-west, Guyana to the south-east, and Venezuela
to the south and west. Trinidad and Tobago obtained independence from
Britain in 1962, becoming a republic in 1976.

Trinidad and Tobago is the third richest country by GDP per capita in the
Americas, after the United States and Canada, and it is recognized as a high-
income economy by the World Bank. Unlike most of the English-speaking
Caribbean, the country's economy is primarily industrial, with an emphasis
on petroleum and petrochemicals. The country's wealth is attributed to its
large reserves and exploitation of oil and natural gas. While the government
publicly stated its commitment to quality healthcare delivery to all its citizens
in 2008, this wealth has not yet been translated into an effective public healthcare system. In light of this, the government of Trinidad and Tobago has
recently embarked on a program of comprehensive health system reforms.

One key objective of these reforms is to establish new administrative and
employment structures, which encourage accountability, increased autonomy, and appropriate incentives to improve productivity and efficiency.
Healthcare in Trinidad and Tobago is organized into five regional health
authorities, each of which has responsibility for acute hospital care, primary
care, mental healthcare, and social care within its geographic boundaries.
There are 117.5 physicians per 100,000 of population, and primary care physicians make up 72.85% of the workforce. Care is paid for by a health surcharge
withheld from the public and private earnings of all employed, and there is
universal access. There is no fee at point of delivery in the public institutions.

Are Nurses the Problem?

In a 2015 newspaper article, *Trinidad Express* columnist Raffique Shah claimed that the public medical institutions in this country are in crisis (Shah, 2015). Responding to a deluge of health system "horror stories," Shah argued that Trinidad and Tobago's network of district health centers and handful of hospitals have failed to fulfill their mandate.

The district health centers Shah referred to are basic clinics that provide pharmaceuticals, primary care, and continued treatment for those suffering from chronic lifestyle diseases. While Shah conceded that service is usually prompt and professional at these centers, the nation's main hospitals, especially the big three—San Fernando, Port of Spain, and Mount Hope—are another matter. It is common to hear stories of unacceptably long waiting times for beds, and patients forced to endure unnecessary pain and discomfort. While there are probably multiple challenges, many of the hospitals' problems have been attributed to inadequate nursing care.

Shah's was not the first high-profile complaint. In 2014, the CEO of the South-West Regional Health Authority (SWRHA) publicly admonished the region's nursing staff for excessive absenteeism (*Trinidad Express*, 2014). The CEO also criticized the nurses' attitude toward their patients, claiming that many nurses seemed to lack the necessary compassion. While San Fernando Hospital's nursing manager defended her staff, citing shortages and maintaining that there were very few complaints from patients (*Trinidad Express*, 2014), the claims echo those made a few years previously by the Minister of Health (Clyne, 2012). While such critiques can be dismissed as anecdotal, the general view of the population, judging from letters to newspapers and comments on call-in radio programs (Mathura, 2013; Rigsby, 2011), is that nursing care in Trinidad and Tobago requires significant reforms if the quality of care is to be improved.

Deficiencies in Current Nurse Training

As nurses from Trinidad and Tobago are in great demand internationally because of their technical competence, the deficiency seems to be in their supervision.

A review of the four institutions that train nurses in Trinidad and Tobago reveals that their curriculum is primarily clinical and technical, and includes little or no training in human interaction skills, supervision, management, or leadership. Thus, nurses are often unprepared for responsibilities that require more than technical competence. This has far-reaching consequences when it comes to supervisory and management positions.

The new nurse manager, for instance, who is often promoted from within the hospital, has to make a complete transition. One day, she is a peer with her colleagues on the nursing unit, seeing them socially, sharing unit stories. The next day, she is a nurse manager with entirely new responsibilities and accountable for outcomes. Major adjustments are required and the learning curve is steep. These adjustments include

- Developing alliances at all levels across the organization
- Recognizing, understanding, and dealing with staff who may have coveted the management position
- Being scrupulously fair and equitable, regardless of previous relationships, in scheduling time, allocating holidays and overtime, and showing consideration and flexibility in meeting staff scheduling requests
- Demonstrating a commitment to 24-hour accountability and availability by regularly visiting all shifts, giving employees timely annual performance appraisals, and working closely with off-shift supervisors to ensure that patient care standards are maintained
- Dealing with unions (if applicable) as a manager, not a union member
- Doing what it takes to get results, regardless of shift requirements or assigned tasks
- Gaining a new perspective that extends beyond the nursing unit to include the overall plan for the hospital
- Developing an attitude of "ownership" toward the organization, and having some understanding of hospital finances
- Getting on the "same side of the table" as senior leaders in the hospital, and being a positive, professional voice for the organization

Without relevant education and training, it is far more challenging for nurses to become good nursing supervisors or managers. The development and use of competencies has proven to be an efficient and effective strategy for skill and talent development and could be successfully used in the training of nursing supervisors and managers.

Competency Models

The development and application of competency models is widely considered to be one of the most effective ways to improve organizational performance. Competencies identify the observable behaviors of top performers—not just what they do, but how they do it. Once the competencies required for a specific role are identified, a clear blueprint for success becomes evident (Human Resource Services Group, 2017).

Competencies are tailored to each specific job. The competencies required for an administrative employee, for instance, differ from those required in front-line nursing. A 2010 study into the development of a competency model at a major hospital in Trinidad ranked the four most desirable competencies for the position of head nurse as follows (Sheppard, 2010):

1. Establishing focus
2. Providing motivational support
3. Building collaborative relationships
4. Personal credibility

These can easily be classified as leadership competencies in which the leader establishes the focus and direction, while motivating and supporting staff.

While the top four competencies emphasize direction and leadership, the next four focus on organization, teamwork, and client orientation. These four competencies are

1. Customer orientation
2. Results orientation
3. Fostering teamwork
4. Managing performance

These were referred to as the performance cluster of competencies.

The following four competencies could to be referred to as the people management cluster:

1. Attention to communication
2. Empowering others
3. Influencing others
4. Developing others

The implementation of a competency-based approach will improve the supervision and management of nurses and should ensure that the public is less likely to be critical of nurses in the future. It is, however, critical that the proposed competency model be validated, that the curriculum of the various nurse training institutions be modified to include training in the selected competencies, and that existing supervisors and managers be trained to enable them to develop the required competencies.

Validation of Competencies

Validation is the process of verifying that a chosen competency model is appropriate for a specific environment. It involves checking the model to ensure that it adequately reflects the knowledge, skills, and abilities that staff must demonstrate. Competency models may be validated using questionnaires, focus groups, or expert panel reviews. Any selected competency model must be defensible; therefore, it is critical that they accurately reflect required on-the-job behaviors.

Enhancing the Curriculum

Once the supervisory and managerial competencies have been verified, the next step is to modify the curriculum of the institutes that train nurses. These competencies are not needed for basic clinical nurse training but should be included as modules for advanced training, preferably in a postgraduate diploma or master's degree, for those nurses who aspire to senior positions. It would be ideal if assessments were based on competency-type methodology rather than using traditional educational methods. For instance, assessments could be carried out in assessment centers, giving the candidate an opportunity to demonstrate a wider range of skills than is possible during a traditional written examination or face-to-face interview (Swain, 2017). Candidates will work both individually and as part of a group on a variety of exercises, including

- Case studies
- Group discussions
- In-tray exercises
- Presentations
- Psychometric tests
- Role play
- Social events
- Written tests

Traditional examinations gauge nurses' ability to recall facts and structure responses to examination questions, basically measuring theoretical knowledge and academic prowess. Actual patients on a ward, however, need clinical care delivered in a humane, customer-friendly manner, while the junior nurses who care for them need strong technical leadership, guidance, coaching, mentoring, and feedback. Competency-based education (CBE) programs provide an alternative approach to advanced education, enabling students to progress at their own pace, ensuring that they gain relevant and demonstrable skills, and providing a clearer signal of what graduates know and are able to do (McClarty and Gaertner, 2015). The CBE model has proven particularly attractive for non-traditional students who are juggling work and family commitments that make conventional higher education class schedules unrealistic (McClarty and Gaertner, 2015).

Implementation and Transferability

The successful development of competency-based models will require more than curriculum modification, however. Educational institutions will also have to train existing staff to deliver the new supervisory and management modules or hire staff with the relevant qualifications and experience.

In Trinidad and Tobago, the University of the West Indies (UWI) School of Nursing is the only institution to have taken this approach, with supervisory, management, and nursing leadership modules in its bachelor's and master's programs. The School of Nursing's mission is to prepare nursing professionals for leadership in nursing education, management, and clinical and other professional roles within the healthcare delivery service. Since its establishment in 2004, the School of Nursing has launched two undergraduate degree programs, the bachelor of science (BSc) nursing in 2005 and the BSc nursing (oncology) in 2007, and has continued to position itself as a major hub of academic development and advancement for nursing and midwifery personnel in Trinidad and Tobago and the wider Caribbean region (*UWI Campus News*, 2009). In 2009, the UWI School of Advanced Nursing Education introduced the master of science in advanced nursing, which is designed to equip a new generation of nurses to assume leadership roles in the areas of nursing education, management, clinical specialties, and other professional positions within the healthcare system. This is the only accredited degree of its type in the country, and it will help meet the Caribbean region's need for high-quality healthcare.

Prospects

In the last 8 years, 54 master of science in advanced nursing students have graduated from UWI. They are now working as head nurses, nursing administrators, and nursing supervisors. Further success will hinge on the willingness of other nursing training schools to modify their curricula and prepare graduates for leadership positions.

Conclusion

Clinical and technical training, while vital, do not guarantee excellent nursing care. Given the increasingly complex nature of both medical institutions and nursing itself, nurses must be trained in leadership, management, and supervisory skills in order to deliver excellent patient care. The nurse leader in first-line patient care management fulfills a crucial role in planning, implementing, and evaluating patient care outcomes. This complex role requires flexible leadership skills, the ability to recognize both short-term and long-term goals, and the ability to design and implement strategic plans (Carrol, 2006). These skills are not learned or developed by chance, but need to be carefully nurtured and developed. If transferable competencies are to be formally developed, training in supervision, management, and leadership must be provided in all nurse training curricula.

8

The United States of America

The U.S. Healthcare System: A Vision for the Future

Robert H. Brook and Mary E. Vaiana

CONTENTS

American Data

- Population: 323,127,513
- GDP per capita, PPP: $57,466.8
- Life expectancy at birth (both sexes): 79.3 years
- Expenditure on health as proportion of GDP: 17.14%
- Estimated inequity, Gini coefficient: 41.1%

Source: All data are from the World Health Organization and World Bank. Latest available data
used as at October 2017.
GDP = Gross Domestic Product
PPP = Purchasing Power Parity

Introduction

During his campaign, Mr. Trump had consistently pledged to repeal the
Affordable Care Act (ACA), and as soon as he was inaugurated on January
10, 2017, the Republican-controlled Congress announced its intention to
repeal and perhaps replace all or part of the ACA with a better plan. While
their initial attempt to do so failed, healthcare remains an important policy
issue and is sure to be revisited.

It is therefore perhaps a propitious time to articulate a vision for the future
U.S. healthcare system.

Essential components of that vision already exist, thanks to major achievements in the field of health services research over the past five decades. We
know how to measure health status and the appropriateness and quality of
care. We understand the link between how much people pay for care and
how much they use. We know that how physicians are paid for care influences the way they practice. We know that depression is one of the leading
causes of morbidity in the world. We realize that the most powerful determinants of health are not the newest drugs or surgical techniques—rather, they
are social determinants, such as education and income.

But how might this science manifest itself in a future healthcare system?

Health Status Measurement Will Be Essential

The ability to measure health status is arguably the most important scientific
development in the last 50 years of health services research. We have reliable

measures of health status across multiple dimensions—social, mental, physical, and physiological (Brook et al., 1979).

In the healthcare system of the future, reliable measures of health status will be the beginning point in any medical history, physical examination, or community assessment; any changes in this status would be used to trigger diagnostic workups, clinical interventions, or community actions.

Consumers Will Know When and How to Use the System

Health services research demonstrates that giving people free care is more expensive than care requiring copayments because, as one might expect, people consume more care. However, free care does not make them healthier: some of the additional care can cause more harm than good (Brook et al., 1983; Chernew and Newhouse, 2008; Newhouse and the Insurance Experiment Group, 1993).

In an ideal system, people would know how and when to use the healthcare system. They would know when an emergency department visit is necessary and when wait and see is the best approach. They would know how to ensure that they receive only appropriate care from their physicians.

Quality of Care Will Have a Central Role

In our future system, patients and doctors will use information about quality on a real-time basis to improve care and to correct policies that are not having their intended effects (McGlynn et al., 2003).

Patients and providers in our future system will also use real-time information about quality to be savvy consumers. For example, is a particular managed care organization, health maintenance organization, or accountable care organization a good buy? If individuals have a chronic disease, such as diabetes or hypertension, they will be able to get a monthly report on their smartphone that describes the quality of care they are receiving, based on broadly accepted standards.

There Will Be Benefits from Updated Information About Appropriateness of Care

We currently have no up-to-date information about the proportion of care that is less than appropriate, that is, when the potential health benefit does not exceed the health risk. Instead, we use findings from studies done decades ago, which show that as much as one-third of care is less than appropriate (Bernstein et al., 1993; McGlynn et al., 1994).

In our future system, the studies will be updated. People will use the results to assess the effectiveness and efficacy of services they are receiving and to identify services that are inappropriate for them (Brook et al., 1986).

Geography Will Not Affect the Amount or Quality of Care a Patient Receives

In our system of the future, decision makers will work to sever the relationship between where one lives and the level of care received because they will want the policies they implement to be viable everywhere (Chassin et al., 1986; Leape et al., 1990; Newhouse and Garber, 2013). Living in Boston instead of Des Moines will no longer predict (everything else being equal—e.g., age and weight) how many tests for diabetes an individual receives.

Care for Physical and Mental Health Conditions Will Be Integrated

In our future system, policies will be evaluated according to the way they integrate care and information about mental health, such as depression and substance abuse, with traditional medical care. Indeed, physical and behavioral healthcare should be integrated not just with each other, but with many other aspects of social and political life (Sherbourne et al., 2008; Wells et al., 1996). In our future system, leaders will strive to merge the roles of medical care, public health, and social determinants of health into one model. Perhaps even intolerance of others will be addressed successfully by the medical system (Brook, 2017).

Health System Leaders Will Understand and Address Social Determinants of Health

In our ideal system, policies will be put in place to reinforce the role of the healthcare system in helping people achieve health even when aspects of health depend on factors outside the traditional medical system (Goldman and Smith, 2011; Marmot, 2004; Smith, 2007). Healthcare professionals will also try to communicate important health-related dimensions of their environment—for example, the opportunities to walk that grid cities offer (Marshall and Garrick, 2010; Marshall et al., 2014).

Decision makers in our ideal system will understand the importance of effective communication between those who are responsible for implementing public health and medical care and those responsible for eliminating social determinants—such as failure to complete high school—that affect a person's health. They will agree that, in principle, policies to control costs that are directed at improving health should be judged by how effectively they promote integration in these three areas, and by the degree to which they alter the physical environment to make it easier for people to maintain health.

Prospects

The difference between the ideal healthcare system just described and the current U.S. healthcare system is vast. It will require comprehensive, disruptive change to bridge the gap.

What ideas would society and physicians contribute to the vision of our future healthcare system? Perhaps something like the following:

- All communities will have a health plan that promotes an environment in which all people can thrive, and provides an integrated set of social and health services.
- All high school graduates will understand how health is produced.
- Education and health policies will be replaced by people policies that target the interaction between health and education as the way to improve a community's health.
- Many face-to-face physician visits will be replaced by video encounters, encounters with computers and people in the community, or self-directed care— approaches that will lower costs, but still be as effective as traditional patient–clinician interactions.
- Medical expertise will be shared so that all people have immediate access, when needed, to world experts via the Internet.
- Expensive equipment will be widely shared so that it can be used before it becomes medically obsolete.

Examples of disruptive change in healthcare might include globalization of labor, making it possible for healthcare clinicians from other countries to offer care in the United States at a cost lower than care offered by U.S.-based clinicians. Disruptive change might also involve providing Medicaid benefits, valid only in Mexico, to American citizens to cover long-term nursing home care.

Disruptive innovations are risky, but we face immense challenges. Our solutions to them need to be commensurately big. The key findings from health services research can keep us honest.

9

Venezuela

Learning from Failure and Leveraging Technology: Innovations for Better Care

Pedro Delgado, Luis Azpurua, and Tomás J. Sanabria

CONTENTS

Venezuelan Data

- Population: 31,568,179
- GDP per capita, PPP: $18,281.2
- Life expectancy at birth (both sexes): 74.1 years
- Expenditure on health as proportion of GDP: 5.26%
- Estimated inequity, Gini coefficient: Unavailable

Source: All data are from the World Health Organization and World Bank. Latest available data used as at October 2017.
GDP = Gross Domestic Product
PPP = Purchasing Power Parity

Background

Venezuela's current humanitarian crisis (Human Rights Watch, 2016) presents both challenges and opportunities. Challenges include the shortage of medications (Brocchetto, 2017), the massive emigration of a large number of medical professionals (Ramones, 2016), the precarious physical infrastructure of public institutions (Watts, 2016), and shifting epidemiological and demographic tendencies, such as the growth in the rate of non-communicable diseases and the significant increase in the number of people over 70 years of age (Institute for Health Metrics and Evaluation, 2017). These, alongside a financial collapse (*The Economist*, 2017) that has seen annual inflation rates of up to 800% and gross domestic product shrinking by almost 20% (Pons, 2017), have put unprecedented pressure on the country's public health and healthcare system (Fraser and Willer, 2016).

The constant churn in healthcare governance in Venezuela—with 16 ministers of health since 1998 (Flores, 2014)—has resulted in the lack of a coherent healthcare strategy to address urgent challenges (Observatorio Nacional de Salud, 2016). Therefore, accessible quality healthcare is the exception rather than the norm, particularly in remote areas, and millions are suffering as a consequence (Fraser and Willer, 2016). Over the years, and particularly since the start of the leftist leadership of Hugo Chávez in the late 1990s, the government has tried to adopt Cuban models and ideas to address these challenges, including two failed innovations that we will briefly describe (Aponte-Moreno and Lattig, 2012).

In 2003, a primary healthcare program, *Barrio Adentro*, was established in the poorer areas of the major cities. This program involved the establishment of small clinics, run by more than 20,000 Cuban health personnel who were also residents in the *barrios*. These personnel were identified as doctors, despite frequently lacking proper medical training. The goal of the *Barrio Adentro* program was to improve access and care quality by providing care close to people's homes. However, over the years many of the facilities have been abandoned and are now derelict (Delgado and Azpurua, 2017), and the program has continued to fail to meet its goals (Jones, 2008).

The second innovation, commenced in 2005, is a new local program of accelerated training to develop holistic community doctors or *medicos integrales comunitarios* (MICs) in non-autonomous Venezuelan universities controlled by the government, without the endorsement of the Venezuelan Medical Federation (Ascencio, 2017). The government's ambition is to develop 50,000 MICs by 2019, largely to work in *Barrio Adentro* facilities (Ministerio del Poder Popular para la Comunicación y la Información [Venezuela], 2016). However, the relative skills of the MICs have been called into question due to the low levels of admission criteria, lack of medical research during training, and the graduate doctors' poor performance in standardized testing (Ascencio, 2017; Matiz Cortes, 2015). The emigration of many of the MICs—including some of the Cuban doctors originally brought to Venezuela to teach—due to poor working conditions, low wages, lack of basic personal safety, and better opportunities abroad (Ramones, 2016; NTN24, 2017) has further reduced access to quality healthcare.

Success out of Failure?

Unprecedented challenges in Venezuela's healthcare system require innovative solutions. With some adjustments and improvements, these two failed efforts to implement well-meaning innovations have the potential to be turned into assets in the future, given the physical and human infrastructures already developed.

With modifications to the current approach, we believe that the role of MICs can be potentiated to a higher level so that they have an increased capacity to implement primary care programs designed for disadvantaged communities, or those who lack access to healthcare.

MICs have the potential to become the lynchpins in the creation of networks in which there is a permanent contact between the primary levels of healthcare, located in rural or suburban areas, and the secondary and tertiary levels, located in cities where most resources are allocated. We suggest that a basic retraining of these doctors to refine core clinical skills, along with a formal revalidation program, will strengthen their professional impact and status, and serve as a guarantee to those who are assessed or treated by MICs. In addition, a targeted investment to improve the nursing curriculum and enhance the working conditions of nursing professionals will be essential in the future.

The *Barrio Adentro* buildings have the potential to provide adequate physical infrastructures for care delivery in the community, either in person or virtually. We suggest that reclaiming these facilities and updating their basic structures will be more cost-efficient than building new health-care centers. Furthermore, with the continued expansion of mobile technology infrastructure and increased accessibility throughout the country, individuals' homes can also be used as venues for healthcare assessment or treatment.

A powerful approach to improving health and healthcare in Venezuela could emerge if the Maniapure Project, a successful non-governmental initiative that leverages technology to provide access to healthcare for rural communities, were to be expanded in combination with these proposed reforms to the MIC and *Barrio Adentro* programs.

Centro La Milagrosa, a tiny clinic in a mostly rural community built beside the Maniapure River in the Venezuelan state of Bolivar—the nation's gateway to the Amazon rainforest—provided the seed for what is now the Maniapure Project (Sanabria and Orta, 2012). In 1995, a joint venture was originated by a non-governmental organization (NGO) and a group of health-related professionals with the objective of providing access to care to the Maniapure community by leveraging the power of telecommunications to connect healthcare needs (of the population) with remote knowledge (of healthcare professionals from large urban areas). The use of technology evolved rapidly in Maniapure: from two-way radio use initially, to satellite phone, then fax, to satellite Internet, which was introduced in 1999.

The arrival of the Internet in the region presented a brilliant opportunity to harness technology to improve rural medical practice, and thus two additional clinics (La Urbana and Guarray) were created using the same model. These clinics covered 10,000 inhabitants in an area of approximately 1500 km² (579 square miles). Healthcare students from national universities became interested in the project, and alliances were made with international medical schools. The Maniapure Foundation was created, taking advantage of the 2005 Science, Technology and Innovation Law (LOCTI) issued by the National Assembly (Ministerio del Poder Popular para Educación Universitaria, Ciencia y Tecnología [Venezuela], 2010), which provided the basis for private corporations to invest in information technology hardware and software, hence making it feasible for the foundation to access much-needed infrastructure resources. The law also opened the door for financing the training and coaching of healthcare professionals from government-funded rural clinics, which were usually underutilized.

With the success of the Maniapure Project, in 2010 the approach expanded to other rural clinics in Venezuela, as well as to more than 50 small clinics in Colombia, Ecuador, and Bolivia. The Maniapure Model, in which telemedicine encompasses the three levels of healthcare delivery, has become a "leapfrogging" experience—that is, the model used new technology to accelerate system-level improvement, thereby skipping over previously unavoidable development stages (World Economic Forum, 2014). The virtual triage center (VTC) is the core of the model. VTCs, which are set up at assigned medical centers in urban areas and staffed by highly skilled professionals, receive voice, data, and images through a limited bandwidth from rural clinics and provide timely healthcare advice and guidance. This approach employs the redesign principle of "moving knowledge, not people," described by Berwick et al. (2015).

Impact

To illustrate the potential impact of the Maniapure approach, let us assume that a typical primary care center covers a population of between 3000 and 7000 people and the usual clinic workload is between 200 and 400 patients per month. The in-house primary care physician can typically meet the healthcare needs of 95% of cases; 5% may need a second opinion or quick referral. If these cases are consulted, 80% can be resolved locally with guidance from the VTC described above (Figure 9.1).

Maniapure's data shows that only one in 100 patients (1%) need to travel from their community to receive specialized medical attention elsewhere. The Maniapure Model increases access to both primary care and subspecialty expertise, which would otherwise be absent, directly within the community, thereby improving the quality of healthcare delivery for these rural populations. Although we cannot accurately describe the potential impact of the suggested combination of MIC expertise, *Barrio Adentro* facilities, and Maniapure's telemedicine approach, we can confidently predict that the combination would lead to a significant improvement to the current healthcare system and provide care to many who, at present, do not have the possibility of accessing it.

FIGURE 9.1
The Maniapure Project, Venezuela (Tomas Sanabria).

Implementation and Transferability

The Maniapure Project appears to be highly transferable, given its expansion in Venezuela and other South American countries. If this model can be used in combination with existing MICs and the *Barrio Adentro* infrastructure, and is proven effective, it too will provide a transferable prototype.

Other challenges to transferability include the availability of the information technology (IT) infrastructure needed to support the approach, the need to rebuild some of the derelict *Barrio Adentro* structures, inadequate numbers of primary and community care professionals, and the reluctance of some professionals to modify their work practices to leverage technology and partner with communities.

Prospects

The success of the proposed approach relies on bringing together key stakeholders—including the government, healthcare professional societies, Maniapure representatives, existing MIC academic institutions, and communities in need—in order to start building a unified vision of what is possible. In the process, setup tasks described in the scale-up framework outlined by Barker and colleagues (2016) should be tackled, such as a clear articulation of what needs to be scaled up, plans made for a full-scale model, and identification of initial test sites, early adopters, and potential champions of the intervention.

Conclusion

This chapter describes how Venezuela might learn from the failure of previous endeavors and use information technology to improve its healthcare system, particularly for individuals living in rural areas. By combining the potential of MICs, the *Barrio Adentro* infrastructure, and the Maniapure Model, high-quality care could be provided in sustainable ways to populations that are currently neglected. We are under no illusions that the task of moving from the current state to the envisioned improved future state is simple. Nevertheless, our commitment to sustainably achieving better care and better health for all Venezuelans remains steadfast, and our ambition is that the content of this chapter will provide optimism and a potential path forward.

Part II

Africa

Stuart Whittaker

Africa is the world's second-largest and second most populous continent (the first being Asia). It covers 20.4% of the planet's total land area. With 1.2 billion people as of 2016, Africa accounts for about 16% of the world's human population.

The continent varies greatly with regard to environments, economics, historical ties, and systems of government. Present political states in Africa originated from a process of decolonization by European countries in the twentieth century.

Most of Africa's 54 sovereign countries have borders that were drawn during the era of European colonialism. Since colonialism, African states have frequently been hampered by instability, corruption, violence, and authoritarianism. The vast majority of African states are republics that operate under some form of the presidential system of rule. However, few of them have been able to sustain democratic governments on a permanent basis, and many have instead cycled through a series of coups, producing military dictatorships.

While healthcare may not be the first priority of many governments in Africa, much is being done to improve population health and system efficacy. Part II consists of four individual country contributions and one overarching regional chapter. These five chapters demonstrate the most common problems encountered by African countries, their challenges, and the efforts being made to address them.

Bruce Agins, Joshua Bardfield, Margaret K. Brown, Daniel Tietz, Apollo Basenero, Christine S. Gordon, Ndapewa Hamunime, and Julie Taleni Neidel discuss Namibia's management of the ongoing HIV epidemic, showing how patient and community involvement can transform the traditional boundaries of the doctor–patient relationship. They describe a national quality program developed at all levels, and creating partnerships in clinical care where patients' voices are given value and applied in care decisions.

The overall efficiency of the Nigerian health system requires further enhancement if the growing and diversified needs of the Nigerian population are to be met, and their health outcomes improved. Emmanuel Aiyenigba explains that to achieve this, the fundamental design of the system needs to be changed and explores the potential of Lean thinking principles in showing the way.

Writing from South Africa, Stuart Whittaker, Lizo Mazwai, Grace Labadarios, and Bafana Msibi investigate the many challenges involved in developing, promulgating, and implementing nationally mandated health system standards. The difficulties of ensuring health establishment compliance and the limitations of external evaluation are discussed, as are programs developed by the Office of Health Standards Compliance to reduce such deficiencies in the system.

Roger Bayingana and Edward Chappy discuss the potential of Rwanda's One Health initiative to combat emerging and reemerging infectious diseases. The initiative promotes collaboration between multiple disciplines and sectors, integrating disease surveillance and response systems to attain optimal health for people, animals, and the environment.

Finally, Jacqui Stewart and Shivani Ranchod provide a comprehensive overview of African healthcare, discussing the commonalities across the diverse sociopolitical, economic, and anthropological landscape of the continent. Their research identifies shared epidemiological and demographic themes, including population dynamics, disease burden, financial exclusion, and supply-side constraints, and looks at efforts to achieve healthcare equity across the continent.

10

Namibia

Lessons from Patient Involvement in HIV Care: A Paradigm for Patient Activation and Involvement across Health Systems

Bruce Agins, Joshua Bardfield, Margaret K. Brown, Daniel Tietz, Apollo Basenero, Christine S. Gordon, Ndapewa Hamunime, and Julie Taleni Neidel

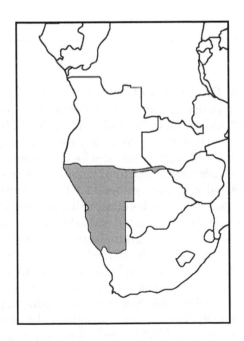

CONTENTS

Namibian Data

- Population: 2,479,713
- GDP per capita, PPP: $10,585.0
- Life expectancy at birth (both sexes): 65.8 years
- Expenditure on health as proportion of GDP: 8.93%
- Estimated inequity, Gini coefficient: 61.0%

Source: All data are from the World Health Organization and World Bank. Latest available data
used as at October 2017.
GDP = Gross Domestic Product
PPP = Purchasing Power Parity

Background

The New York State Department of Health (NYSDOH) AIDS Institute was created in 1983 in response to the AIDS epidemic. The mission of the AIDS Institute is to protect and promote the health of New York State's diverse population through disease surveillance and the provision of quality healthcare, preventative measures, and support services for those impacted by HIV, AIDS, sexually transmitted infections, and viral hepatitis; intravenous drug users; the lesbian, gay, bisexual, and transgender (LGBT) community; and those with related health concerns. This includes eliminating new infections and improving the health and well-being of infected and affected communities. The AIDS Institute's HIV Quality of Care Program, launched in 1992, is responsible for systematically monitoring the quality of medical care and supportive services for people living with HIV (PLWH) in New York State to achieve patient-centered care and positive health outcomes (Agins et al., 1995).

At the epicenter of the AIDS crisis, New York became the birthplace of a vibrant culture of activism among PLWH, driven by a groundbreaking document entitled the "Denver Principles" (UNAIDS, 1983). A central recommendation of the Denver Principles is to involve PLWH at every level of decision-making, setting the stage for consumers to provide recommendations in national HIV/AIDS policy and program development, service delivery, and evaluation. Involving consumers in these capacities also provided a basis for the Joint United Nations Programme on HIV and AIDS (UNAIDS) principles of greater involvement of people living with HIV (GIPA) (UNAIDS, 2007) (Figure 10.1).

Figure 10.1 presents the fundamental actions required to realize the rights and responsibilities of PLWH, formalized at the 1994 Paris AIDS Summit and endorsed by the United Nations in 2001.

Patient and Community Involvement

Community engagement and capturing the patient experience is key to understanding priorities for improving care. In low- and middle-income countries (LMICs), patients often have priorities that are very different from those in wealthier nations. For example, patients who prioritize mental health needs or food provision will not adhere to treatment until these needs are addressed. Gaining buy-in from communities and PLWH networks at local, provincial, and national levels or representing key populations is key to the success of improvement activities. Involving the community in code-signing measures and coproducing healthcare enhances participation and contributes important dimensions of quality and related experiences that are not accounted for otherwise.

In the context of LMICs, where patient and community involvement continues to evolve, it is important to examine the core elements of the NYSDOH HIV Quality of Care Program to understand how this model can be adapted—not only to stretch scarce resources to optimize treatment delivery, but also as a core component of a national quality program.

On a global scale, the World Health Organization (WHO) and others are adopting frameworks to ensure that public health systems promote and champion patient involvement as part of the people and family-centered care framework to improve patient outcomes and public health systems (Figure 10.1). As patient and community groups evolve from loosely organized structures into more formal designs, and as their role is integrated into health systems, resources can be stretched even further to bridge gaps in patient care and local capacity, as dedicated disease-specific funding wanes. As HIV-specific patient involvement programs advance, these models can subsequently be leveraged to build patient involvement across the health system. Today, the process of integration is underway in many LMICs, representing a unique opportunity to accelerate patient and community involvement in care.

The benefits and ultimate success of patient and community involvement in healthcare delivery are numerous, not only influencing patients and their families, but also providers and the system (Brett et al., 2014; Groene and Sunol, 2015). Advantages of this approach include better care coordination, improved health literacy, improved flow and self-management, better access to services, improved efficiencies and effectiveness, and improved patient safety (WHO, 2016e) (Figure 10.2).

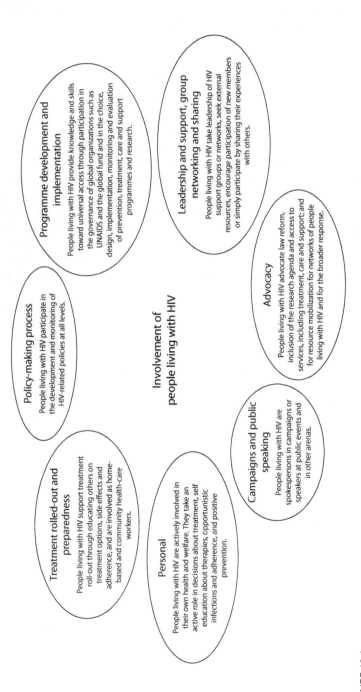

FIGURE 10.1

The greater involvement of people living with HIV (GIPA). (From UNAIDS, UNAIDS policy brief: The greater involvement of people living with HIV (GIPA), 2007. Retrieved from http://data.unaids.org/pub/briefingnote/2007/jc1299_policy_brief_gipa.pdf.)

The potential benefits of integrated people-centered health services	
To individuals and their families: • Improved access and timeliness of care • Better coordination of care across different care settings • Shared decision making with professionals with increased involvement in care planning • Improved health literacy and decision-making skills that promote independence • Increased ability to self-manage and control long-term health conditions • Increased satisfaction with care and better relationships with care providers	*To health professionals and community health workers:* • Improved job satisfaction • Improved workloads and reduced burnout • Role enhancement that expands workforce skills so they can assume a wider range of responsibilities • Education and training opportunities to learn new skills, such as working in team-based healthcare environments
To communities: • Improved access to care, particularly for marginalized groups • Improved health outcomes and healthier communities, including greater levels of health-seeking behavior • Better ability of communities to manage and control infectious diseases and respond to crises • Greater influence and better relationships with care providers that build community awareness and trust it care services • Greater engagement and participatory representation in decision-making about the use of health resources • Clarification of the rights and responsibilities of citizens toward health care • Care that is more responsive to community needs	*To health systems:* • Enables a shift in the balance of care so resources are allocated where really needed • Improved equity and enhanced access to care for all • Improved patient safety through reduced medical errors and adverse events • Increased uptake of screening and preventive programmes • Improved diagnostic accuracy and appropriateness and timeliness of referrals • Reduced hospitalizations and lengths of stay through stronger primary and community care services and a better management and coordination of care • Reduced unnecessary use of health care facilities and waiting times for care • Reduced duplication of health investments and services • Reduced overall costs of care per capita • Reduced mortality and morbidity from both infectious and non-communicable disease

FIGURE 10.2
The potential benefits of integrated people-centered health services. (From WHO, Strengthening people-centered health systems in the WHO European region: Framework for action on integrated health services delivery, 2016. Retrieved from http://www.who.int/servicedeliverysafety/areas/people-centred-care/framework/en/.)

The Namibian Experience

The experience of Namibia provides an appropriate and insightful illustration of how components of patient and community involvement through a disease-specific program can be integrated into a national quality program at all health systems levels—national, regional, and local—even as implementation continues to evolve and spread.

With a population of approximately 2.4 million, Namibia is an expansive country located in southwestern Africa bordering the Atlantic coast. Namibia's HIV epidemic is generalized: transmission primarily occurs through heterosexual and mother-to-child transmission. The HIV prevalence is 13%, representing approximately 210,000 people over the age of 15 (Ministry of Health and Social Services [Namibia], 2015).

The Ministry of Health and Social Services (MoHSS) conducted a national assessment in 2012 to review the entire quality management (QM) system at health facilities across the country. As part of that assessment, healthcare workers, patients, and community members were interviewed. One specific objective of this review was to explore how patients and the community were involved in quality improvement (QI) activities and to capture their perspectives about quality of care in health facilities. The results of this assessment provided critical insight and facilitated relevant planning for integration of patient and community engagement into the broader health system.

National Level

Recognizing the importance of consumer and community engagement as a part of the new QM directorate, the MoHSS established a subdivision for patient and community engagement. This subdivision is driven by four primary objectives:

1. To establish mechanisms for active involvement of patients in their care
2. To encourage respect and dignified treatment of patients
3. To coordinate efforts to ensure continuity of care, reduction in delays, and access to the package of services
4. To disseminate information related to patient care, patient safety, and community engagement

At the national level, patients are represented in the National QM Steering Committee, where they attend monthly meetings. They have assisted in the development of terms of reference for the program and in updating the annual Quality Management Plan. This work has included establishing a subcommittee to design a consumer satisfaction questionnaire to be tested at the health facility level, as well as developing guidelines and standard operating procedures around consumer and community involvement.

In January 2016, the MoHSS held QM training sessions for consumers at eight pilot sites, covering basic principles of QI and techniques for engagement. Consumers attended with providers from their agencies and developed plans to activate consumer participation at their respective sites. A formal consumer training curriculum is under review, which will be adapted to the Namibian context. Additional plans include training local PLWH to adapt course content and engaging community-based organizations to facilitate future training sessions.

Consumers are invited to attend regional HIV peer-learning sessions every 6 months with the goal of accelerating their QI knowledge and skills to function at the same level as other health workforce members. A patient charter was recently revised and nationwide patient education on consumer rights and expectations is in the planning stages.

The MoHSS has also engaged a public relations officer who provides health-related information to the public and responds to consumer questions and concerns that emerge either individually or through the media.

Regional Level

Program officers have received training on patient involvement and have attended regional HIV peer-learning sessions together with patients from healthcare facilities. This approach was designed to sensitize and prepare healthcare workers to more fully engage with patients, and to involve patients during regional meetings, such as the regional AIDS Coordinating Committee.

District and Facility Levels

Applying the regional approach at the district and facility levels, healthcare workers have been trained in patient involvement as part of the national QM curriculum. Patients have also gained representation within key administrative and management structures, such as hospital advisory committees, facility QI teams, and constituency coordinating committees.

Patient feedback is routinely solicited using suggestion boxes and through a customer care desk in referral hospitals. Some HIV clinics also engage patient experts to help PLWH navigate the clinical experience.

Namibia exemplifies the way in which a number of LMICs have begun the process of formally engaging patients and the community into routine improvement activities, and as noted, the process of integration is already underway elsewhere. Over the next 10–15 years, it will be critical to harness adaptable components of HIV quality programs, for example, by identifying key elements and considering how they can be integrated, especially for HIV patients who require general medical care and management of non-communicable diseases. In the context of patient and community involvement, this includes a list of core components previously described. In LMIC specifically, adaptation and integration will take time and require government structures and regulatory systems to support sustainable implementation. Namibia has made these steps possible through the recently approved QM directorate structure, which governs the national quality program. Patient and community involvement is embedded in administrative structures, technical working groups, and/or committees to ensure adequate management of these groups at various levels of government—local, regional, and national—and alignment of their work with national policies, frameworks, and priorities.

Prospects

Involvement of patients and the community in healthcare delivery is directly linked to the prospect of success of an improvement culture that encompasses the entire public health system, one that can truly improve quality of care and achieve patient outcomes in Namibia and beyond. This continues to be most evident in HIV care, and it will be critical to harness systems in that sector to build patient involvement beyond these disease-specific programs to ensure that similar successes can be achieved in the broader public health system. Namibia is now addressing this need by focusing on building the capacity of consumers to participate in the national QM program through the QM directorate.

Through a combination of leadership, the sensitization of healthcare providers, and the active engagement of consumers, Namibia's health systems can move beyond involvement, creating true partnerships in clinical care where the patient's voice receives equal consideration and is valued and applied in care decisions. Equality would also extend to relevant committees and governing bodies, allowing patient and community concerns to duly influence not only practice, but also policy.

Conclusion

True patient activation and patient-centeredness can only be accomplished by addressing existing challenges, which include transforming the traditional boundaries of the doctor–patient relationship and sensitization to the needs of patients and their role in care beyond token participation. In Namibia, the MoHSS is designing a well-structured approach, which is embedded in an overall strategy of quality in the health sector. By building the capacity of and empowering consumer representatives to ensure that they understand their roles and responsibilities in care delivery, the MoHSS has committed to reinforcing systems and processes that will facilitate success.

Continued involvement of patients and communities in the improvement process will encourage ownership and further the spread of patient-centered strategies, while increasing the long-term sustainability of changes. Achieving sustainability will require multidisciplinary partnerships within future healthcare systems; adequate structures to support patient-centered care through committees and other formal mechanisms, as previously described; and opportunities to scale-up and spread the concepts of patient involvement through training, recognition, and the promotion of success.

11

Nigeria

Doing More with Less: Lean Thinking in the Health System

Emmanuel Aiyenigba

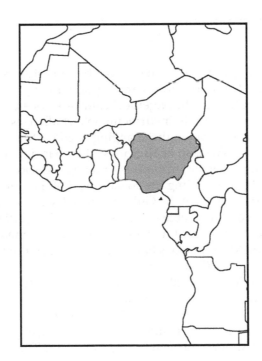

CONTENTS

Nigerian Data

- Population: 185,989,640
- GDP per capita, PPP: $5,867.1
- Life expectancy at birth (both sexes): 54.5 years
- Expenditure on health as proportion of GDP: 3.67%
- Estimated inequity, Gini coefficient: 43.0%

Source: All data are from the World Health Organization and World Bank. Latest available data
    used as at October 2017.
GDP = Gross Domestic Product
PPP = Purchasing Power Parity

Background

Nigeria, the most populous African country, has an estimated population of 193 million people (2016 projection) and a population growth rate of 2.6%. From independence to date, the Nigerian healthcare system has faced numerous and sometimes seemingly insurmountable challenges and must evolve continuously to stay relevant to the changing needs of the population. In the last two decades, the Nigerian healthcare system has developed some noteworthy systems improvement initiatives, and enjoyed successes in a number of areas (WHO, 2016c), including the containment of the spread of the deadly Ebola virus, the eradication of the Guinea-worm disease, and improvements in key indices for communicable diseases, such as HIV/AIDS, malaria, and tuberculosis. The development and implementation of maternal, newborn, and child health (MNCH) policies have seen a significant reduction in maternal and child mortalities (Kana et al., 2015).

The National Health Bill of 2014 established a Basic Health Care Provision Fund for the provision of a basic minimum package of health services, the provision and maintenance of primary healthcare facilities, the provision of drugs for primary healthcare facilities, and the training of primary healthcare personnel (Federal Ministry of Health [Nigeria], 2014a). The implementation of the fund has led to improved access to primary healthcare across the regions.

Current Efforts to Enhance Care

The lack of centralized quality data to support evidence-based policy and planning has been an ongoing problem in the Nigerian healthcare system. In 2014, the Federal Ministry of Health approved the establishment of a central database—the National Health Workforce Registry (NHWR)—becoming the first country in the world to mandate the registration of all health workers and other health workforce production data (WHO, 2014). In addition to providing a series of checks and balances, the NHWR monitors recruitment and training of required health personnel, and provides real-time data on the availability of personnel at each level of the healthcare service delivery system. The registry helps to optimize the deployment, utilization, and retention of skilled healthcare workers at all levels in an equitable manner. The Nigerian Public Health Training Initiative (NPHTI) was also implemented by the federal government, in collaboration with the U.S.-based Carter Foundation, with the aim of improving the quality and quantity of health professionals serving in the country (Federal Ministry of Health [Nigeria], 2014b).

The Primary Health Care under One Roof (PHCUOR) initiative, modeled on guidelines developed by the World Health Organization (WHO) for integrated district-based service delivery, became policy in 2011. The PHCUOR has integrated all primary healthcare services, which are now delivered under one authority and provide for health education and promotion, maternal and child health, family planning, immunization, disease control, essential drugs, nutrition, and the treatment of common ailments. This framework has created a single management body with decentralized, clearly defined roles and responsibilities that can adequately control services and resources, especially human and financial resources. Effective referral systems between/across the different levels of care, along with enabling legislation and associated regulations which incorporate these key principles have also been established (National Primary Health Care Development Agency [Nigeria], 2016).

The adoption of the Program for Results (PforR) by the Federal Ministry of Health in the Saving One Million Lives Initiative (SOML) has had a positive impact on primary healthcare and the wider health system. The SOML initiative was launched in 2012 to help save the lives of the 900,000 women and children who die yearly in Nigeria from diseases that are largely preventable. In 2016, the World Bank added its support, approving a US$500 million International Development Association (IDA) credit to improve maternal, child, and nutrition health.

While laudable, these improvement efforts still occur within poorly defined boundaries with ineffective connections, rather like independently siloed software applications without the driving stable force of a strong operating system. Despite the efficiency of each subsystem, the overall efficiency of the Nigerian health system requires further enhancement.

Challenges

Nigeria's ongoing political instability and economic crises means that the forces that enhance and inhibit improvement efforts are highly complex. Some influences on the improvement work in Nigeria are listed in Table 11.1.

TABLE 11.1

Analysis of Improvement Efforts in Nigeria

Forces Influencing Improvement Efforts in Nigeria	
Forces Enhancing	Forces Inhibiting
Development agencies' support and technical assistance	Weak sense of system ownership at the subnational levels
Improving technological climate that enhances the available options of care delivery	Continued fragmentation of the system despite the efforts toward integration
Autonomy of states, which creates the space for innovation and creativity in the implementation of policies	Autonomy of states, which potentiates inequalities and unpredictable variation in health outcome across the states
Strong supportive policy climate, which breeds opportunities for collaborative partnerships	Economic instability, which deters and discourages potential investors in the health system
A technical working group (TWG) constituted by the Federal Ministry to drive the process of the operationalization of the National Health Act	

In 2017, the federal government of Nigeria decided to reduce the allocated budget for human resources for health (HRH) management. While such cutbacks seem counterproductive in the face of growing concerns about the state of HRH in the country, this strategy may in fact compel system administrators to become resourceful in the redesign of the health system to meet the growing and diversified needs of the Nigerian population.

Lean Thinking

What is Lean thinking? According to Lean principles, "the obstacle is the way": the components that have made the system weak and volatile will provide the ingredients for a highly integrated and resilient system. The principles of Lean thinking thus focus on the elimination of waste in all its forms from the system. The eight types of waste described in the Lean thinking literature are summarized in Table 11.2.

Implementation

There have already been some improvements that can be regarded as a practical application of Lean principles. Cuts in HRH funding, for example, have led health systems across all levels to seek effective and innovative

TABLE 11.2

Summary of Lean Management

Type of Waste	Present Picture	Steps toward the Future
Defects (rework) Work that contains errors or lacks value	• Errors in diagnosis, surgical intervention, and medication • Unplanned variation in outcomes (especially resulting from special causes)	• System redesigns that understand human factors and provide a supportive workplace • Establish simple and clear definitions of defects, standards of care, and process of elimination of waste within and between the levels of care delivery • Establish clear boundaries for roles and responsibilities between tiers of government, levels of care, and partners in collaboration
Overproduction Redundant work	• Duplicate documentation, redundant patient flow, multiple forms with the same information, and steps in care that are unnecessary • Information overload that does not lead to decision-making or system improvement	• Simplified process maps and clear interpretation of regulations • Streamlined processes that eliminate redundancies in the processes • Systems of data and information management that speak to one another and inform decision-making at all levels
Waiting Idle time created when people, information, equipment, or materials are not at hand or aligned	• Delays in meetings, surgeries, procedures, reports • Interdependencies that lead to inefficiencies • Misallocation of resources within the system on account of poor needs assessments • Patients waiting for appointments, doctor visits, or procedures	• Resource mapping and needs assessment that lead to "right now" scheduling and resourcefulness, as well as efficiencies within the system

(Continued)

TABLE 11.2 (CONTINUED)

Summary of Lean Management

Type of Waste	Present Picture	Steps toward the Future
Not clear (confusion) People doing the work are not confident about the best way to perform tasks	• Lack of standardization, which causes the same activities to be performed in different ways by different people at different times/contexts • Unclear patient safety, quality improvement approaches, and standards goals	• Standardization to ensure that all activities of work are clearly specified • Establish "triggers" that help initiate the uniform execution of tasks on the job • Create standards for both technical and administrative work • Capacity building for all levels and cadres of workers to be able to execute work correctly and according to standard • Articulated job descriptions and job aids to empower workers on the job
Transportation Required relocation or delivery of a patient, materials, or supplies to complete a task	• High fragmentation and poor coordination of subsystems, like lab and pharmacy supplies • Siloed health facilities with little or no referral linkages	• Reduce fragmentation by resource mapping and allocation based on needs • Remove redundant steps between processes • Create efficient flow of both patients and consumables at every level of the system
Inventory More materials on hand than are required to do the work	• Ineffective and inefficient supply chain management • Poor communication between levels of care	• Create effective inventory systems and highly responsive supply chain management systems to supply only what is needed in a timely fashion • Ensure proper and timely use of data for improved inventory management

(*Continued*)

TABLE 11.2 (CONTINUED)

Summary of Lean Management

Type of Waste	Present Picture	Steps toward the Future
Motion Movement of people within the system that does not add value	• Lack of directional pointers, which leads to waste as both caregivers and patients move around searching for this information • Poor patient and provider database; workplaces poorly constructed with no consideration for human factors	• Create functional and reliable information technology (IT) systems that match the demand of work • Establish reliable communication systems between key stakeholders in the health system to ensure timely information dissemination and shared learning and decision-making • Create databases that equip both caregivers and patients with the relevant information needed for decision-making in a timely fashion • Ensure that workplaces are designed paying attention to human factors
Excess processing Processes or activities that do not add value from the patient perspective	• Prolonged time of communication between levels of care • Redundant data management processes • Regulatory paperwork	• System architecture redesigns to create effective communication and continuous flow within and between levels of care • Simplified, consistent, standardized delivery systems for medication, materials, and information • Data collection tools that document only essential and relevant information

work-arounds to the increasing demands of the patient population through process redesign, task shifting, demand-driven health services planning, and increased focus on patient-centered care, among others. A shift to encouraging intrinsic motivation for HRH is also important to the sustainability of Lean thinking to health system improvement. Health professionals of all types and ranks have had to make the mental shift to look inward for reasons to go the extra mile in their job. Lean processes in hospital management and administration have indirectly compelled caregivers to find fulfillment in non-monetary sources and seek non-financial rewards for their labor.

Senior management of health institutions are now compelled to cut costs without compromising quality. Keenly seeking opportunities to apply the

Pareto principle* in selecting projects, these institutions continually improve the processes that produce the services.

Structures and boundaries that will support a continued adherence to Lean thinking at every level of the system should be introduced. These would include

- Incorporating the teaching of Lean thinking and other broad process improvement methodologies in the curriculum of health professionals' education at under- and postgraduate levels
- Restructuring the health system to accommodate a reorientation of the system hierarchy, that is, ensuring patients' needs are central
- Empowering the patient population as well as the designers of the system to request and create high-quality services and products using Lean principles

Prospects

Over the next decades, the embrace of Lean thinking should

- Advance systems thinking and the understanding of human factors. Enhance the ability of the health systems players to understand and leverage the interactions between the system components and how they influence human performance.
- Design the health system at every level with this understanding of how the human interacts with the system components.
- Guide the design of programs with a shift in focus from methodologies to the underpinning of concepts and principles.
- Align methods, programs, and initiatives with clearly defined blueprints based on proved principles adapted to the context.

Conclusion

In order to improve the outcomes of a system rife with waste and inefficiency, and in the face of resource restrictions, the fundamental design of the Nigerian health system needs to be changed. While it might seem

---

* The Pareto principle (or the 80/20 rule) is a management-based principle that states that 80% of effects come from 20% of causes.

counterintuitive that something perceived and received as detrimental and inhibitory could hold the potential for transformation and positive results, the principles of Lean thinking create systems that do more with less, and their application could see the development of a more efficient and effective system. The precarious situation of the Nigerian health system makes this approach not only attractive but also necessary.

12

South Africa

Regulated Standards: Implementation and Compliance

Stuart Whittaker, Lizo Mazwai, Grace Labadarios, and Bafana Msibi

CONTENTS

South African Data

- Population: 55,908,865
- GDP per capita, PPP: $13,225.4
- Life expectancy at birth (both sexes): 62.9 years
- Expenditure on health as proportion of GDP: 8.80%
- Estimated inequity, Gini coefficient: 63.4%

Source: All data are from the World Health Organization and World Bank. Latest available data
            used as at October 2017.
GDP = Gross Domestic Product
PPP = Purchasing Power Parity

Background

The apartheid system, which was established in South Africa in 1948, legislated the homeland policy of "separate development" that perpetuated colonial inequities between rural and urban populations. This inequitable dual system of healthcare has continued along racial, geographic, and political lines, and is now perpetuated by socioeconomic differences that play out in private- and public-sector healthcare inequities.

In 1994, after the first democratic elections, the new democratic government of South Africa had to deal with the task of transforming a fragmented and hugely inequitable healthcare system into a national health system working toward universal coverage, and incorporating both the public and private sectors. To this end, the National Health Act of 2003 (National Department of Health [South Africa], 2004) was amended by the National Health Amendment Act of 2013 (National Department of Health [South Africa], 2013), paving the way for the establishment of the Office of Health Standards Compliance (OHSC) in September 2013. The OHSC has a mission to "act independently, impartially, fairly, and fearlessly on behalf of the people of South Africa in guiding, monitoring and enforcing health care safety and quality standards in health establishments" (Office of Health Standards Compliance [South Africa], 2017).

Roles of the OHSC

In order to achieve its mandate, the OHSC must carry out multiple functions in a coordinated manner (Figure 12.1).

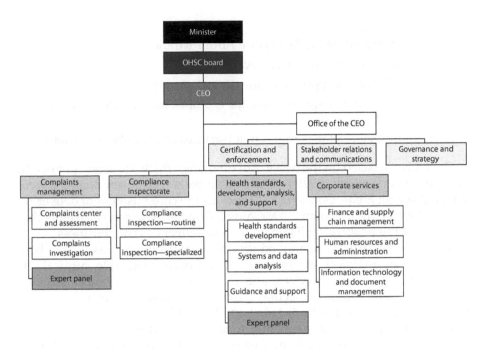

FIGURE 12.1
Organogram summarizing the structure of the OHSC and its related functions. (From Office of Health Standards Compliance [South Africa], Strategic plan 2015/16–2019/20, 2017. Retrieved from http://ohsc.org.za/.)

The Health Standard Development Analysis and Support (HSDAS) unit plays an important role in the OHSC—providing high-level technical, analytical, and educational support in relation to the research, development, and analysis of norms and standards. The HSDAS unit also helps stakeholders establish communication networks with the OHSC and build capacity in relation to quality management within health establishments.

Problems Experienced in the Development of the NCS and Subsequent Regulated Standards

National Core Standards (NCS) were drafted in 2008 as policy initiated by the Minister of Health, and regularly updated as forerunners to a proposed set of regulated standards to be developed in the future and which would become law when promulgated.

The content of the NCS included the following:

1. Legislation and policies from the National Department of Health (NDOH)

2. Requirements of other government departments

- Treasury
- Department of Public Service Administration
- Guidelines for corporate governance
- Guidelines and protocols for clinical governance

3. Clinical systems requirements, for example, medical devices, pharmacy, and diagnostics
4. Support services, for example, hotel services and facility maintenance

The NCS are based on the concept of domains, which are aspects of health service delivery where quality or safety can be at risk. Seven domains were selected as risk areas. Domains 1–4 focus on the core business of delivering quality healthcare to patients. Domains 5–7 focus on support systems that ensure that the health establishment system delivers its core business.

Initially, standards were applied directly to functional areas (FAs) where clinical and non-clinical services were carried out, for example, surgical services, infection control services, and management and technical services. There were numerous inspection problems, however, which frequently resulted in incorrect standards being applied to FAs.

The standards were subsequently modified to ensure that the domains (risk areas) and related standards were relevant to the FAs to which they were allocated.

Regulated Standards

The promulgation of the National Health Amendment Act in 2013 established the OHSC as an independent regulator of health services responsible for ensuring safe, quality health services for all South Africans. Following this, the NCS were reviewed to provide a regulatory framework against which service delivery at health establishments could be evaluated. Because data collected during inspections must withstand legal scrutiny, the framework requires evaluations of a rigorously objective nature.

Regulated health standards are currently being developed by the OHSC with the following aims:

- Ensuring the provision of quality health services
- Developing a common definition of quality care that should be found in all health establishments in South Africa, as a guide to the public and to managers and staff at all levels
- Establishing a benchmark against which health establishments can be assessed, gaps identified, and strengths appraised
- Providing for the national certification of compliance of health establishments with mandatory standards

There is, however, some debate as to which authority is the most appropriate organization to carry the responsibility for development of regulations for health establishments. This has led to delays in their promulgation, and hence finalization and implementation of the regulated standards.

Regulated Standards Compliance

Quality standards alone are not enough—it is their implementation and maintenance that ensures quality patient care. Health establishments must have support and guidance in relation to quality and risk management, as well as comprehensive training in the standard requirements, so that they can meet the essential requirements, even if they are not able to meet the specific details.

Adequate training and support must be provided to help health establishments facilitate compliance, implement effective quality assurance (QA) and quality improvement (QI) services, and achieve certification. While training in QI is the responsibility of individual healthcare establishments—both public and private—and their administrators, government entities such as OHSC and National Health Insurance (NHI) have an important role to play.

The OHSC is responsible for ensuring that healthcare workers understand the requirements of the regulated standards and how to assess them. Currently, however, the NDOH and provincial departments do not have the full capacity to provide adequate support.

Regulated Standard Certification

The OHSC is also responsible for the certification of establishments that meet the regulated standards. To achieve its legislated mandate, the OHSC is required to inspect South Africa's 4010 health establishments every 4 years. This number will increase as additional standards are developed and implemented, for example, family practitioner standards.

During the development phase of the NCS standards, prior to the development and promulgation of the regulated standards, pilot or "mock" inspections were carried out using the various drafts of the NCS as they were developed and improved. The ongoing inspections provide experience for inspectors, and sensitize health establishments to quality management principles in preparation for the implementation of regulated standards, which will lead to the certification of health establishments.

In the 2015–2016 financial year, a total of 627 inspections were carried out across the country. This included four central hospitals, 11 provincial tertiary hospitals, nine regional hospitals, 27 district hospitals, nine community health centers, and 567 clinics. Of these, 132 were reinspections. Improved performance was found in the four central hospitals and 12 provincial or tertiary hospitals that were reinspected. However, although there were some improvements in the 27 reinspected district hospitals, many weaknesses remained.

In contrast to the improvements shown in the hospitals, the reinspected clinics across the country showed little improvement. Problem areas included poor infrastructure, lack of medication, and severe staff shortages. In addition, policies and procedures were either missing or poorly implemented, with little support in relation to training or guidance on how to carry out the necessary processes. Clinics supported by feeder hospitals generally did better than those with district office oversight.

The reinspections indicated that there was lack of guidance and support by the provinces. In particular, there was a lack of encouragement to conduct self-assessments in order to monitor ongoing compliance.

Implementation of Regulated Standards

The delay in promulgation of the legislated standards will impede the implementation of NHI, since health establishments are required to be certified by the OHSC in order to participate in the NHI program (National Department of Health [South Africa], 2017).

Without the promulgation of the regulated standards, the OHSC is unable to legally enforce compliance in health establishments that consistently fail to meet the required standards. However, once the draft regulated standards are completed and ratified, the standards become law and must be met.

Limitations of External Evaluation

Every 4 years, the inspection program produces "snapshots" in time of the degree of compliance to agreed standards. The degree of compliance between inspections is generally unknown. The following two OHSC programs or units have been designed to reduce this deficiency by identifying problems generally related to service provision.

Early Warning System

Section 79(1)(d) of the National Health Amendment Act empowers the OHSC to monitor indicators of risk as an early warning system (EWS) relating to serious breaches of norms and standards and report any breaches to the minister without delay.

The EWS was designed to identify health establishments where patients, staff, or services are highly likely to suffer serious harm. These indicators will be used as part of the system to prioritize inspections and monitor risk in health establishments in the intervening years between inspections.

Data collected via the EWS will contribute to research on areas of risk to patient safety and improve the accuracy in predicting risk to prompt corrective action.

Complaints Management Unit

The mandate of this unit is to receive, investigate, and resolve complaints about health establishments in relation to breaches of regulated standards, where it has not been possible to resolve the complaint locally. The unit also provides support to the minister-appointed health standards ombud. The ombud adjudicates particularly complex matters and oversees all decisions.

The aim of the unit is to provide an easily accessible, responsive, central service whereby individuals can lodge legitimate complaints directly or after exhausting all local processes. Complaints can be reported by phone, by e-mail, in writing, or in person. Efforts are underway for complaints to be reported via electronic media platforms. All complaints are recorded electronically and are then evaluated and dealt with appropriately.

Expert medical, pharmaceutical, and health system knowledge is sometimes necessary in order to fully understand the complexities of complaints. Although the OHSC's investigating team will have healthcare and legal training, members may sometimes lack the specialist expertise required for particular cases. The OHSC and the ombud will have access to an expert panel, comprising specialists who are leaders in their field, to assist them on a part-time basis. Figure 12.2 summarizes the activities of the complaints management unit.

To monitor performance longitudinally, the OHSC will utilize the indicators of the EWS, as well as reports generated via the complaints management unit, which identifies problems in "real time." This will substantially strengthen the risk reduction process provided by the inspection unit.

Next Steps

OHSC IT System

An advanced information technology (IT) system that will be used for the capture, collation, and analysis of inspection data has been procured. The system will also calculate certification status by means of a predetermined

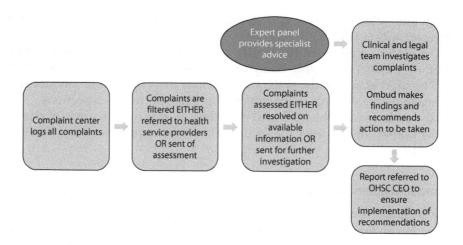

FIGURE 12.2
Activities of the complaints management unit and omsbud. (From Office of Health Standards Compliance [South Africa], Overview on complaints management and the work of the ombud, 2014. Retrieved from http://www.ohsc.org.za.)

algorithm and generation of inspection reports. Unfortunately, its implementation will be delayed until new premises for the OHSC are secured. This requires the approval of the national treasury. Delays in the implementation of the IT system will severely impede the efficiency of the inspection, analytical, reporting, and feedback systems of the compliance inspectorate.

Ongoing Development of the OHSC IT System

The data capture component of the IT system should be user-friendly, and will operate on multiple levels.

Once the system has been installed,

- Health establishments will be able to capture their own self-evaluation requirements.
- Health establishments will be able to efficiently input QI data monthly for certification purposes.
- Stakeholders with appropriate permissions should be able to gain access to the IT system to follow up and encourage health establishments to make the necessary changes to gain certification.
- Reporting systems should be such that it is easy for permissioned users to evaluate FAs in health establishments and the overall certification status of the health establishment.

- Reports will be issued to participating facilities and other appropriate stakeholders.
- National, provincial, and local authorities and the private sector will provide the data necessary to bring about an improved health system delivery.

Conclusion

Although progress in implementing regulations has been slower than anticipated, the delays, along with their associated causes, have provided experience and knowledge that will ultimately enhance the effectiveness and efficiency of the OHSC. Such knowledge, in combination with the collaboration of relevant stakeholders, particularly the NHI, will see the mission of the OHSC accomplished, improving the quality and safety of the South African health services for all of South Africa's people.

13

Rwanda

Embracing One Health as a Strategy to Emerging Infectious Diseases Prevention and Control

Roger Bayingana and Edward Chappy

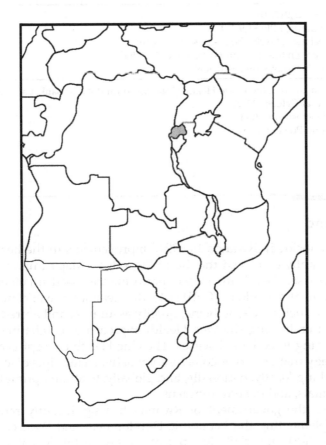

CONTENTS

Rwandan Data

- Population: 11,917,508
- GDP per capita, PPP: $1,913.4
- Life expectancy at birth (both sexes): 66.1 years
- Expenditure on health as proportion of GDP: 7.53%
- Estimated inequity, Gini coefficient: 50.4%

Source: All data are from the World Health Organization and World Bank. Latest available data
used as at October 2017.
GDP = Gross Domestic Product
PPP = Purchasing Power Parity

Background

Since the genocide in Rwanda in 1994, improvements in the health system have significantly enhanced the life of Rwandans. Major initiatives include community-based health insurance; performance-based financing; adding community health workers (CHWs) to the healthcare workforce; eHealth technologies, such as telemedicine, quality assurance, and accreditation initiatives; and embracing the One Health approach to combating emerging and reemerging infectious diseases. The One Health concept is an international strategy that promotes collaboration between multiple disciplines and sectors working locally, nationally, and globally, to attain optimal health for people, animals, and the environment.

Since 1994, the government of Rwanda has significantly reduced poverty by implementing the Economic Development and Poverty Reduction Strategy (EDPRS) 2013–2018 (Republic of Rwanda, 2013). Two of the key components of this strategy are to reduce poverty and to achieve universal health coverage. Data show that since the 2011 implementation of community-based health insurance, covering 90% of the population, there has been a significant improvement in health service utilization and a significant reduction in financial catastrophe and impoverishment caused by out-of-pocket spending (Lu et al., 2012). Other efforts that have been key to improving the health-care system in Rwanda include performance-based initiatives and the use of ehealth technologies to track childhood and maternal morbidities, better monitoring of epidemic outbreaks, and the mapping of vaccination coverage. Most importantly, the country's top leadership has shown the political will to strengthen the health sector through proper coordination of donors and external aid and close monitoring of the effectiveness of this aid.

Future Prospects

Rwanda is one of the most densely populated countries in the world, and has a fast-growing population. Because of its geographical location in the Great Lakes Basin in East Africa, Rwanda is vulnerable to environmental disasters and epidemics of a zoonotic nature. Rwanda is also undergoing significant demographic and environmental transformations, including rapid urbanization, closer integration of livestock and wildlife, forest encroachment, changes in ecosystems, and globalization of trade in animal and animal products. In combination, these factors have created conditions that are highly conducive to the spread of infectious diseases at the interface between animals and humans. The potential catastrophe that these infectious diseases would cause cannot be taken lightly. Rwanda has embraced the One Health approach both to reduce the threat of such epidemics and to ensure that any outbreaks are effectively contained and managed.

Research shows that 61% of human pathogens (Cunningham, 2005) and 75% of emerging diseases (Blancou et al., 2005) are zoonotic. In the last two decades, we have seen multiple factors contributing to the occurrence and spread of diseases, such as severe acute respiratory syndrome (SARS), highly pathogenic avian influenza H5N1 (HPAI H5N1), the H1N1 pandemic that originated from wildlife, and cholera (which is climate related). Rwanda's geographic location and physical environment make it susceptible to most of these emerging infections. While these diseases may originate from wildlife and domestic animals, thus far the lessons learned regarding the prevention and control of these diseases have highlighted the fact that no single discipline, no single sector, no single ministry, and no single country can handle these public health events independently (Republic of Rwanda Ministry of Health, 2014). However, the One Health initiative, which has been embraced by the Rwandan government, takes a holistic approach to addressing human, animal, and ecosystem health. It emphasizes multisector, transdisciplinary action across professions to ensure well-being within human, animal, and ecosystem interfaces. One Health has gained international attention as an approach to control infectious disease outbreaks and to address interconnected health threats affecting animals, humans, and the ecosystem domains (Gibbs, 2014).

Rwanda's health system has improved significantly because of the government's investment in the health of its people over the last two decades. During this period, Rwanda was one of the few countries to achieve most of the Millennium Development Goals, and has now agreed to work toward achieving the Sustainable Development Goals. The political will to improve the health of all Rwandans, and the coordination of all health sector players and development partners, are factors that are leading to success within the Rwandan health system. This success, coupled with the recently approved healthcare quality and accreditation policy and strategy, the human resources for health strategy and sustainability plan, and other health systems development policies and strategies, has prepared the way for Rwanda to achieve its vision of becoming a healthcare tourism destination in the future. This vision has been strengthened by the validation of the One Health strategic plan. As emerging diseases and health priorities evolve into global and multisectoral issues, public health professionals—from interventionists to advocates to researchers—must step outside their silos and work collaboratively, pooling resources and setting priorities before interventions begin (Yang, 2011).

For example, the Rwandan Ministry of Health (MoH) has developed an integrated disease surveillance and response (IDSR) system based on the 2010 version of the World Health Organization, Africa Regional Office (WHO-AFRO) integrated disease surveillance and response guidelines (WHO-AFRO, 2010), which provides a platform for all activities in disease surveillance and response at all levels of the health system in Rwanda. This enables health teams and communities to determine priorities, plan interventions, anticipate and detect epidemic outbreaks, and mobilize and distribute available resources. Such a surveillance system can be upgraded to capture data for other health sectors (domestic, wildlife, and environment), thereby creating a system that provides for rational utilization of resources devoted to the prevention and control of diseases.

Impact

One Health will be a success because of its open approach to solving health issues. As professionals in different sectors come to realize that they cannot work alone to combat infectious diseases, their combined efforts will lead to improvement in the health of all Rwandans. Okello et al. (2011) state that such collaborative efforts fill vital knowledge gaps and provide a strong evidence base to support policy decisions at the international, regional, and national levels in developing countries.

It is imperative that all those working in the health sector (human health, animal health, and environmental health) understand the principles of One

Health. Promoting interdisciplinary collaboration requires a workforce trained in its principles and application, if it is to be successful. Providing One Health education to those already working in the relevant professional disciplines, and to students seeking professional qualifications in these areas, is key to a One Health–oriented workforce. The One Health curriculum at the undergraduate and postgraduate levels will integrate collaborative problem-solving approaches with elements of infectious disease control, epidemiology, ecology, environment, finances, food safety, and leadership.

As globalization continues, One Health will become increasingly relevant, and there will be an even greater impetus to ensure that all graduates in the medical, veterinary, and environmental science fields are appropriately trained.

Implementation and Transferability

Rwanda is one of the few countries to implement a One Health strategic plan (Nyatanyi et al., 2017), underpinning collaboration between the various sectors and systems governing environmental, animal (wildlife and domestic), and human health. In the past 2 years, collaboration between the three sectors has increased with the appointment of the One Health Steering Committee, made up of members from different sectors. Ideally, such collaboration should involve a willingness to share emerging infections data for the good of the Rwandan population. It is expected that the merging of both animal and human surveillance and response systems into one infectious disease surveillance system, accessible to both sectors, will lead to smoother collaboration between the different health stakeholders.

An intersectoral team approach will be necessary if One Health is to be effectively integrated into the Rwandan health system. According to Coddington et al. (2001), barriers to team collaboration are plentiful. Central to these barriers is the tendency of many professionals to work independently, and to resist collaboration. It is important to nurture One Health–minded professionals in preservice education so upon entry into the workforce, they are clear on the strategies, aims, and benefits of collaboration. Second, intersectoral collaboration will be enhanced by developing a clear communication strategy to avoid duplication and ensure efficient transfer of information across sectoral boundaries. Central to this will be the merging of the human and animal infectious disease surveillance and response systems. Implementation and operation of an integrated health system requires leadership and an organizational culture that is congruent with a common vision—something that Rwanda possesses.

In terms of transferability, Rwanda is advocating for the institutionalization of One Health in the health systems of fellow East African countries.

Although the International Health Regulations (WHO, 2005) provide guidelines related to any event detected by a national surveillance system, there is a consensus at the global level of the need to address zoonotic diseases and other public health events using a multinational, multisectoral, multidisciplinary approach, and to enhance local, national, regional, and global collaboration to prevent and control zoonotic diseases and other public health events. Thus, in April 2014, the permanent secretary of the Rwandan MoH presented the concept of One Health to the East African Community (EAC) Council of Ministers for the very first time, calling on fellow EAC member states to adopt the One Health approach as a means to implementing most of its activities (East African Community, 2014). The One Health approach was adopted in that same meeting (East African Community, 2014).

Prospects

Rwanda has many CHWs and community-based animal health workers. Once One Health has been integrated into Rwandan health systems, the two groups can be brought together and rebranded "One Health CHWs" (OHCHWs) (Nyatanyi et al., 2017). The two groups of health workers are both well situated to quickly identify unusual events or problems affecting humans, animals, or ecology and agriculture. OHCHWs will routinely collect local information on the health of humans, animals, and crops and report it in real-time.

Conclusion

The incorporation of One Health principles in Rwanda's health strategic plans and policies is a recognition of the realities of a global community in which humans, animals, and the environment impact each other and do not always respect geopolitical boundaries. Rwanda's One Health approach provides innovations that are important to emerging infections prevention and control, and offers synergy across systems, resulting in improved communication, development of a new generation of systems thinkers, improved surveillance, decreased response lag time, and improved health and economic savings.

14

Africa

Equity for All: A Global Health Perspective for the Continent

Jacqui Stewart and Shivani Ranchod

CONTENTS

African Data

- Population: 1,204,859,690
- GDP per capita, PPP: $5,686.1[a]
- Life expectancy at birth (both sexes): 62.7 years
- Expenditure on health as proportion of GDP: 6%[b]
- Estimated inequity, Gini coefficient: 51.8%[c]

Source: All data are from the World Health Organization and World Bank. Latest available data
      used as at October 2017.
[a]   Data unavailable for Somalia.
[b]   Data unavailable for Somalia, South Sudan, and Zimbabwe.
[c]   Data unavailable for Equatorial Guinea, Eritrea, Libya, and Somalia.
GDP = Gross Domestic Product
PPP = Purchasing Power Parity

Background

Africa is a vast continent: it contains more than a quarter of the world's countries (Worldometers, 2017) and 16.2% of the world's population (Population Reference Bureau, 2017). Seeking commonalities across the diverse sociopolitical, economic, and anthropological landscape of the continent is challenging. When it comes to healthcare, however, there are some epidemiological and demographic themes that are common. There are also shared challenges of financial exclusion and supply-side constraints—or put differently, shared opportunities centered on financial inclusion and supply-side innovation.

Demographic and Epidemiological Commonalities

While much of the world faces the challenge of aging populations and decreasing rates of reproduction (Lutz et al., 2008), the African continent, which has a growing working-age population, is the youngest in the world (Drummond et al., 2014). By 2050, Africa is projected to be home to more than a quarter of the world's population (Population Reference Bureau, 2017).

Disease burden across the continent is strongly related to demography and poverty (Figure 14.1). Maternal and child health, nutritional deficiencies, and infectious disease are specific problem areas; however, non-communicable diseases associated with an aging population are less common (Institute for Health Metrics and Evaluation, 2017). This is not to say that non-communicable diseases are not present. Health systems across the continent have to deal with a complex mix of communicable and

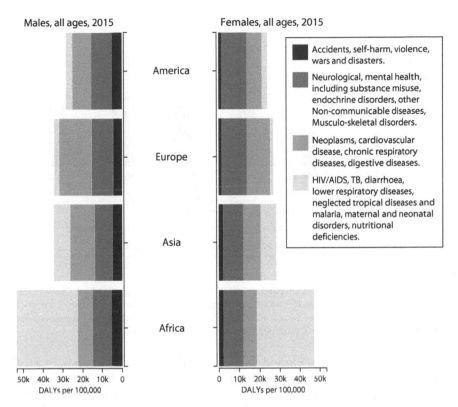

FIGURE 14.1
Disability-adjusted life years per 100,000 for Africa, Asia, Europe, and America. (From Institute for Health Metrics and Evaluation, Global burden of disease, n.d. Retrieved from http://www.healthdata.org/gbd.)

non-communicable disease, as well as the unique challenges associated with a high prevalence of HIV, tuberculosis (TB), and malaria.

Supply-Side Constraints and Innovations

The high burden of disease across the continent is compounded by supply-side constraints—issues including stock, skills mix, and geographical distribution (Chen et al., 2004). The continent also has both high-density urban and rural contexts to deal with, each of which poses a different set of challenges. In Figure 14.2, we can see that the density of healthcare workers is lower on the African continent than on any other continent. This comparison does not adjust for quality and productivity, and hence may be understating the disparity (Hongoro and McPake, 2004).

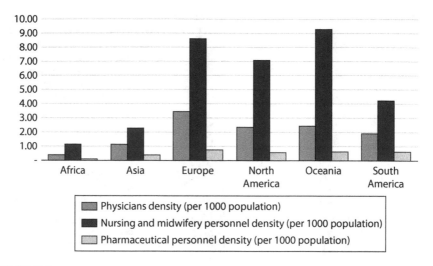

FIGURE 14.2

Health worker density per continent. (From WHO [World Health Organization], Global Health Data Observatory, Density per 1000: Data by country, 2017. Retrieved from http://apps.who. int/gho/data/node.main.A1444?lang=en. World Bank, Data: Population, n.d. Retrieved from http://data.worldbank.org/indicator/SP.POP.TOTL?name_desc=false.)

Initiatives such as the United Nations' Millennium Development Goals (MDGs) and more recent Sustainable Development Goals (SDGs), along with new targets to improve access to HIV and TB services, while well intentioned, have increased the demands on the systems (Hongoro and McPake, 2004). Three of the MDGs (a set of quantified targets related to addressing poverty by 2015), namely, reducing child mortality, improving maternal health, and combating HIV, malaria, and other diseases, could not be met by many African countries without health systems strengthening (United Nations, 2017b). The SDGs follow on from the MDGs and are similarly time-bound. They are, however, significantly more ambitious. The health targets for 2030 include *inter alia* further reductions to maternal and child mortality, achieving universal health coverage, and an end to the AIDS, TB, and malaria epidemics (United Nations, 2017a). Disease-specific drives, such as those for HIV and TB, create further obligations to increase access, and are most often related to donor funding (Hongoro and McPake, 2004).

As a means of strengthening delivery with limited resources, task shifting* holds some promise, although implementation requires sustained effort (Lehmann et al., 2009; Zachariah et al., 2009). While much of the attention has been directed toward HIV (Callaghan et al., 2010), there are other positive

---

* "Task shifting is the name given to a process of delegation whereby tasks are moved, where appropriate, to less specialized health workers. By reorganizing the workforce in this way, task shifting presents a viable solution for improving healthcare coverage by making more efficient use of the human resources already available and by quickly increasing capacity while training and retention programmes are expanded" (WHO, 2008b).

examples. Mentor mother programs across the continent engage community members in a relatively low-cost manner to support new mothers, and address the critical area of maternal and child health (Rotheram-Borus et al., 2011). More recent examples involve tasks such as vital signs monitoring and screening (Haac et al., 2016; Sanyahumbi et al., 2017).

Technological developments have also enabled some resource constraints to be loosened. Sema-doc in Kenya is an example of how mobile technology has been leveraged to enable remote access to doctors (Ranchod et al., 2016). This development makes use of Kenya's high level of mobile penetration (Communications Authority of Kenya, 2015) and has been integrated with a healthcare financing solution, which includes insurance, credit, and savings components (Ranchod et al., 2016). The Rwandan government has partnered with international firm Babylon Health to build mobileHealth capability at scale—the service has 450,000 users and makes use of artificial intelligence to drive referral pathways and limited diagnosis (Asaba, 2016; Shead, 2017).

Equity and Healthcare Financing

Issues of equity are of critical importance for the African continent. If we accept that access to health services should be provided according to need, it is obvious that the continent should be a global priority, based purely on the burden of disease. Rather, the "inverse care law," initially described in the *Lancet* more than 30 years ago, remains alive and well: "The availability of good medical care tends to vary inversely with the need for it in the population served" (Hart, 1971: 405).

Healthcare financing reforms are a key component of addressing inequity. Effective financing mechanisms have the capacity to improve household cash flow, facilitate access, and reduce the risk of financial destitution as a result of healthcare events.

In 2001, African heads of state committed their governments to spending at least 15% of public expenditure on health (the Abuja declaration) (Witter et al., 2014). Based on 2014 data from the World Bank (2017b), the average across the continent remains below 10%. The Making Access Possible (MAP) studies that have been rolled out in 11 African countries to date indicate that even where public healthcare is free of charge at the point of service, people still incur health-related expenses (Ranchod et al., 2016). These include transport costs, medicine costs, and opportunity costs in terms of lost earnings (Ranchod et al., 2016).

The MAP studies also reveal that very few low-income households in the region make use of prefunding mechanisms to meet their health needs. There is considerable variation between countries in terms of financial regulation and the development of insurance markets. Low insurance penetration is deemed to be due to a combination of a lack of appropriate products and limited services being offered (Ranchod et al., 2016).

Efforts to implement universal health coverage via public insurance mechanisms have varied across the continent, in terms of both the reforms implemented and the consequences (Lagomarsino et al., 2012). Affordability remains a serious concern for many countries. In addition, the large, informally employed populations in many countries present challenges for both public and private insurance in terms of premium or tax collection (Lagomarsino et al., 2012)—hence coverage is frequently limited to the formally employed sector (Ranchod et al., 2016). Thus, the poorest members of society are frequently required to fund their out-of-pocket health needs from credit, savings, or alternative resources. This may increase the poverty cycle and extend it across generations. The high level of out-of-pocket financing across the region can be seen in Figure 14.3.

Pluralistic approaches to healthcare financing reforms seem to hold some promise for the continent, as they allow countries to make use of wide-ranging solutions (Sakunphanit, 2006). This is increasingly relevant as fintech* developments become more common. As with supply-side innovation, high levels of mobile penetration could facilitate health systems strengthening. Pluralistic approaches also allow for path dependency, and are less reliant on complex "big-bang" reforms (ILO Social Security Department, 2008).

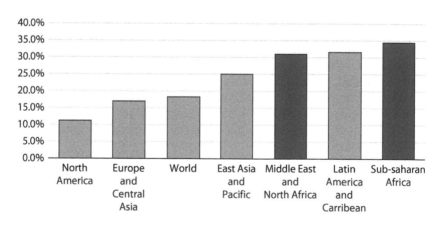

FIGURE 14.3

Out-of-pocket expenditure by region (2014). (From World Bank, Out-of-pocket health expenditure, n.d.a Retrieved from http://data.worldbank.org/indicator/SH.XPD.OOPC.TO.ZS.)

---

* Fintech is a broad term referring to technologies that have the potential to disrupt more traditional financial businesses. Examples include robo-advice, blockchain, mobile banking, and peer-to-peer lending.

Health System Strengthening and Resilience

Donor funding remains a substantial component of funding across the continent. Where there are limited resources and weak systems, donor-funded programs tend to be vertical and focused on specific diseases or patient groups. According to Victora et al. (2004: 1543), "vertical approaches seem to be more attractive to donors, who want rapid and hard-hitting results to feed back to their constituencies, whereas investment in health-systems strengthening needed for delivering comprehensive programmes is less appealing."

Through context-specific strategic planning, it is possible to shift from the vertical to the horizontal, and hence impact systems in a more equitable and sustainable way (Victora et al., 2004). After the Ebola epidemic in West Africa in 2014, the most widespread outbreak of the virus in history, those on site concluded that robust and resilient health systems could have contained the epidemic more efficiently (Kieny et al., 2014; Kruk et al., 2015). Building resilient health systems that are sensitive to Africa's health burden and that are cost-effective needs to be at the driving edge of planning. The cornerstone of resilient health systems must be the implementation of improvements to deliver high-quality care at scale.

A global health perspective is useful when considering health systems developments in Africa. Koplan et al. (2009: 1995) define *global health* as "an area for study, research, and practice that places a priority on improving health and achieving equity in health for all people worldwide. Global health emphasizes transnational health issues, determinants, and solutions; involves many disciplines within and beyond the health sciences and promotes interdisciplinary collaboration; and is a synthesis of population-based prevention with individual-level clinical care." The goal of achieving health equity between countries, and not just within countries, is deeply resonant for the African continent.

Part III

Europe

Russell Mannion

In our continuing journey across the globe, we now arrive in the World Health Organization (WHO) European region. The 47 countries covered in Part III are widely diverse in terms of their topographies, demographics, per capita income, sociopolitical systems, and the organization, structure, and financing of their respective health and social care systems. They range from Russia, the most populous country in Europe (144 million), to the smallest, Montenegro (622,000). The topics discussed in this section relating to predictions for the future development of health systems are wide ranging and diverse, spanning policies and programs initiated and implemented at the macro-, meso-, and micro-levels, with some already established, while others are more nascent, emergent, and evolving.

We begin our pan-European tour in Austria, where Maria M. Hofmarcher, Susanne Mayer, Nataša Perić, and Thomas E. Dorner outline the development of new primary health centers (PHC) aimed at reducing demand for hospital outpatient services through improved access to primary care staff and services, including extended opening hours. Moving to Denmark, Liv Dørflinger, Jesper Eriksen, Janne Lehmann Knudsen, and Carsten Engel discuss attempts to create a more patient-oriented health system by moving away from a focus on measuring, analyzing, and acting on healthcare process measures toward the development of more patient-reported outcome measures (PROs).

Shining a spotlight on England, a country that has witnessed continuous health reform over many decades, Martin Powell and Russell Mannion chart the development of the controversial policy of awarding personal health budgets (PHBs)—an individualized form of funding designed to encourage greater self-directed support, particularly for those people requiring long-term care.

In a theme common to many other countries, authors Ruth Kalda, Kaja Põlluste, and Margus Lember from Estonia in the next chapter analyze an e-consultation service developed by a collaboration of that country's peak medical and insurance groups.

Then onward to Finland's highly decentralized health system, where Persephone Doupi discusses the development potential of a range of precision medicine initiatives, including those relating to health and social care processes and developments in data collection.

Shifting the attention toward recent health reforms in France, Catherine Grenier, René Amalberti, Laetitia May-Michelangeli, and Anne-Marie Armanteras-de-Saxcé outline a new "global-local" approach to enhancing patient safety that offers the potential to radically improve the delivery of safer care to local populations by 2030. Continuing on our way, we arrive in Germany, the second most populous country in Europe, where Wolfgang Hoffmann, Angelika Beyer, Holger Pfaff, and Neeltje van den Berg set out a number of promising current initiatives for improving health services research and future planning in relation to the coordination of integrated care, including increased cooperation and joint partnerships of hospitals involved in delivering acute outpatient pediatric care.

Next, we journey to Greenland, the world's largest non-continental island, where Tine Aagaard and Lise Hounsgaard elaborate a number of key common principles and a range of practical methods for engaging citizens in professional practice with the aim of facilitating more democratic and culturally sensitive healthcare services. We then travel to Italy, where Americo Cicchetti, Valentina Iacopino, Silvia Coretti, and Marcella Marletta outline the development of a new national health technology assessment plan that should accelerate the adoption and procurement of new medical devices.

Sandra C. Buttigieg, Kenneth Grech, and Natasha Azzopardi-Muscat outline how a new national cancer plan in Malta has resulted in significant achievements in fighting cancer. There have already been remarkable improvements in survival rates, particularly for patients with malignant melanoma, and breast, testicular, thyroid, and prostate cancers. We then take in the Netherlands, where Madelon Kroneman, Cordula Wagner, and Roland Bal set out recent long-term care reforms aimed to keep people living in their own homes as long as possible. Key changes include the implementation of population health management programs, which hold the promise of establishing better integrated and networked (social) care arrangements.

Journeying to Northern Ireland, Gavin Lavery, Cathy McCusker, and Charlotte McArdle discuss the Attributes Framework (AF) for Health and Social Care. This program aims to help health professionals assess the attributes they currently possess (and those they require) to lead quality improvement and facilitate the building of workforce capability and capacity. In Norway, Ånen Ringard and Ellen Tveter Deilkås outline how already established clinical ethics committees (CECs) have the potential to fulfill an important role in advising and deciding on important topics related to the use and allocation of scarce resources at both the clinical and national levels.

Moving to Portugal, José-Artur Paiva, Paulo André Fernandes, and Paulo Sousa outline how a nationwide plan designed by the Portuguese National Program for the Prevention of Infection and Antimicrobial Resistance has been set up with the aim of reducing and hopefully eliminating the unnecessary use of antibiotics. Next, we arrive in Russia, where Vasiliy V. Vlassov and Mark Swaim discuss the challenges of attenuating some of the adverse side effects of the growing trend toward specialization among physicians. They also identify a number of possible solutions to this phenomenon, including the provision of generalist training and the nurturing of a more collegial, team-based approach to medical training and practice. We then journey to Scotland, where Richard Norris, Andrew Thompson, and David R. Steel set out the principles of the Our Voice framework. This involves a collaborative attempt to transform approaches to health and social care improvement by encouraging citizen and stakeholder involvement in health and social care planning and decision-making.

Moving on to Spain, Laura Fernández-Maldonado, Sergi Blancafort Alias, Marta Ballester Santiago, Lilisbeth Perestelo-Pérez, and Antoni Salvà Casanovas assess current policies to promote citizen engagement and involvement, including the Patients' University initiative. This initiative is an attempt to meet patient information needs and promote educational programs and research projects for staff and patients. We then travel to Sweden, where John Øvretveit and Camilla Björk discuss the learning health system (LHS) concept, which is an initiative to better integrate digital data and technologies into healthcare provision. This should be beneficial for all stakeholders, including patients, clinicians, and service delivery managers.

Anthony Staines and Adriana Degiorgi then discuss the prospects for new initiatives around teamwork and simulation in Switzerland. These seek to combine two fundamental approaches in patient safety: multiprofessional cooperation (teamwork) and individual accountability (acquiring and using teamwork skills). Onward to Turkey, Mustafa Berktaş and İbrahim H. Kayral discuss a range of initiatives to promote standardization and accreditation of health services outside the hospital system as a means of enhancing the delivery of high-quality healthcare, including accredited school health services, to promote the health and well-being of citizens in a number of contexts.

In Wales, Andrew Carson-Stevens, Jamie Hayes, Andrew Evans, and Sir Liam Donaldson report on a number of new policies and programs designed to improve patient safety in primary care based on the development and use of better data-driven learning systems, including devising more effective measures for safety incident recording and reporting. Our next destination is Central and Eastern Europe (CEE)—the bloc of 26 countries that make up the most easterly part of Europe—where Jeffrey Braithwaite, Wendy James, Kristiana Ludlow, and Russell Mannion assess reforms related to community-based family care and efforts to reduce health inequalities. Overall, they report slow progress in both these areas, although they identify pockets of CEE where more effective primary care and family medicine have been provided.

Finally, Jeffrey Braithwaite, Wendy James, Kristiana Ludlow, and Yukihiro Matsuyama discuss post-Soviet health reform in the countries that make up "the Stans" of Central Asia: Kazakhstan, Kyrgyzstan, Tajikistan, Turkmenistan, and Uzbekistan. While there is significant variation in terms of expenditure, access, and outcomes, all five nations have made substantial health system improvements in the last 25 years.

15

Austria

Primary Healthcare Centers: A Silver Bullet?

Maria M. Hofmarcher, Susanne Mayer, Nataša
Perić, and Thomas E. Dorner

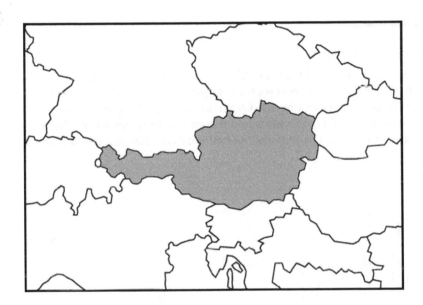

CONTENTS

Austrian Data

- Population: 8,747,358
- GDP per capita, PPP: $50,077.8
- Life expectancy at birth (both sexes): 81.5 years
- Expenditure on health as proportion of GDP: 11.2%
- Estimated inequity, Gini coefficient: 30.5%

Source: All data are from the World Health Organization and World Bank. Latest available data
used as at October 2017.
GDP = Gross Domestic Product
PPP = Purchasing Power Parity

Background

By international standards, healthcare provision in Austria is unbalanced.
While the Austrian hospital sector is large and appears to be expanding, with
Austria ranked first among the Organisation for Economic Co-operation and
Development (OECD) nations when it comes to the number of admissions per
capita, activity in outpatient and ambulatory care has stagnated (Figure 15.1).

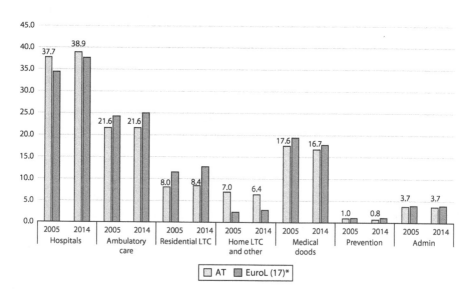

FIGURE 15.1
Healthcare spending components as a percentage of total current expenditure on health. The
latest year from which data are available is 2014. Data for Greece, Ireland, and Italy are missing
for 2005. EuroL = Euroland countries; LTC = long-term care. (Authors' own calculation based on
OECD Health Data, https://data.oecd.org/health.htm.)

Even though health outcomes have improved in many important areas, for example, stroke (Hofmarcher et al., 2017), there is evidence that increased life expectancy and better health could be achieved, even without additional investment (Gönenç et al., 2011; Joumard et al., 2010).

Underlying health determinants, such as smoking, weight, and alcohol intake, need greater attention. Many "survivors" need further health and social care; however, this is often poorly coordinated across care sectors. This may have a direct adverse impact on health and well-being—in particular for chronically ill people—and it puts additional pressure on financial sustainability (Kringos et al., 2013).

The legislative foundation of health services planning in Austria is a public contract between the central government and federal states, or Länder (Article 15a B-VG, 2016). Together with legislation about fiscal equalization across Länder, this contract defines the content and scope of mutual responsibilities for financing and providing healthcare. While most public hospitals are owned and governed by the Länder, primary care is the responsibility of social health insurance (SHI).

Primary ambulatory care is mostly provided per the benefit-in-kind principle, which gives priority to provider reimbursement. Insured persons can obtain benefits in kind from

- Contracted doctors who are self-employed and mainly work in solo practices. About 65% of practicing self-employed doctors have a contract with one or more SHI branches (Hofmarcher, 2013).* All licensed physicians are registered with the Chamber of Physicians. Placement and contract terms are jointly planned and administered by SHI and the chambers at regional levels.
- Hospital outpatient care departments, which are included in hospital plans.
- Outpatient clinics that are owned and run by either social insurance institutions or private individuals who may have a contract with one or more social insurance institutions. Many of these outpatient clinics are registered as businesses in the chamber of commerce.

Primary healthcare (PHC) centers provide an additional pillar of support in the Austrian healthcare system. Multidisciplinary PHC teams are scarce in Austria; 85% of Austrian general practitioners (GPs) are still practicing in solo practices for an average of 20.5 opening hours per week, while seeing on average 50 patients per workday (Hoffmann et al., 2015).

---

* The rest may offer services privately where patients are required to pay the bill directly at the point of service and are partly reimbursed by their respective insurer.

Success Story

In 2013, legislation introduced new primary care models to ensure better coordination of care at the "best point of service." It was agreed that by the end of 2016 at least 1% of the population would be served by such enhanced primary care models, and that at least two multidisciplinary ambulatory outpatient care or PHC centers should be established (Hofmarcher, 2014). Currently, only two PHC centers are being piloted—Maria Hilf (http://www.medizinmariahilf.at/) and GHZ Enns (https://www.ghz-enns.at/). Both PHC centers have extended opening hours (50 and 64 hours per week, respectively) and offer an extended range of services.

The current reform foresees the setup of a minimum of 75 PHC centers by 2020 (GRUG, 2017). These numbers should be achieved by the establishment of either group practices, freestanding outpatient clinics at one location, or a network of providers in different locations across Austria.

PHC centers aim to both improve patient access through increased proximity and extended opening hours, and increase the attractiveness of such models to healthcare professionals. Legislation passed in August 2017 requires that closely linked multidisciplinary cooperative practice teams provide and coordinate services. A core team comprising GPs, nurses, and assistants will be linked to professionals in other areas, such as child and elderly care, physiotherapy, health promotion and prevention, psychosocial healthcare, and palliative care.

Traditionally, the planning of ambulatory care capacity across different providers is fragmented. The new legislation widens the options of joint planning of ambulatory care by SHI and regional government, which is novel in the Austrian context (Table 15.1). Finally, regional plans have defined terms of references of ambulatory care levels to accommodate care provision at the best point of service. While current contracts with physicians will remain valid, the law ensures that contracted doctors will be given priority when they decide to create or participate in a PHC center.

Sixteen rounds of negotiations preceded the adoption of the law, and the draft received more than 365 reviews. There has been considerable controversy over the decision to allow commercial companies to run PHC centers. This led to the requirement that new centers can only be run by non-profit providers, for example, municipal entities.

Impact

Better health outcomes are generally associated with strong PHC sectors (Shi, 2012), and in those countries where strong PHC sectors exist, healthcare expenditure is found to be growing at a slower pace (Kringos et al., 2013). In

TABLE 15.1

Current and Future Market Authorization in Ambulatory Care

Regulatory Instruments	Ambulatory Care Settings	Market Authorization		Payment Schemes	
		Current	Future	Current	Future
"Public laws," for example, hospital acts	• Social Security Act (ASVG) • Supraregional (ÖSG) and regional capacity plans (RSG)	Outpatient departments in hospitals Outpatient clinics	Regional market authorization in the context of hospital plans Regional market authorization	Better-coordinated regional procedure with approval coming from ÖSG and RSG	Flat rate per case plus subsidies from the hospital budget Budget or flat rate per case	
"Professional laws," for example, Ärztegesetz		Primary care centers	—	ASVG/GRUG* and RSG as either • Group practice • Outpatient clinic • Provider networks	—	Combination of flat rates, fee-for-service, and bonus payments; not clear yet, but always with obligatory electronic reporting of diagnosis and services
		Contracted doctors	Regional "location plans" established through negotiations between SHI and the chamber of physicians		Primary care: Flat rate and fee-for-service Primary specialist care: Largely fee-for-service	

Source: Authors compilation based on Hofmarcher, M. M., Ambulatory care reforms fail to face the facts? *Health Policy Monitor*, 2010. Retrieved from http://hpm.org/en/Surveys/GOEG_-_Austria/15/Ambulatory_care_reforms_fail_to_face_the_facts_.html.

* GesundheitsReformUmsetzungsGesetz.

the Austrian healthcare system, a strong PHC sector is expected to replace hospital outpatient department utilization (Czypionka et al., 2017). A recent evaluation shows that PHC centers help shift utilization from costly inpatient care to outpatient care outside hospitals (Fröschl and Antony, 2017). Care delivery becomes more responsive and the satisfaction of patients and providers is high (Fröschl and Antony, 2017). Finally, the law provides the means to strengthen the organizational capacity in the outpatient care sector, especially in rural areas, and aims to enhance the quality of healthcare with improved coordination and access to care (GRUG, 2017).

PHC centers are generally expected to ensure better balanced utilization and improved coordination of inpatient and outpatient care. In contrast to earlier efforts, the current approach seeks to ensure the best point of service. The policy is therefore more attractive to other providers, rather than single contracted physicians entering the market.

Implementation and Transferability

Successful implementation of PHC centers in Austria will be supported by several key factors. In the current funding framework (2017–2020), a total of 200 million euros (US$234.7 million) is earmarked for the development of the PHC sector. It is hoped that workforce changes will enhance cooperation, with the upcoming generation of practitioners more likely to be open to change (Sensor Market Research, 2014). As work conditions at PHC centers are generally family-friendly, they may be attractive to physicians with children. While physicians are not obliged to enter PHC centers, successful best-practice examples, in addition to planned financial incentives, will also operate as pull factors (Ministry of Health [Austria], 2014). Importantly, the Austrian population supports the PHC reform plans and clearly requests extended opening hours and increased coordination and cooperation between healthcare professions (Hauptverband, 2015).

Professionalization of management in primary care was named one of the most important success factors in the PHC pilot. Trust, transparency, and financial incentives are also considered crucial. These issues are echoed in a recent evaluation of the Medizin Mariahilf PHC (Fröschl and Antony, 2017), where, in addition, a high level of provider satisfaction was found.

While the Austrian approach to enhancing high-quality primary care is path dependent, there are several lessons to learn. It may be practical to complement existing pillars of ambulatory care to improve supply. The best point of service approach may compel regional stakeholders to improve their collaboration through enforcing joint planning. The law prevents for-profit companies from entering the market; this may increase the odds of success because established providers may be less resistant to change.

Prospects

PHC centers promise to be successful in the future. However, major obstacles exist.

For example, there is, in principle, free choice when it comes to any level of care, and many patients use primary care for administrative issues, for example, certification of work absence due to sickness and prescription renewals (Pieber et al., 2015).

Important obstacles may arise from traditional task sharing, especially concerning doctors and nurses. While the education of GPs in Austria is based on adequate curricula, the postgraduate training is mostly carried out in hospital settings. At the same time, well-educated nurses with diplomas often carry out auxiliary nursing tasks, and there are very few community nurses. Task shifting is prevented because nurses would be required to work independently in PHC centers; this is currently not permitted, however. Both medical doctors and nurses lack adequate incentives to better collaborate in caring for patients.

The possibility of giving doctors employment contracts in a PHC center is very limited. This may inhibit market concentration and crowding out of contracted doctors in solo practice. At the same time, it weakens PHC to create optimally sized centers to improve both patient care and cost-effectiveness.

Conclusion

Austrian efforts to improve PHC date back decades. While earlier legislation proved to be insufficient, health reform in 2013 defined PHC centers as the best point of service for the provision of PHC. Recent legislation implementing PHC centers paves the way to achieving reform goals. While in the past the creation of more and better primary care capacity has always stalled because of flaws in legislation and slow uptake of providers, current efforts represent a renewed attempt to improve patient care and health system sustainability through a much-needed rebalancing of care provision. The success of this will largely depend on how payers and providers respond to fundamental changes in light of growing and specialized demand and current financial constraints.

16

Denmark

Patient-Reported Outcomes: Putting the Patient First

Liv Dørflinger, Jesper Eriksen, Janne Lehmann Knudsen, and Carsten Engel

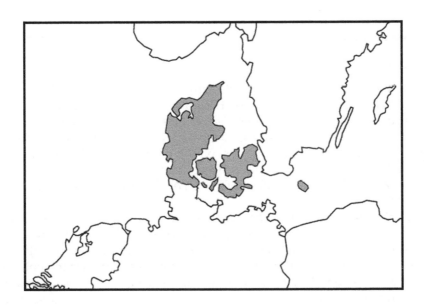

CONTENTS

Danish Data

- Population: 5,731,118
- GDP per capita, PPP: $49,696.0
- Life expectancy at birth (both sexes): 80.6 years
- Expenditure on health as proportion of GDP: 10.8%
- Estimated inequity, Gini coefficient: 29.1%

Source: All data are from the World Health Organization and World Bank. Latest available data used as at October 2017.
GDP = Gross Domestic Product
PPP = Purchasing Power Parity

Danish Healthcare

In Denmark, as in many other Western European countries, the quality of healthcare is high (Ministry of Health [Denmark], 2017). Healthcare is financed by income tax and there is free and equal access for the 5.4 million Danish residents and visiting European Union citizens. The foundations of the Danish healthcare system rest on the provision of efficient and high-quality healthcare, with providers freely chosen by users (OECD, 2015a). Denmark spends just above 10% of GDP on healthcare (OECD, 2015a; Danish Regions, 2015a).

Since the millennium, high productivity has been a crucial goal for Danish healthcare providers. The political goal has been, and still is, the reduction of waiting times; however, as these have been reduced to acceptable levels for most patients, other priorities, including patient engagement, have received more attention. Ambitious national strategies and local initiatives are strengthening the Danish healthcare system's focus on consumer needs and preferences (Danish Regions, 2015b).

A Greater Focus on Patient-Centered Healthcare

The core of patient-centeredness is to focus on patient needs and preferences when decisions about care are being made in a partnership between patients and caregivers; this is a cornerstone of quality care. As the population ages, many more people suffer from chronic and multiple diseases, and the demand for patient engagement is increasing (Freil and Knudsen, 2009). Each patient must be better equipped to engage as an equal healthcare partner, through co-decision, enhanced self-care, and self-treatment.

The Danish healthcare system is one of the best in Europe when it comes to patient engagement (Health Consumer Powerhouse, 2017). Nevertheless, one out of five patients in Denmark report not being sufficiently involved in treatment decisions (Unit for Evaluation and User Involvement, 2016). In repeated national population-based surveys, no substantial improvements have been found since 2000 (Unit for Evaluation and User Involvement, 2016). These surveys utilize patient-reported experiences (PREs) to monitor patients' experience and satisfaction levels, and have been an essential element in quality improvement measures in Denmark for almost two decades. Health profiles, which list various health status indicators, such as whether patients smoke cigarettes or engage in exercise, have also been used for some time.

More recently, the introduction of patient-reported outcomes (PROs) in Danish healthcare has resulted in a shift toward a more systematic and clinically-based approach to patient involvement. PROs have the potential to further improve healthcare quality, particularly when treatment is provided across different healthcare sectors (Figure 16.1).

What Is a Patient-Reported Outcome?

A PRO measure is the patient's own report of his or her health status. The American Food and Drug Administration (FDA) defines PROs as "a measurement based on a report that comes directly from the patient about the status of a patient's health condition without interpretation of the patient's response by a clinician or anyone else" (Food and Drug Administration, 2009: 2). Patients report on their symptoms; treatment outcomes, such as pain and health-related quality of life; and functional status.

Over the past 18 months, a group of 29 experts have worked to develop a Danish PRO data strategy, Program PRO. The program provides a blueprint

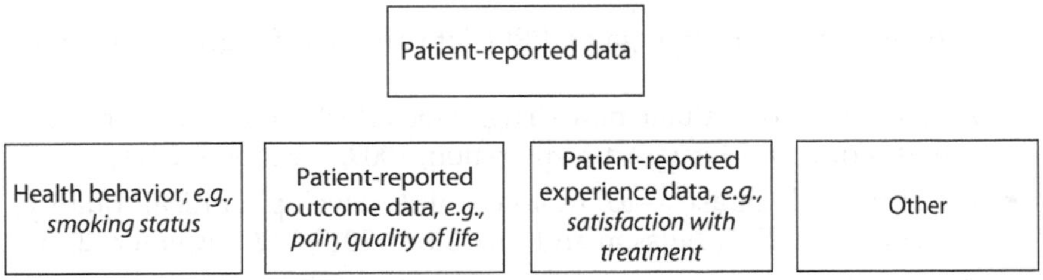

FIGURE 16.1
Types of patient-reported data: health behavior, experiences with care, and health outcome. (From Program PRO, Use of PRO-data in quality improvement in Danish healthcare: Recommendations and knowledge base, 2016. Retrieved from https://danskepatienter.dk/files/media/Publikationer%20-%20Egne/B_ViBIS/A_Rapporter%20og%20unders%C3%B8gelser/program_pro-rapport.pdf)

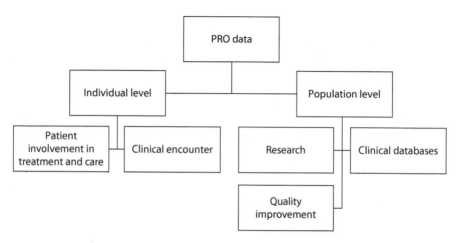

FIGURE 16.2
Use of PRO data on an individual and population level. (From Program PRO, Use of PRO data in quality improvement in Danish healthcare: Recommendations and knowledge base, 2016. Retrieved from https://danskepatienter.dk/files/media/Publikationer%20-%20Egne/B_ ViBIS/A_Rapporter%20og%20unders%C3%B8gelser/program_pro-rapport.pdf.)

for the systematic integration of PRO data in quality improvement across all sectors of the Danish healthcare system (Program PRO, 2016). PRO data can be collected and used at both the individual and the population level (Figure 16.2). On an individual level, PRO data is used

- By clinicians—to target diagnostic evaluation and improve treatment, for example, measuring the side effects of chemotherapy treatment (Chen et al., 2013; Nelson et al., 2012)
- By patients—to achieve greater satisfaction and better communication between patient and health professional, for example, using real-time PRO data in the consultation to focus the dialogue on the issues most bothersome to the patient (Chen et al., 2013)

At a population level, aggregated PRO data are used for groups of patients

- In research—to evaluate new drugs, side effects, and treatment outcomes (Food and Drug Administration, 2009; Meadows, 2011).
- In clinical databases—to monitor and develop clinical quality (Nelson et al., 2012; Nilsson and Lindblom, 2015; PROMCenter, 2015).
- As a quality indicator—to ensure systematic quality of treatment. This approach is currently restricted to a few countries, such as England, Sweden, and to a lesser extent, the United States (Black, 2013).

The most obvious gain from the introduction of PROs is the development of a more precise assessment of health status than is possible through physician assessment and clinical process and outcome data, such as blood tests and mortality rates. PROs demand and encourage both patients and clinicians to endorse a more patient centered care.

PROs in Danish Healthcare: How Far Are We?

PROs were introduced in Denmark and internationally more than 20 years ago (Sprangers et al., 1993). During this time, many Danish projects have been carried out with the aim of ensuring quality care, facilitating value-based management, and providing solutions to regulatory challenges. There are many examples of PRO use in hospitals, municipalities, and general practice, but efforts are currently underway to consolidate the databases at a national level (Government of Denmark and Danish Regions, 2015, 2016).

A national initiative has recently been established to implement PROs across all sectors and disease groups (Government of Denmark and Danish Regions, 2016). The initiative, a world first, satisfies both financial and quality care imperatives, ensuring a more patient-centered approach to healthcare (Government of Denmark and Danish Regions, 2015).

The 2016 budget agreement between the Ministry of Finance and the five hospital regions states that patients must be at the center of treatment decisions. PROs provide a key tool to support this (Government of Denmark and Danish Regions, 2015).

In late 2016, a national PRO working group was established to lead the implementation of PROs across all diseases: "The Parties agree to establish a working group to support a standardized and broad use of patient-reported information (PRO) in all sectors in healthcare. The group has a particular responsibility for the standardizing of PRO questionnaires and guidelines, and should encourage the use of PROs in daily clinical practice and quality development" (Government of Denmark and Danish Regions, 2016: 7). Between 2017 and 2020, the national working group aims to

- Standardize PRO questionnaires to patients
- Determine guidelines for standardized use of PRO data across geographies, sectors, and treatment areas
- Contribute to systematic knowledge sharing

The foundation of all PRO-related work in Denmark is based on the detailed work carried out by the 29 experts in Program PRO. These experts have recommended 12 steps (Table 16.1) for "good practice" when implementing PRO measures across sectors (Program PRO, 2016). These 12 steps have been developed using the experiences and evaluations of many smaller

TABLE 16.1

Twelve Steps for Good Practice

Choose a relevant patient group.
Map all relevant stakeholders in the patient trajectory.
Establish a steering group.
Choose quality improvement systems for the PRO quality improvement work.
Collect PRO data on the basis of clinical practice.
Use PRO data as a quality indicator for quality improvement work.
Plan methods for collecting data.
Choose a suitable PRO measure.
Inform and train patients and healthcare professionals—and carry out pilot studies.
Analyze PRO data from a quality improvement perspective.
Evaluate the results of PRO data analysis and adjust practice.
Evaluate the use of PRO data for a quality improvement system.

Source: Program PRO, Use of PRO-data in quality improvement in Danish healthcare: Recommendations and knowledge base, 2016. Retrieved from https://danskepatienter.dk/files/media/Publikationer%20-%20Egne/B_ViBIS/A_Rapporter%20og%20unders%C3%B8gelser/program_pro-rapport.pdf.

Danish PRO initiatives—including the Danish Rheumatology Database (DANBIO, 2015), the generic web-based system AmbuFlex, and the initiative for patients with arthritis, GLA:D (GLA:D, 2016; Hjøllund et al., 2014).

Perspective on PROs: Changing Healthcare in Denmark

The Danish health system is moving from measuring process to covering those outcomes that matter to patients. The aim is to improve clinical care and patient satisfaction. The success of the healthcare system can not only be measured by its output of services, but also must include the value it provides to patients. This value can only be measured by asking the patients.

While PROs can be used as tools in clinical encounters with individual patients, they also provide aggregated data to assess the performance of the entire healthcare system. Healthcare professionals sometimes consider the collection of quality data burdensome; this burden should be minimized when the same data, reported by patients, can be used for clinical purposes as well as quality monitoring.

PROs may also have a role to play in breaking down the silos of hospitals, general practice, and community-based care. The discourse could be shifted from distributing tasks *a priori*, with each silo focusing on those assigned to itself, toward a common discussion on how to achieve the desired outcomes through collaborative actions. Further, research potential is increased significantly when PROs and data from the clinical databases are linked.

Conclusion

PROs, with their focus on patient involvement and optimal use of resources, have the potential to optimize Danish healthcare by making it more patient oriented. However, while there is considerable merit in assessing success by outcomes, rather than processes, process data are still needed in order to understand and influence the outcomes. PRO data have the potential to change the encounter with individual patients, the way different actors cooperate, and the way in which we measure the performance of the healthcare system and assess its success.

17

England

Getting Personal? Personal Health Budgets

Martin Powell and Russell Mannion

CONTENTS

United Kingdom Data

- Population: 65,637,239
- GDP per capita, PPP: $42,608.9
- Life expectancy at birth (both sexes): 81.2 years
- Expenditure on health as proportion of GDP: 9.1%
- Estimated inequity, Gini coefficient: 32.6%

Source: All data are from the World Health Organization and World Bank. Latest available data
used as at October 2017.
Note: Data from England, Scotland, Wales, and Northern Ireland.
GDP = Gross Domestic Product
PPP = Purchasing Power Parity

Background

The English National Health Service (NHS) is a single-payer health system that is directly funded through general taxation and national insurance contributions and is (largely) free at the point of use. It caters to a population of 54.3 million and employs more than 1.2 million equivalent staff. In terms of structure, the NHS is made up of a complex array of organizations with different responsibilities and reporting arrangements. NHS England is an executive non-departmental public body of the Department of Health that oversees the budget, planning, delivery, and day-to-day operation of the commissioning side of the NHS and allocates funding to clinical commissioning groups that purchase care for their populations from providers (which may be run by the NHS, or by private- or third-sector organizations).

The 2016 Autumn Statement confirmed that the period from 2009–2010 to 2020–2021 will be the most austere decade that the NHS has ever experienced (HM Treasury, 2016). Funding will rise by an average of 1.0% a year, which is less than half the previous lowest growth over a comparable period (2.1% a year between 1975–1976 and 1986–1987). Taking population growth into account, health spending per head in 2020–2021 will be almost the same in real terms as it was in 2009–2010. Thus, additional pressures from rising costs, an aging population, and increasing numbers of people with chronic conditions will need to be met by efficiency growth, with the NHS in England mandated to deliver efficiency improvements of £22 billion (US$27.6 billion) by 2020–2021. Every health and social care system in England now produces a multiyear sustainability and transformation plan (STP), detailing how local services will evolve and become sustainable over the next 5 years. In a major policy volte-face, NHS organizations are now expected to collaborate rather than compete with each other, with a new focus on integrating and coordinating services between health and social care.

Success Story

Our success story involves personal health budgets (PHBs), which are defined as "an amount of money provided to support health and well-being needs. Money is spent according to a plan set out between an individual and health professionals" (Williams and Dickinson, 2016: 151). In our view, PHBs are the "best of a bad bunch": a policy option chosen reluctantly and unenthusiastically from an uninspiring field. Given the significant problems generated by austerity in the NHS, perhaps it is time to make this type of radical choice, which comes with significant potential, along with great risks. As it appears so difficult to change the "producers" from top-down diktats, perhaps increasing bottom-up consumer power through PHBs is worth trying.

Personal budgets are an international phenomenon, with many countries in Europe and North America experimenting with comparable methods of individualized funding and encouraging greater self-directed support, particularly for long-term care (Gadsby, 2013; Williams and Dickinson, 2016). PHBs come in a variety of shapes and sizes across a number of countries and may involve a direct payment, a notional budget, or a real budget held by a third party (Gadsby, 2013).

In the United Kingdom, PHBs build on the earlier introduction of direct payments (a cash payment made in lieu of service delivery, with individuals able to choose how this resource should be allocated) in social care (Forder et al., 2012; Williams and Dickinson, 2016). The PHB initiative was proposed in the NHS Next Stage (or Darzi) Review in 2008, as a way to encourage the NHS to become more responsive to the needs of patients, and was reaffirmed in the 2010 white paper "Equity and Excellence" (Department of Health [England], 2008, 2010; Forder et al., 2012). In 2009, a three-year pilot program for PHBs began (Forder et al., 2012).

At the center of a PHB is a care plan that is developed by the individual in conjunction with his or her clinical team and signed off by the NHS from a clinical and financial standpoint (Alakeson et al., 2016; Gadsby, 2013). Budget holders are theoretically able to use their budget for a wide range of health services, including complementary therapies and personal care, or to purchase one-off items for enhancing health and fitness, such as a computer or a Wii Fit. Budgets cannot be used to pay for emergency care or care normally received from a general practitioner, or for gambling, debt repayment, alcohol or tobacco, or anything unlawful. Individuals can choose to manage their PHBs in different ways depending on the level of financial responsibility they wish to take on (Alakeson et al., 2016; Gadsby, 2013). In 2015, an article in *Pulse*, a magazine for UK general practitioners, estimated that the overall level of spending on PHBs was £123 million (US$154.4 million) for 2015–2016, which was just over 0.1% of NHS spending (Price, 2015). During the same period, 4800 people were using PHBs, which suggests an average package cost for individuals of £25,600 (US$32,140).

Impact

International systematic reviews broadly suggest that while PHBs may not lead to better health, they bring positive outcomes in terms of patient satisfaction, feelings of well-being, and quality of life for the majority of users. There are, however, some methodological limitations in the evidence base (Gadsby, 2013; Webber et al., 2014).

The evaluation of the PHB pilot in England by Forder et al. (2012) reports that PHBs did not appear to have an impact on health status, mortality rates, health-related quality of life, or costs over a 12-month period, but were associated with significant improvement in patients' care-related quality of life and psychological well-being over the same period. They note that due to the relatively short time period examined, it is unsurprising that the underlying health status of patients was unaffected. Moreover, a PHB could be seen as a vehicle to effectively manage the health condition rather than improve clinical health status. It seems that the more experience people have of their long-term conditions (LTCs) and associated needs, the more they can use a PHB to improve their well-being. Mays (2013) notes that although improved care planning was not the main aim of the PHB pilots, the study suggests that care planning in the context of PHBs may well be worthwhile for some patients with LTCs. Forder et al. (2012) suggest that the findings provide support for the further implementation of PHBs, which were provided as an option to all people in receipt of NHS Continuing Healthcare from April 2014, and available to anyone with an LTC who could benefit after April 2015. The chief executive of NHS England stated that 5 million people could be receiving PHBs by 2018 (Scott-Samuel, 2015).

Other analyses of the impact of the PHBs are less positive, with some objecting to the recipients' spending decisions (see, e.g., Scott-Samuel, 2015). While the pilot PHBs were most commonly used to pay for physical exercise, alternative therapies, and carers, Scott-Samuel (2015) is critical of the "very wide range" of services accessed: these included neurolinguistic sessions, acupuncture, personal trainers, gym memberships, reiki, manicures, driving lessons, mobile phones, and theater and football tickets. The *Pulse* article, too, is disparaging of the PHB recipients' choices, claiming that NHS funding had been "splashed on holidays, games consoles and summer houses" (Price, 2015). Scott-Samuel (2015) further reports that a review of Dutch PHBs found substantial use of ineffective therapies and inappropriate consumer spending, as well as widespread fraud. He argues that patients are not necessarily capable of making the correct choices: "diabetes care is not like shopping for baked beans: I want my care prescribed by my diabetologist and my primary care team, not by my own subjective whims and prejudices" (Scott-Samuel, 2015: 76). This paternalistic view appears to discount the possibility of any patient control, and to deny the idea of the "expert patient," particularly where LTCs are concerned (see Alakeson et al., 2016; Duffy, 2015; Gadsby, 2013).

Implementation and Transferability

It is unclear how the program might roll out from its current low-patient-numbers base. There has been significant opposition to the initiative by the public service union Unison, the Royal College of General Practitioners, the Royal College of Nursing, and the British Medical Association (but then the British Medical Association was against the introduction of the NHS in the 1940s, so perhaps the "doctor does not always know best"). Duffy (2015) suggests two possible reasons why many in the medical establishment are not convinced of the benefits of PHBs, speculating that senior doctors and others are genuinely concerned about the threats that they perceive PHBs pose to the welfare state, and that senior doctors do not want to share power with patients in issues of decision-making or coproduction.

Notwithstanding these reservations, PHBs can be used to meet identified health and well-being needs in possibly new and innovative ways (Alakeson, 2013; Alakeson et al., 2016). For example, Alakeson (2013) notes that PHBs lead to the individual taking on the role of "service integrator," which presents an alternative to structural integration and (professionally led) care coordination. According to then health minister Lord Darzi, "the main aim of introducing personal health budgets is to support the cultural change that is needed to create a more personalised NHS" (Gadsby, 2013: 2).

Prospects

Simon Duffy, who was credited—or blamed—for inventing the personal budget, responds to the ideological question that was raised in *Pulse*, namely, the notion that personalization is just a right-wing neoliberal idea. Duffy is puzzled by this criticism, and struggles "to understand how shifting socioeconomic power to disabled people or to people with long-term health conditions is right-wing. I still don't understand why leaving power with the NHS bureaucracy or medical establishment is left-wing" (Duffy, 2015).

Williams and Dickinson (2016) point to the dangers of concentrating on individual service provision, and ignoring the impact on wider considerations of equity or structural inequality.

However, this is to view consumerism through a narrow "consumer as chooser" prism rather than as part of the wider vision of social democratic citizenship as envisioned by Duffy (2015). This model was originally advanced by Michael Young, who helped to draft the Labour Party manifesto in 1945, and founded the Consumers' (note plural) Association in 1956. Such ideas also formed the core of the organized consumer protection movement associated with Ralph Nader in the United States (Powell et al., 2010; Simmons et al., 2009). If one of the architects of the 1945 Labour Party manifesto was comfortable with the notion of consumerism, we should not dismiss it out of hand. It is important to place consumerism in the wider context of Simon Duffy's original broad objectives, with an emphasis on involvement, empowerment, coproduction, placing patients at the center of decisions about their care, and learning lessons from the "wisdom of crowds" about what works.

Conclusion

In the United Kingdom, the concept of PHB, both in theory and practice, divides opinion (Alakeson et al., 2016; Gadsby, 2013). Williams and Dickinson (2016) point to an explosion of claims and counterclaims over both the benefits of personalization and the nature of its ethical underpinnings: regarded from one perspective as a radical, user-led program of reform, and from the other, as a symbol of a neoliberal agenda to dismantle the welfare state. Perhaps PHBs will not live up to some of the grander claims made for them, such as "turning the welfare state upside down" or having "the potential for the most fundamental reorganisation of welfare for half a century" (Williams and Dickinson, 2016: 150), but if they contribute to a better quality of life for patients, especially in the area of LTCs, that may be a good start.

18

Estonia

e-Consultation Services: Cooperation between Family Doctors and Hospital Specialists

Ruth Kalda, Kaja Põlluste, and Margus Lember

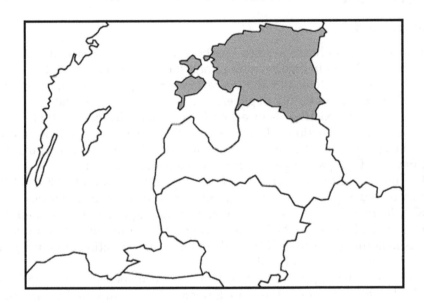

CONTENTS

Estonian Data

- Population: 1,316,481
- GDP per capita, PPP: $29,364.7
- Life expectancy at birth (both sexes): 77.6 years
- Expenditure on health as proportion of GDP: 6.4%
- Estimated inequity, Gini coefficient: 33.2%

Source: All data are from the World Health Organization and World Bank. Latest available data used as at October 2017.

GDP = Gross Domestic Product
PPP = Purchasing Power Parity

Background

The Estonian healthcare system has been extensively reformed since the early 1990s. The main objectives of reform include the reorganization of the public funding system and the overextended hospital system, improving the quality and accessibility of general medical care services, and increasing the efficiency of existing services, such as primary healthcare (PHC). The training and introduction of family doctors (FDs) was central to PHC reform, which was supported by necessary organizational changes, such as instituting new funding, partial gatekeeping, and a patient list system.

The reforms also supported the widening of the scope of PHC and the increased participation of nurses in individual consultations. Various other PHC initiatives have been implemented, including opening a 24-hour primary care call center in 2005, introducing financial incentives as a quality bonus in family practice in 2006, and voluntary accreditation of family practices in 2009 (Põlluste et al., 2017).

A centrally managed eHealth system was developed between 2005 and 2008. The eHealth system has transformed infrastructure and information exchange by linking health service websites and databases, including health service e-registration, e-prescriptions, electronic health information records, imaging databases, and pathology. This improves the continuity and integration of care by providing patient information to healthcare workers at different levels or institutions of the health system (Lai et al., 2013).

Today, all hospitals and FDs use electronic health records (eHRs); unfortunately, however, hospital software varies, as does the software used in family practices. All healthcare providers are therefore obliged to send each patient's clinical history to the national eHR system. According to official statistics, while hospitals provide histories in almost 100% of cases, FDs provide significantly fewer (Anderson and Panov, 2015). The ePrescribing system was launched in January 2010, and currently almost 100% of prescriptions are created digitally and sent to the central ePrescribing system (Widen and Haseltine, 2015).

Success Story

While there are some concerns about extended waiting times for outpatient specialist services and overall access to healthcare services, Estonia's annual population satisfaction surveys have shown high satisfaction with health-care provider services since the 1990s (Kantar Emor, 2016). FDs, however, are somewhat less satisfied: the aging population, with a growing number of chronically ill patients, has increased their workload considerably. Although there is no published official data, surveys of specialists and FDs reveal that both parties are unhappy with the current referral system due to perceived communications problems. Although trust in FDs is rising, there is still a greater "belief" in specialist doctors' expertise.

In response to this difficult situation, the Estonian Health Insurance Fund (EHIF) and the Estonian Society of Family Doctors (ESFD) worked together to develop an e-consultation* system, based on the system used by Finnish doctors for some years. This system has impressed the many Estonian FDs who have worked in neighboring Finland (Doupi and Ruotsalainen, 2004).

There are a number of steps in the e-consultation process:

1. The patient visits the FD with a health problem.
2. The FD determines whether specialist advice is required.
3. Instead of referring the patient directly, the FD sends an e-consultation request to a specialist doctor. The request describes the situation or diagnostic problem of the patient and includes relevant patient health data from the eHR.
4. The specialist should respond in 3–4 days, either with advice to the FD on how to continue the treatment himself or herself or by recommending a specialist consultation. If a consultation is needed, the specialist books the appointment and informs the patient.

The system has multiple benefits: FDs are able to communicate with specialty doctors more efficiently, hospital specialists can use their time more effectively, and FDs are provided with the information required to solve most health problems at the primary care level. To guarantee that the patients who really need referrals are seen by specialists, priorities are identified and first examinations are done at the PHC level.

It should be emphasized that although the ESFD played an important role in the development of the e-consultation system, the willing collaboration

---

* By e-consultation, we mean electronic bidirectional communication between clinicians about patient-specific questions. e-Consultations utilize the nationwide eHR system, which includes an electronic referral system.

of FDs and hospital specialists was essential. The specialists were particularly motivated by the need to get the "right patients" at the "right time" and to reduce the workload of the hospital emergency departments. In 2011, a pilot project of e-consultation services was carried out by the ESFD and the North Estonian Regional Hospital, and in 2013 the EHIF began financing the e-consultation service.

Another important aspect of the e-consultation system is that service standards for e-referral and e-consultation have been agreed on by FDs and specialty doctors. The service standards correspond to the recommendations of existing clinical guidelines, list the cases when an e-consultation request is relevant, and describe what information should be presented.

By 2014, e-consultation services were available in seven specialties in North Estonia, and 1358 e-consultation services were provided. By 2015, the number of e-consultations had almost doubled to 2354. The service was extended to other parts of Estonia in 2016. By the first half of 2017, the e-consultation service had been used 6023 times. Currently, the e-consultation service is available in 13 specialties and is most often used by neurology, endocrinology, urology, and ear, nose, and throat physicians.

A number of important preconditions facilitated the implementation of the e-consultation process in Estonia. The already existing central eHealth platform, in combination with well-established quality standards for e-services, provided solid infrastructural and technical foundations. In addition, interested stakeholders, including the ESFD, regional hospitals, and the EHIF, have taken an active role in the implementation of the system. Financial incentives appear to have had a positive impact on utilization: while there is no specific financial incentive for FDs to use e-consultation, the e-consultation reimbursement for specialist institutions is 68% of the reimbursement cost of an ordinary initial face-to-face visit. FDs are currently funded via a capitation arrangement, but discussions about adding e-consultation services to their quality payment system are currently underway.

In April 2016, a web-based national survey was undertaken by 912 healthcare providers (including FDs, hospital specialists, and hospital management) to determine the most serious obstacles to using the e-consultation services and to look for possible solutions. The survey found that almost 47% of the FDs who responded had used the e-consultation services, and 86% of those who had not yet used the service said that they planned to do so in the near future. The most common reasons for not using the service were lack of information about availability, lack of time, and the fact that traditional consultation referrals are still possible. About 41% of the hospital specialists who responded to the study found that e-consultation takes the same or even more time as face-to face consultations and is therefore not time-efficient. Allocating appropriate time for e-consultation services was therefore seen as necessary. The lower reimbursement for the e-service was also reported to be a disincentive (Saar Poll, 2016).

Impact

Initial studies indicate that e-consultation programs can alleviate pressure on limited health system resources by improving access to specialty care at relatively low cost, and that they are feasible in a variety of settings and flexible in their application (Liddy et al., 2013; Vimalananda et al., 2015). In the 2016 survey, healthcare specialists reported many positive features of e-consultation, among which faster access to specialist care, time efficiency, and better triage of patients according to their real needs were most frequently mentioned. About 17% of the respondents stated that e-consultation is for those patients who are unwilling to travel to the specialist, either because they are elderly and frail or they are unable to leave work (Saar Poll, 2016).

E-consultations are not a substitute for face-to-face specialist visits, which are really needed, but they can reduce unnecessary referrals and allow specialists to see more urgent cases. Liddy et al. (2013) found that the e-consult program significantly reduced waiting times from months to an average of 5 days—with 25% of e-consultations taking place the following day. The study found that in 40% of cases, an unnecessary referral was avoided. The first survey of the e-consultation process indicates that in 2015, face-to-face specialist visits at the North Estonian Regional Hospital were not needed in 3–25% of cases, depending on the specialty (with the lowest rate of unnecessary visits in endocrinology and the highest in rheumatology).

E-consultation systems may improve the quality and documentation of consultations because they use standardized formats that help structure questions and expectations for care, reducing opportunities for miscommunicated clinical information (Reichman, 2007).

In order to improve the system, regular meetings and round-table discussions are held so that a range of stakeholders (FDs, hospital specialists, hospital management, health insurers, and information technology [IT] experts) can share opinions and ideas and give feedback. The main purpose is to work together to make the system satisfactory for all participants.

Implementation and Transferability

Currently, the use of the e-consultation service is not compulsory for FDs, who can choose to use paper-based referrals, e-referrals, or e-consultation; however, the trend is toward high use of e-consultation. At the same time, the potential of e-consultation varies by specialty. Most likely, it is appropriate in communication with specialists who provide verbal and academic advice (e.g., in endocrinology or nephrology) rather than performing procedures. In Estonia, patients can consult dermatologists, psychiatrists, and ophthalmologists without a referral from an FD; however, this is currently being reconsidered. Many FDs already use teledermatology services to screen potential melanomas, and because photographs of skin lesions can be attached to e-consultation requests, it is likely that dermatologists will soon require initial e-consults. e-Consultation services can even be used also in ophthalmology in some cases, for instance, in the screening of diabetic retinopathy. This will be particularly useful for those with type 1 and 2 diabetes who miss their advised annual retina exam due to long waiting lists or lack of access to specialist services.

Prospects

If the e-consultation system is to continue to expand and improve, it must meet the needs of both the patients and providers. To ensure this, assessments of the following indicators should be made regularly:

- Referral volume
- Time from referral to treatment of first visit
- Clinical reason for consultation
- Quality of the referrals
- Reasons for rejecting the e-consultation
- Providers and patient satisfaction

Conclusion

In contrast to e-referrals, e-consultation services involve bilateral communication and thus have the potential to improve interactions between FDs and hospital specialists. Such interaction may also have an educational value for FDs. The service has already improved patient access to appropriate specialist care. The e-consultation service will provide support to FDs in rural areas, treating patients who are not able to travel to specialized care. Establishing satisfactory provider payments to incentivize utilization, and the development of integrated and easy-to-use software are essential to the system's continued success.

19

Finland

A Real-Life Experiment in Precision Medicine

Persephone Doupi

CONTENTS

Finnish Data

- Population: 5,495,096
- GDP per capita, PPP: $43,052.7
- Life expectancy at birth (both sexes): 81.1 years
- Expenditure on health as proportion of GDP: 9.7%
- Estimated inequity, Gini coefficient: 27.1%

Source: All data are from the World Health Organization and World Bank. Latest available data used as at October 2017.
GDP = Gross Domestic Product
PPP = Purchasing Power Parity

Background

Finland is in the process of designing and implementing the biggest health and social care reform in its history, with the double aim of reducing system inequalities, as well as containing costs (Finnish Government, 2015). The Finnish healthcare service sector's current focus is on achieving better accessibility and equity of services through the digitization and unification of the fragmented financing system, as well as on providing better-quality care. Freeing up patient and client choices when it comes to health service providers is also a priority.

Many of the reforms have faced setbacks in terms of acceptance and alignment with constitutional requirements (European Social Policy Network, 2016). Nevertheless, large-scale national projects are ongoing in key areas, including child and family services, elderly home care, and informal care; support for those with reduced work ability; and the reduction of inequalities in the health and well-being of vulnerable population groups (Ministry of Social Affairs and Health [Finland], n.d.)

The healthcare and social care sector reform is complemented by a strong business counterpart in health-related research and innovation (R&I). The drivers behind both are familiar: an aging population with increasing service needs, in combination with limited resources and an insecure global economy. The tandem initiatives form part of a strategy to guide the country through the difficult times it has been experiencing and into a promising future. The common denominators supporting both efforts are data, technology, and digitization, which in the health R&I field should pave the path to precision medicine.

Precision Medicine in Finland

Precision medicine is one of the high-end health technology initiatives currently being rolled out globally through strongly publicized projects, such as NHS England's 100,000 Genomes Project and the Precision Medicine

Initiative in the United States (Genomics England, n.d.; National Institutes of Health, 2017).

Precision medicine is a term for which no generally accepted definition exists (Feero, 2017). It is nevertheless closely connected to the trend of personalized medicine (European Commission, 2013), and involves the use of molecular data and genomics. The National Institutes of Health described precision medicine as "an emerging approach for disease treatment and prevention that takes into account individual variability in environment, lifestyle and genes for each person" (National Institutes of Health, 2017).

The roots of the precision medicine paradigm lie in pharmacogenomics, the use of genomic information to tailor drug prescribing and use. Effective screening methods and other preventive interventions are also in the scope of precision medicine, but they are dependent on certain preconditions being met, including the ability to connect genomic, medical record, and lifestyle data across large cohorts (Ashley, 2015; Joyner and Paneth, 2015).

A number of barriers must be overcome if precision medicine is to become the standard practice of future care. These include

- Massive investments in infrastructure for data collection, analysis, and interpretation
- Policy, legal, and regulatory reforms
- Privacy and ethical challenges
- Collaborative efforts from a variety of stakeholders, with patients and citizens playing a key role (Aronson and Rehm, 2015; Ashley, 2015)

Finland has boldly decided to join the gradually increasing number of countries who commit to the somewhat daunting task of realizing the precision medicine model. Having already been at the forefront of global eHealth implementation (Ministry of Social Affairs and Health [Finland], 2013), Finland moved its healthcare digitization game to a new level by announcing the establishment of an IBM Watson Health Center of Excellence in Helsinki (Tekes, 2016).

Health technology has traditionally been a priority domain for Finland's key funding players, such as the Finnish Innovation Fund, Sitra; the Finnish Funding Agency for Innovation, Tekes; FinPro; and the Team Finland network. Despite successful investment in health-related research and the high international standing of Finnish researchers and health professionals, the main conclusion of a 2012 report assessing the growth potential of the sector was crushing: Finland had not succeeded in wide-scale commercialization and was a net importer rather than exporter in the health sector (Ministry of Employment and the Economy [Finland], 2012). A new goal and vision were set: private research activities in the pharmaceutical and health technology research sectors were to grow by 2.5 times by 2020. Finland's target was to become a pioneer in personalized healthcare.

Between 2012 and 2016, three strategic ministerial-level documents were published, outlining the actions required to ensure the economic growth and success of the healthcare sector (Ministry of Employment and the Economy [Finland], 2014, 2016; Ministry of Social Affairs and Health [Finland], 2016). The focus is on multistakeholder collaboration in health-related R&I activities.

Building the Precision Medicine Ecosystem

Systematic long-term investments in utilizing digital technologies in the health and social service and research sectors have created a developed infrastructure featuring most of the essential components the precision medicine vision relies on (Aronson and Rehm, 2015). Existing assets span fully implemented electronic health records (EHRs), health and social care service data and registries, biobanks and collections of biological samples, and computing centers and solutions (Figure 19.1). The Institute for Molecular Medicine Finland's (FIMM) world-class Technology Centre and Biobanking platform is involved in several European infrastructures, such as the European Advanced Translational Research Infrastructure in Medicine (EATRIS) and Biobanking and Biomolecular Resources Research Infrastructure (BBMRI). The Finnish IT Center for Science (CSC), the country's supercomputing center, is central to both BBMRI and the Partnership for Advanced Computing in Europe (PRACE), the European high-performance computing infrastructure.

Services in the Future Healthcare System

In accordance with policy targets set in 2015 (Ministry of Social Affairs and Health [Finland], 2015), there should be a stronger role for patients and citizens in the future, both in terms of assuming responsibility for their own health and taking care of minor ailments, and in the form of increased freedom in service provider selection. Complementing this, the service provider landscape should also become far more variable than today, allowing for a bigger share for the private and voluntary and community sectors. Changes in modes of service delivery are already visible and are expected to evolve further, with a shift in remote service provision, such as various forms of telehealth (Keränen, 2017; Seppänen, 2016). In the future, remote services will play an increasingly important role in healthcare, with citizens and patients given tools, including genetic testing, to self-diagnose (at least partially or initially) and self-manage their conditions (Figure 19.2). Significant shifts will take place in the division of professional roles and tasks, and laypeople will play a more consequential role. For example, nurses are already being allocated broader responsibilities.

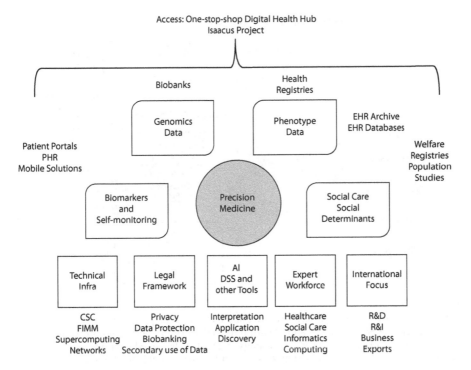

FIGURE 19.1
Components of the Finnish precision medicine infrastructure. (PHR = Personal Health Record, CSC = Finnish IT Center for Science, AI = Artificial Intelligence, DSS = Decision Support Systems.)

The operational environment required for the realization of the precision medicine scenario is starting to emerge. Heterogeneous data sources are currently pooled in large data warehouses, which are tested for their use potential in care and research in the context of the Isaacus project (https://www.sitra.fi/en/projects/isaacus-pre-production-projects/#contact-us), a digital health hub focused on gathering and coordinating access to health and well-being data (Sitra, n.d.). Genomic data and its utilization are gradually progressing toward integration with healthcare practices—as seen, for example, in pharmacogenomics tests currently being offered through private service providers to assess the suitability of certain drugs to individual patients (Abomics, 2017).

Much healthcare and lifestyle data will be captured by patients and eventually combined with the data provided by professionals and recorded in EHRs. Measures of functional ability and quality of life will increasingly be used in the assessment of patients and care planning (Kunnamo, 2015). When combined with genomic data, these assessments will enable treatments to be streamlined to individual patients and/or used to prevent diseases of genetic origin.

Because the data on which health-related decisions will be based will change dramatically, both new skills and improved decision support tools

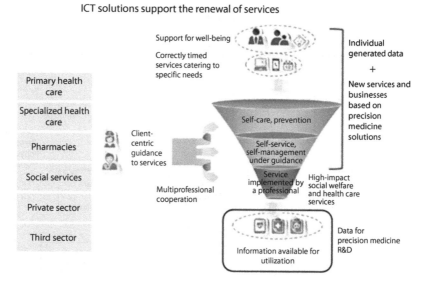

FIGURE 19.2
Future healthcare system: services structure and interplay with precision medicine. (Modified from Ministry of Social Affairs and Health [Finland], Information to support well-being and service renewal: eHealth and eSocial strategy 2020, 2016. Retrieved from http://julkaisut.val-tioneuvosto.fi/bitstream/handle/10024/74459/URN_ISBN_978-952-00-3575-4.pdf.)

will be required, for both professionals and citizens. So far, some steps have been taken with regard to education of professionals, but citizens are still lagging behind.

Transferability and Prospects

One of the most profound changes in the way we view health and disease is the stronger emphasis on prediction and prevention, rather than just care. However, what is understood as prevention is not unequivocally agreed on. Traditionally, prevention has been regarded as a public health issue, and population-level interventions have primarily aimed at reducing gaps caused by socioeconomic differences. Precision medicine focuses on the individual and utilizes high-end technology to achieve its goals. Most likely the strength lies in utilizing a good mix of both worlds (Khoury et al., 2016).

The building blocks required by precision medicine initiatives can be seen as elements that should be present in any modern digital healthcare system: user-friendly software, access to the broad spectrum of data available on each patient, and internationally standardized coding enabling automatic data processing and multiple applications (Learning Healthcare Project, 2015). On the other hand, the promise of new technologies needs to be weighed against each healthcare and social care system's needs and priorities. Some reports of the progress of precision medicine efforts and tools, both internationally and in Finland, have been less positive recently, and skepticism around the feasibility and worth of the undertaking has been growing (Freedman, 2017; Mediuutiset, 2017; Ross and Swetlitz, 2017).

It has been conjectured that in the coming decades the rise of technology in healthcare will radically change healthcare delivery, lead to health profession reforms, and generate significant economic impact (Ailisto, 2017). As yet, however, it is unclear how and whether possible gains will be achieved.

The unfolding developments in data analysis and its utilization raise new ethical and security problems that need to be addressed (Temperton, 2016). In Finland, the legal framework needed for the infrastructure to function as envisaged is still under revision, although commitment to change has been made at the policy level. There is a possibility that all relevant data will one day be made available to patients, too, for direct use in their care, but this will require considerable further changes to both legislation and culture, and a significant overhaul of the data collection systems and repositories.

Conclusion

The potential of precision medicine initiatives has created extensive interest around the globe, both within scientific circles and among the wider society, but the jury is still out on their value (Joyner and Paneth, 2015; McCartney, 2017). Finland offers a unique example where internationally proposed implementation steps and priorities are being tested in real life. Whether the experiment will benefit both citizens and the health industry still remains to be seen.

20

France

Horizon 2030: Adopting a Global-Local Approach to Patient Safety

Catherine Grenier, René Amalberti, Laetitia May-Michelangeli, and Anne-Marie Armanteras-de-Saxcé

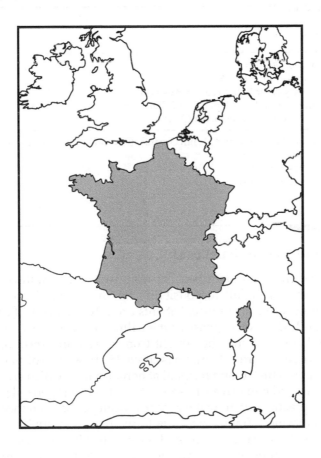

CONTENTS

French Data

- Population: 66,896,109
- GDP per capita, PPP: $41,466.3
- Life expectancy at birth (both sexes): 82.4 years
- Expenditure on health as proportion of GDP: 11.5%
- Estimated inequity, Gini coefficient: 33.1%

Source: All data are from the World Health Organization and World Bank. Latest available data
used as at October 2017.
GDP = Gross Domestic Product
PPP = Purchasing Power Parity

Challenge of the Global-Local

The French healthcare system benefits from universal health coverage. The government has taken full responsibility for the financial and operational management of health insurance. Patients are refunded 70% of most healthcare costs, and 100% of expenses when illnesses are costly or long-term. Supplemental coverage may be bought from private insurers, most of them non-profit, mutual insurers. France has about 2500 hospitals of all types (acute, rehabilitation, psychiatry, and hospital at home), half of which are public non-profit hospitals, and half private for-profit hospitals. Most general practitioners are in private practice but draw their income from public insurance funds. The system is increasingly fragmented due to the discontinuity between private and public hospitals, primary care, and social care. However, due to the financially secure and independent status of its hospitals, primary care, and social care, healthcare in France has long been regarded as one of the best systems in the world (Degos et al., 2008; Nolte and McKee, 2008; WHO, 2000).

The challenge now for the French system is to maintain this excellence in the face of current difficulties. And there are many. France, like most other countries, has a rapidly aging population and significant public debt. Urban centers and medical settings are expanding at the expense of the regions, which in turn have become unappealing to medical practitioners while retaining a population that has a disproportionate number of retired and chronically unwell. These social changes, combined with the fragmentation of the system, threaten the stability of the healthcare system.

The French healthcare system needs to create a balance between a global centralized vision and rapidly growing regional disparity. Achieving such a balance will require both a significant increase in regional autonomy and cooperation between medical professions and care systems, while still maintaining the historical imperatives of equality, equity, and fraternity. A global system is necessary to maintain uniformity of access, quality, and cost, and a local emphasis is needed to ensure that regional demographic differences are catered to. Health systems must also be reoriented toward the promotion of healthy living rather than just healing. France will have to consider revised safety and quality standards to reflect the large number of end-of-life and frail aged citizens living at home, and those who are likely to be living with cognitive impairment and dementia by 2030 (estimated to be more than 10% of the population).

Identifying Obstacles

The centralized vision and one-size-fits-all attitude that prevails in France, in combination with the strict demarcation that exists between private and public sectors, hospitals, primary care, and social care, are likely to prove major obstacles to reform.

France has a long history of centralization, and of imposing rules and regulations in an effort to eradicate or suppress regional differences. In the 1970s, the regions were given a degree of financial and administrative autonomy, via regional councils of elected officials. Regional autonomy, however, is still quite limited, and there continues to be very little variation between the laws and codes at the regional level regardless of local specificities.

Despite the creation of a regional health and social care authority, the Agence Régionale de Santé (ARS), in 2010, the provision of healthcare remains highly centralized. Each ARS is bound by the same rules and regulations despite administering to regions that are demographically diverse and that vary widely when it comes to healthcare access and practice.

Similar difficulties arise when it comes to aligning the planning of medical services, resources, and professionals within the regions. Since 2005, the allocation of resources to healthcare organizations has been based on an activity payment rate (*Tarification à l'activité-T2A*). While this system has had a positive effect in that it has given individual hospitals greater control over their activities and medical offerings, it has also encouraged fierce competition between hospitals and sectors, which in turn has increased efficiency to the detriment of quality and safety, and led to a decline in interhospital cooperation.

The integration of hospitals, primary care, and home care in France is also historically problematic; in fact, France has been ranked among the worst-performing Western countries in this area (Commonwealth Fund, 2016). Not only do individual hospitals, both private and public, compete, but hospitals and primary care, which operate independently of one another, are competing for the same limited resources. French primary care and hospital sectors are also contending with rapidly decreasing staff numbers in remote and socially deprived regions.

Priorities

The reform process rests on five pillars:

1. More regionalized structure and regional autonomy
2. More collective and cooperative work among medical settings and with primary care
3. Better care continuity through the organization of regional medical sectors (emergency, cancer, heart disease, etc.)
4. Reduced burden of care for patients when accessing professionals and exams, and increased proximity and coordination
5. A global-local vision for the process of surveillance and accreditation

France has already implemented two acts that herald the future system. The first dates back to 2009 with the *Hôpital, patient, santé et territoires* (Hospital, Patient, Health and Territory [HPST]) reform act, which made continuity of care a priority. In 2016, the *Loi de modernisation de santé* (Modernizing the Healthcare System) act was passed. This act definitively reinforced the need for a structured regional approach to healthcare delivery.

In 2016, building on the 2009s reform act, around 130 health territories, the *Groupement Hospitalier de Territoire* (GHT) were established. Authority has been delegated to the district hospital of each territory, under supervision of the ARS, to reorganize and rationalize services, to ensure that staffing is commensurate with requirements and that all citizens have access to specialist care.

There is some flexibility within the system, with territories able to share resources. The largest public hospital in each GHT is also responsible for increasing inter-hospital coordination of patient care, and for coordinating joint evaluation and accreditation of all public hospitals of the territory.

Other improvements are underway, with the introduction of digital technology a priority. Such improvements include the implementation of telemedicine for remote areas, the installation of new monitoring and alert systems for remote medicine, and standardizing and improving the compatibility of medical software. President Macron has pledged to increase public health spending by an annual 2–3% over his 5-year term (which is slightly more than in recent years); to invest €5 billion (US$5.8 billion) in hospitals, primary care, and innovation; to cut existing costs by €15 billion (US$17.5 billion); and to balance the deficit-ridden social security budget. He has also promised to speed up the merger of hospitals into regional groups, ensure that all effective drugs and devices will continue to be reimbursed, and phase in 100% reimbursement for spectacles, dental treatment, and hearing aids by 2022 (Casassus, 2017).

Three Priorities for Improving Quality and Safety

When it comes to the future of patient safety, the global-local approach that is emphasized in the reforms flows into three complementary priorities.

Engaging All Local Professionals in Mutual Cooperative Work

The customary barriers between medical castes must be crossed, and a culture of cooperation established, if the global-local approach is to succeed. In 2014, the Haute Autorité de Santé (HAS) launched the *Programme d'Amélioration Continue du Travail en Equipe* (Continuous Improvement Program for Teamwork) (PACTE), in order to encourage regional medical teams to develop better cooperative and teamwork skills. The program is voluntary, goes toward accreditation, and is based around a diagnosis of cooperation problems using medical team training (MTT), safety culture questionnaires, and process and outcomes indicators. The program is currently underway in hospitals and medical homes, and will soon be undertaken by professionals within the GHT. The ultimate aim is to ensure that the patient journey between hospitals, medical homes, and social care is as seamless as possible. The accreditation process is also addressing the GHT, and studying the possibility of bridging the gap between hospital, primary care, and social care through territorial accreditation.

Managing Quality and Safety Solutions in Context

Accreditation has highlighted the many French establishments in which one or more areas of care are substandard. The HAS have negotiated to give hospitals that fail to meet accreditation standards 3–12 months to comply. However, regional disadvantages, such as isolation, lack of medical and technical resources, low numbers of professionals, low-density populations, and disproportionately large aged populations, mean that at least 50% of medical establishments will be unable to achieve compliance by the due date. In addition, there is no way to guarantee that hospitals meet the requirements before the end of the compliance period.

Because it is simply impossible for some disadvantaged hospitals to comply with the required standards, it may be better to ensure that effective risk management procedures are put in place. These hospitals could instead use proven safety strategies to compensate for non-standard care and maintain safety when local circumstances are not ideal. Such strategies might include interventions to support staff, reorganizing, risk control, monitoring, education, and mitigation (Vincent and Amalberti, 2016).

In settings where care can be precisely delineated, strategies to control exposure to risk and maintain standards will predominate, hopefully accompanied by concurrent strategies to improve working conditions and support staff. By contrast, in more demanding environments, strategies to improve monitoring adaptation may be more to the fore, although all environments require solid procedural foundations. We need to consider how these strategies and associated interventions might be combined and in what proportions.

The HAS is considering a more flexible accreditation process at the territory level—which represents a significant transformation of the one-size-fits-all mentality. Such a transformation will have a flow-on effect on the way authorities share and manage risks. The ARS will be given the authority to contract with each territory regarding compliance exemptions, and the ways in which any identified risks will be managed. Needless to say, this new partnership cannot work unless all stakeholders cooperate.

Hospital and primary care pay-for-performance for quality and safety could also be tailored to the specifics of each territory or region.

Indicators, Public Information, and Accessibility Are Key Processes for Guaranteeing Medical Quality in the Territories

Over the last decade, France has made tremendous progress with informing the public about "voice and choice" issues in the French public healthcare system—giving service users greater choice in the services they use, and more say about the way that services are provided. Public process and outcome indicators, as well as accreditation results, are posted on the national website *Scope Santé* (Health Scope) (https://www.scopesante.fr/#/). It is even possible for citizens to benchmark medical settings on the website.

Although this information is useful, it does not cover everything. More is expected, and more is in development, especially in terms of patient-reported outcome measures (PROMs) and patient-reported experience measures

(PREMs). See, for example, the result of the PREMs-e-SATIS indicator on http://www.atih.sante.fr/qualite-des-soins/e-satis.

The *Scope Santé* website also lacks a global-local focus. Regional citizens need to know how to physically access the various medical professionals and sectors: where, when, and how to make an appointment in their local area; how to coordinate appointments and exams; how to benefit from social care in complementary care; and so forth. This level of information depends more on developing a local integrated strategy for each territory, incorporating cooperative web tools, and coordinating media and local government campaigns. This element of the reform is still in its infancy.

An Attitudinal Transformation

If the coming demographic shift is to be supported by an effective global-local system—a flexible regionalized system tailored to individual needs—an attitudinal transformation regarding the very purpose of healthcare is required. The current focus on healing and restoration should be replaced by the longer-term aim of supporting well-being in the face of chronic disease and increasing frailty. Helping citizens stay in their own homes as long as possible should also be a primary objective. Success will require a transformation of decades-long thinking about healthcare.

Conclusion: A Long Way to Go

To paraphrase President Macron, healthcare in France is *en march* (Casassus, 2017). The goal is to deliver the best and safest service to the local population by 2030. The paradigm shift calls for a revolution of mind—thinking locally rather than nationally, thinking adaptation and personalization rather than common ideal, thinking of the entire patient journey rather than isolated actions, and last but not least, envisioning healthy aging rather than healing.

21

Germany

Health Services Research and Future Planning in Pediatric Care

Wolfgang Hoffmann, Angelika Beyer, Holger Pfaff, and Neeltje van den Berg

CONTENTS

German Data

- Population: 82,667,685
- GDP per capita, PPP: $48,729.6
- Life expectancy at birth (both sexes): 81.0 years
- Expenditure on health as proportion of GDP: 11.3%
- Estimated inequity, Gini coefficient: 30.1%

Source: All data are from the World Health Organization and World Bank. Latest available data used as at October 2017.
GDP = Gross Domestic Product
PPP = Purchasing Power Parity

The Pediatric Care System in Germany

The German outpatient healthcare system is planned around the ratio of physicians to residents. The ratios vary over medical specialties and the demographics of the planning region. This planning system has two objectives: to provide all necessary outpatient healthcare near patients' homes and to ensure the economic base of the medical practices (Kassenärztliche Bundesvereinigung, 2013).

In the German healthcare system, outpatient and inpatient sectors are kept strictly separate, as are the tasks and competencies of the different healthcare specialties. This division means that separate planning systems are required for different sectors and professions, a complicating factor in the organization of healthcare in regions where there are gaps in the geographical distribution of healthcare providers. Hospitals in regions with few outpatient pediatric practices, for instance, may find it difficult to receive authorization to deliver outpatient care.

Pediatricians in outpatient practices have a special position in the health-care system. While their main clinical focus involves the provision of primary and preventative care for children (Fegeler et al., 2014), pediatricians are also involved in multiprofessional networks, such as child protection, child development, and early education, although better integration is needed (Siebolds et al., 2016; Thyen, 2010). Pediatricians are treated as a medical specialty in the healthcare planning system, however, and the planning regions for pediatricians are larger than the planning regions for general practitioners (GPs). In rural regions without additional supply from nearby cities, one pediatric practice may be responsible for around 4000 children (<18 years), while a nearby GP may be responsible for only 1700 persons, regardless of the degree of rurality or urbanity of the planning region.

The current healthcare planning system was introduced in 1992 to control the number of physicians. Since that time, it has been developed to allow more flexibility regarding the size of the planning regions and the integration of demographic parameters. At present, a reconsideration of a number of socioeconomic and infrastructural particularities is underway (Kassenärztliche Bundesvereinigung, 2013).

Challenges in Pediatric Healthcare Delivery in Rural Regions

Sustaining economically efficient, close-to-home pediatric healthcare is increasingly difficult in rural and remote parts of Germany. In many rural regions in Germany, the number of children and their proportion of the total population has decreased over the last 25 years. For instance, only 16% of the population of the rural federal state of Mecklenburg–Western Pomerania (the state with the lowest population density) is aged under 20 years, significantly lower than the German average of 18% (Statistisches Bundesamt, 2017). Such demographic variation means that in some rural regions in Mecklenburg–Western Pomerania, pediatric outpatient practices can no longer operate in an economically efficient way.

Pediatric hospital care faces similar problems: small pediatric units in rural regions with a low population density may not be economically efficient but are nevertheless necessary. It is also increasingly a struggle to recruit enough young physicians and nurses to fill staff vacancies in both inpatient and outpatient pediatric care in rural and remote regions.

Children living in close proximity to pediatric healthcare providers report more visits than children living farther away (van den Berg et al., 2017).

In order to develop a sustainable model of pediatric care in rural areas of Germany, a more flexible approach is required.

Innovative Healthcare Concepts in Pediatric Care

To ensure high-quality medical care in rural regions, innovative demographic-specific, patient-oriented care models should be developed, implemented, and evaluated. Possible elements of innovative care concepts include

- Defined indication-specific transsectoral treatment pathways
- Cooperation and compensation models within and between different medical specialties and sectors
- Delegation and substitution models involving nurses and other medical professions, based on new qualifications
- Various types of telemedical functions, for example, teleconsultation or telemonitoring

These elements need to be implemented in defined regions or patient groups, either separately or combined in multidimensional solutions.

The Mecklenburg–Western Pomerania Innovative Care Projects

A series of research projects have recently been run in Mecklenburg–Western Pomerania to examine the possibilities and limitations of innovative healthcare concepts in pediatric care. The focus of these projects includes

- Integration of hospitals in acute outpatient pediatric care
- Efficacy of telemedical triage in those hospitals without a pediatric unit
- Division of tasks and cooperation between GPs, pediatricians, and other specialties
- Delegation of medical tasks to non-physician professions (nurses and physician assistants)

Integration of Hospitals in Acute Outpatient Pediatric Care

Researchers from the University Medicine Greifswald recently examined data from a small hospital in the village of Anklam, in the disadvantaged rural region of Western Pomerania, to test the hypothesis that hospitals already participate significantly in the delivery of acute outpatient pediatric care. Data from the pediatric emergency unit of the hospital from 2007 to

2010 were analyzed. Over this time, the number of ambulant visits in the emergency department continuously increased from about 1350 in 2007 to 1600 in 2010, while at the same time, the number of inpatient cases decreased. The distribution of the visits showed that the highest numbers of visits occurred on weekends, between 8:00 and 12:00 a.m. Furthermore, a significantly higher number of visits occurred on days with reduced consultation hours in the local pediatric and general medicine practices, especially on Wednesdays (most practices in Germany close on Wednesday afternoons) and, to a lesser extent, on Fridays. In the observation period (2007–2010), the majority of outpatient consultations (75–78%) were conducted outside the office hours of pediatric and general practices.

A second analysis concerned the appropriateness of pediatric emergency unit visits. All unscheduled visits at the emergency unit (outpatient visits and visits with a subsequent hospital treatment) were analyzed with respect to their urgency as based on diagnosis. According to the evaluation by the pediatricians involved, the hospital admission of 85% of the patients with an inpatient treatment subsequent to the visit of the emergency unit was appropriate. However, in about 50% of these cases no immediate special ist intervention was necessary; rather, a scheduled admission on the basis of a referral from an outpatient physician would have been sufficient. With respect to ambulant visits to the pediatric emergency unit, while a visit to a pediatrician was justified in about 75% of the cases, in 80–85% of these cases a visit to the pediatric emergency unit was not necessary (van den Berg et al., 2014).

The Efficacy of Telemedical Triage

Not all hospitals in rural regions run a specialized pediatric department. To compensate for the absence of pediatric specialists, these hospitals can be supported by hospitals with a pediatric department using telemedical functionalities.

The possibilities and limitations of a telemedical pediatric triage are being examined in an ongoing study (Warren et al., 2008). Telemedical triage is performed via a videoconferencing system equipped with a remote-controlled camera with a high-level optical zoom function operated by a telemedicine physician in a larger hospital with a pediatric department. The triage is conducted first by a physician on site in the small hospital and then by the telemedicine physician. Both physicians suggest an urgency level on the basis of the standardized triage; these are then compared in a non-inferior analysis.

The main research question of this project is whether the assessments of both physicians concur. Possible barriers in the utilization of the video-conferencing system, as well as technical and organizational problems, are assessed after each use.

Division of Tasks and Cooperation between GPs, Pediatricians, and other Specialties

In a study funded by the umbrella organization of German scientific pediatric societies and pediatric professional associations (DAKJ), possibilities of cooperation between pediatricians and other physicians in outpatient care were assessed. Cooperation is necessary to support pediatric medical care near the patients' residences. A standardized questionnaire was answered by 30 DAKJ experts. The results showed the experts' preference for a reduction of sectoral limits, which means that hospitals with pediatric departments or wards should also treat ambulatory patients. In other contexts, outpatient pediatric practices cooperate with inpatient pediatric departments or wards to ensure an around-the-clock supply of pediatric care. However, in rural regions with insufficient medical care, other solutions can be developed, depending on regional resources. In Western Pomerania, for example, inpatient pediatric rehabilitation facilities have been integrated with the supply of acute pediatric care.

Surveyed experts also supported intensified cooperation between outpatient pediatricians and GPs. GPs are more widely dispersed, even in rural regions. Cooperation concepts may vary, but they all need to be based on defined agreements with respect to tasks, patient groups, and financing. For instance,

- The distribution of tasks must be based on the severity of the pediatric cases.
- The distribution of tasks must be based on the urgency of medical treatment.
- All treatment responsibilities should be clearly delineated (e.g., for chronic patients, a specialist pediatrician provides the initial diagnostic process and treatment, while a GP provides therapy and symptom monitoring).

Logistically, the pediatrician can offer regular practice hours on site in one or more GP practices. A video connection between the main practice of the pediatrician and the GP practice can support consultations during the remaining time. One prerequisite of all these regional solutions is the implementation of joint patient records and standardized joint documentation (van den Berg et al., 2016).

Delegation of Medical Tasks to Non-Physician Professionals

The delegation of general medical tasks to non-physician professionals can now be reimbursed and has become an integrated part of GP outpatient care in rural regions (Dini et al., 2012; van den Berg et al., 2009, 2010).

In a project funded by the Federal Ministry of Health of Mecklenburg–Western Pomerania, options to implement delegation in outpatient pediatric care were assessed on the basis of surveys among different healthcare professionals (inpatient and outpatient pediatricians, nurses, and physician assistants) and the children's parents.

The results of the surveys, which involved 206 healthcare professionals and 411 parents, show that there is a high degree of acceptance of delegation among both healthcare providers and parents, particularly when it comes to preventative counseling and health promotion, the care of chronically ill children, and support of the transition process between pediatric and adult medicine.

Although tasks in pediatric practices can be delegated to nurses and physician assistants, appropriate additional training is crucial. A targeted curriculum should be developed, and the necessary qualifications should be built as far as possible on already existing training modules.

Delegation with respect to pediatric care in rural regions will have a significant impact outside pediatric practices. In cases where no pediatrician is present and high levels of expertise are required, for instance, in a pediatric branch practice or in GP practices, delegated tasks should be conducted by qualified nurses (van den Berg et al., 2017).

Perspectives

Innovations in the organization and delivery of healthcare are often received skeptically by stakeholders who have diverse and sometimes competing interests (e.g., ambulant vs. hospital care, specialists vs. GPs, and physicians vs. nurses). However, the younger generation of physicians is considerably less competitive and more interested in integration and cooperation in healthcare planning and delivery. Changing work–life balance priorities and increasing digital competence offer new opportunities for collaborative models of pediatric care delivery.

According to population forecasts and predicted demographic changes in rural regions, the challenges to the pediatric care system will increase over the coming decade. Future planning should include innovative options developed in a variety of individual model projects. Improved concepts should be re-evaluated in larger-scale implementations in all pediatric settings. Successful approaches should be implemented into comprehensive regional healthcare planning.

22

Greenland

Everyday Life with Chronic Illness: Developing a Democratic and Culture-Sensitive Healthcare Practice

Tine Aagaard and Lise Hounsgaard

CONTENTS

Greenland Data

- Population: 56,186
- GDP per capita, PPP: $37,900
- Life expectancy at birth (both sexes): 72.4 years
- Expenditure on health as proportion of GDP: 8.8%
- Estimated inequity, Gini coefficient: Unavailable

Source: All data are from the World Health Organization and World Bank. Latest available data
used as at October 2017.
GDP = Gross Domestic Product
PPP = Purchasing Power Parity

Background

The healthcare system in Greenland faces many challenges (Deloitte, 2010),
for many reasons. Greenland was a Danish colony from 1721 until 1953,
when it was incorporated into the Danish state as a county on par with the
other counties. A healthcare system was developed, based on the model of
the Danish system, with some adjustments for Greenlandic conditions. For
example, Greenlandic midwives educated by Danish medical doctors represented the healthcare system in towns and settlements because of the lack of
medical doctors. This system contributed to, among other things, the control
of a severe tuberculosis epidemic. In 1978, Greenland achieved home rule,
and in 2009 self-rule.* Jurisdiction over public healthcare was taken over
from Denmark in 1992.

Throughout the period since colonization, Danish medical doctors
have conducted health research among the Greenlandic population

---

* Greenland is a welfare society that builds on the so-called "Scandinavian model" with some
modifications (Skatte-og Velfærdskommissionen, 2010). This means that the state secures
social rights for all citizens, not only the poorest. The system was established to realize the
common good by the redistribution of economic wealth through taxes. Healthcare is not
defined as a private, personal problem, but as a common good for the benefit of both the
individual and the community, and as such, health services are free for all citizens.

(Bjerregaard, 2005). Since 1993, health status has been surveyed and monitored regularly by epidemiologists to inform the development of health policies (Dahl-Petersen et al., 2014). In addition to this, there is current biomedical research into diseases prevalent within the Greenlandic population (Homøe et al., 2009). Greenland's health services have also received scholarly attention (Pedersen, 2012). A great deal of the research has been conducted by Danish institutions. In 2008, an independent institution for health research in Greenland, Greenland Centre for Health Research, was established under the auspices of the University of Greenland, Ilisimatusarfik. Since 2009, a humanistic health research environment has been developed at the Institute for Nursing and Health Science, Ilisimatusarfik (Hounsgaard et al., 2013; Aagaard, 2015; Olesen, 2016).

Current discussions among healthcare researchers and professionals indicate a degree of frustration over the fact that epidemiological and biomedical knowledge about disease and health appears to have little impact on the health of the population. For example, the incidence of lifestyle diseases, such as diabetes and lifestyle-influenced cancer, is growing (Dahl-Petersen et al., 2014), and as in Western countries, more and more Greenlanders are living with chronic disease.

The healthcare budget in Greenland is tight, and this is made explicit in the nation's healthcare strategies. There is a growing focus on low-tech interventions, like prevention, health promotion, and lately, rehabilitation, to prevent people from having to access expensive hospital treatment and care. Reducing hospital care is of particular significance in Greenland, where a population of 56,000 is spread over an area of 2.2 km², and the options for transportation—by ship, helicopter, or airplane—are all extremely expensive. Telemedicine is being established to underpin local interventions, but the technology is still being developed and professionals must be trained to use it. Because the Greenlandic healthcare system lacks sufficiently educated personnel, many health professionals come from Denmark and other countries, and the turnover is consequently quite high.

Chronic Illness in Everyday Life

The growing number of people who live with chronic illness calls for a new professional focus, not only on health, but also on well-being. Dealing with the psychosocial problems that chronic disease brings about in everyday life is a huge challenge. Empirically, ignoring psychosocial consequences of disease often causes new health-related problems that demand further interventions from the healthcare system. The issue is actually well documented, with a considerable body of research in Denmark and other countries regarding the experiences of both those living with chronic illness and

their families (Hounsgaard et al., 2011). However, according to a recent qualitative study, "Everyday Life with Illness—Patients' Cultural Perspectives
on Healthcare Practice in Greenland" (Aagaard, 2015, 2017), there are few
formal strategies or practical procedures that professionals can use to ascer-
tain patients' health-related worries in their everyday lives. The interviewed
patients and relatives in the study praised the healthcare system for its excellent treatment of physical disease, but all noted the lack of follow-up.

A conclusion of the study is that the existing disease-oriented healthcare
system is not geared to manage the psychosocial problems of chronically ill
patients or citizens. Because academic and scientific understandings about
disease and treatment are given precedence over patients' perspectives and
knowledge about their own lives, professionals often have no insight into the
circumstances under which patients have to conduct their lives when suffering from a chronic illness. In addition, professionals often embrace prevailing concepts of disease and health, not realizing that patients and citizens
may have differing, culturally formed views of what constitutes a good and
meaningful life. There is an urgent need, therefore, to develop means and
procedures to incorporate patients' knowledge and perspectives into professional practice. Combining this patient-oriented information with professional academic and experiential clinical knowledge will create professional
interventions that are efficient, relevant, and meaningful for patients. It is
crucial that professionals develop an awareness of their own values, and the
ways in which these are formed by their own particular way of life (e.g., as
educated professionals and operating within the disease-oriented healthcare
system), and that they never work from the assumption that their values are
in any way superior to those of their patients (Aagaard, 2015, 2017).

Home Rehabilitation

In response to the findings of the study, research into home care
(Kommuneqarfik Sermersooq, 2016) has recently begun in the town of
Nuuk,* the capital of Greenland. The research project will evaluate the pro-
cess of establishing a municipality home care project, Training Before Care.†
In line with the strategies for low-tech initiatives for prevention, rehabilita-
tion, and health promotion, "home rehabilitation" is now regarded as a way
to avoid expensive treatment in the secondary healthcare sector. Training
Before Care aims to strengthen the physical functioning of elderly and

---

* The project is expected to cover the municipality of Sermersooq, which includes three towns
  and a number of settlements. Initially, however, the project is concentrated in the town of
  Nuuk.
† The working title of the research project is "Cross-Professional Cooperation on Rehabilitation
  in Everyday Life: A Process Evaluation of Training before Care."

chronically ill people in order to help them have "an active and meaningful life" (Kommuneqarfik Sermersooq, 2016: 9). As in healthcare strategies, while the citizens' everyday quality of life is given as an objective of the project, the actual purpose can be seen as an expression of the values of policy-makers: for instance, the inclusion of positive health markers, such as an "active life," may conveniently reflect the values of the system rather than those of the patients themselves. Therefore, there is no guarantee that the aim of the project is in accordance with the needs and wishes of the citizens.

Municipal home care in Nuuk provides physical care and practical assistance by care assistants (skilled and unskilled), nurses, occupational therapists, and physiotherapists.* The teams work 7 days a week, and night care is to be introduced, so that the system will be operational 24/7. The home care system also provides physical aids, activity centers, and seniors' clubs. Thus, there is a relatively well-developed structure for home care in Nuuk. A constraint is that the majority of personnel are unskilled and staff turnover is relatively high.

Developing a Framework for Professional Practice Reflection

The aim of the research project is to develop a value-guided and practical framework for care professionals' practice reflection as an integrated, cross-professional, and current activity in home care. This includes the development of tools for practice reflection that build on principles and values relating to the involvement of citizens in their rehabilitation. Here, involvement means interplay between professional knowledge and the citizens' knowledge and perspectives. The project is designed as practice research and builds on a concept of knowledge as distributed among the participants: citizens, healthcare personnel, leaders, administrators, and researchers. Each participant in a practice will have different perspectives on and knowledge about the practice according to his or her location, position, relation to other participants, and so on (Dreier, 2008). Therefore, all perspectives must be involved in developing the practice in question. In the same way that citizens must be involved in professional practice as subjects of their own lives, professionals must be involved in developing their professional practice as subjects of that practice.

The illumination of different perspectives—in research and in professional practice—gives a better picture of the social conditions under which those with chronic illnesses have to make their everyday lives function. This

---

* In the following, we call employees in home care both "professionals" and "healthcare personnel." Many employees do not have formal professional qualifications, but their work involves some kind of professional skill.

is a robust foundation from which professionals can critically discuss means and ends in home care practice in order to create a good practice. Reflections on values always involve reflections on social justice. That is, the reflection on means and ends in a local practice involves more general questions about societal ends (Schwandt, 2005). In other words, practice reflection aims at changing practice for the sake of the common interest (compare the purpose of the welfare state). Therefore, professionals' practice reflections must involve the citizens' social and societal conditions for handling their lives. For instance, many people do want to live an active life but are hindered by their conditions. This must be understood by professionals who want to support citizens; otherwise, they risk deeming them non-compliant or non-cooperative, rather than providing them with appropriate support.

Likewise, researchers' reflections on professional practice in home care must include the circumstances under which the professionals act in order to understand the subjective reasons for their actions. Otherwise, researchers risk interpreting the professionals' actions as immoral. People who work in the "relational professions" are mostly engaged in efforts to improve the well-being of their fellow human beings, but many conditions can weaken this engagement—for example, the prevailing concept of an active life as referring solely to physical activity. These conditions must be examined in cooperation with healthcare personnel. Understanding what the conditions mean in terms of professionals' actions is a prerequisite for changing professional practice in a way that is in the common interest.

Future Perspectives

The Training Before Care project is still in the early stages, so no concrete tools can be presented here; however, it is obvious that many structures in municipal home care in Nuuk are already in place. These include

- Daily meetings within the different personnel groups in which care work is discussed and organized

- Regular cross-professional meetings about citizen-related issues and cooperation to solve them

- Managerial encouragement and practical possibilities for supplementary training and/or education

- Possibilities for organizing seminars and project days on topical subjects

These structures provide an initial framework for professionals to develop home care and rehabilitation in cooperation with citizens. Cross-professional practice reflection following the principles described above will engender new potentials for action. The framework can be adjusted and new tools to incorporate citizens' perspectives into professional practice can be developed in a dialectic interplay between citizens and professionals, and between theory and practice.

Conclusion

In welfare states based on the Scandinavian model, welfare professional contributions are meant to support citizen participation in social living. Frequently, however, the support given reflects the sociocultural norms and ways of life of policy-makers, experts, and professionals. With the growing need for everyday life support for people with chronic illness, there is also a growing need for reciprocity and cooperation between citizens and health-care personnel. This chapter has provided some common principles and practical means for involving the citizens in professional practice, in preparation for creating a democratic and culturally sensitive healthcare practice.

23

Italy

The Introduction of New Medical Devices in an Era of Economic Constraints

Americo Cicchetti, Valentina Iacopino, Silvia Coretti, and Marcella Marletta

CONTENTS

Italian Data

- Population: 60,600,590
- GDP per capita, PPP: $38,160.7
- Life expectancy at birth (both sexes): 82.7 years
- Expenditure on health as proportion of GDP: 9.3%
- Estimated inequity, Gini coefficient: 35.2%

Source: All data are from the World Health Organization and World Bank. Latest available data used as at October 2017.
GDP = Gross Domestic Product
PPP = Purchasing Power Parity

Background

The role of medical devices (MDs) in the healthcare sector is essential to both the enhancement of quality in the provision of healthcare services and the monitoring and treatment of diseases. Given their significance, the implications of costs and benefits associated with the adoption of MDs and their sustainability are of paramount importance. In order to successfully promote innovation while containing costs, multidimensional approaches, such as the health technology assessment (HTA), have been widely used in order to elicit the value of MDs (Jonsson and Banta, 1999).

In Europe, the role of cooperation in HTA has been recognized as having a positive effect on healthcare systems. Expanding such cooperation into technical and scientific realms would involve HTA bodies as well as those involved in the innovation process. To this end, the European Directive for cross-border healthcare (Directive 2011/24/UE) made the establishment of a collaborative HTA network a priority, and proposed a shared governance model based on the European Network for Health Technology Assessment (EUNetHTA) experience.

Since 2000, due to the absence of any national framework, Italian National Health Service (I-NHS) hospitals have been setting up individual hospital-based HTA units to inform managerial and clinical decision-making regarding health technologies.

In 2007, a number of Italian regions developed their own HTA programs based on different approaches and methodologies. In 2009, the National Agency for Regional Healthcare Services (Agenas) created a national HTA network (RIHTA) in an effort to coordinate these units. Thus far, only 12 regions have joined this network (Cicchetti et al., 2016).

Since 2015, some normative law reforms have been issued, all aimed at recognizing the role of HTA in orienting decisions on the uptake of health technologies at a national level, and thus promoting stronger coordination among regions. In order to ensure uniform and equitable access to healthcare services across the nation, a national steering committee, the Cabina di Regia per l'HTA, was established in 2015–2016 to coordinate HTA activities. A technical body directly reporting to the Minister of Health, and charged with setting the priorities of the national HTA plan, the Cabina di Regia per l'HTA is made up of

1. The general director of the Medical Devices and Pharmaceutical Care Department of the Ministry of Health (MoH), who covers the role of president
2. Two representatives of the MoH
3. Representatives identified by regions
4. A representative identified by Agenas
5. A representative identified by the Italian Medicine Agency (AIFA)

Cooperation between the committee and the stakeholders is ensured by the presence of one technical secretary and one organizational secretary.

Success Story

The objective of the national HTA program is to ensure that all HTA activities are successfully coordinated and to support and promote HTA best practices across the country. The program aims to make recommendations about the efficacy of MDs at a national level, which will then be used to inform and accelerate the adoption and procurement of valuable MDs at a local level on the basis of decisions made nationally by the Cabina di Regia per l'HTA. The main activities of the steering committee will result in the following:

- A list of technologies signaled by the different stakeholders to be included in the assessment process
- A list of selected prioritized technologies eligible for assessment
- HTA reports
- Recommendations for the appropriate use within the Italian-National Health Service (I-NHS) (appraisal)
- Policy documents for the coordination of HTA national program activities

This will ensure a stricter regulation when it comes to the introduction and adoption of MDs. Currently, European Conformity (CE) marking is assumed to be a necessary and sufficient condition of market readiness, while clinical indication and the impact on health and clinical practice are not taken into account in the devices' procurement procedures. The national HTA program is expected to guarantee access to the latest technologies in clinical practice, while observing the system's financial constraints. In order to achieve this goal, HTA can exert a supporting and technical role in the purchase process. The Cabina di Regia per l'HTA will also provide regional and hospital procurement departments with informative HTA to consider when making purchasing decisions.

Moreover, the establishment of the national HTA program will reduce variability in technology adoption and prices across healthcare organizations within the I-NHS, ensuring that only the best products are bought and keeping costs to a minimum.

Impact

Proposals for the assessment of innovative technologies within the national HTA program can be advanced by different stakeholders, such as the MoH and related organizations, the regions, healthcare organizations and professionals, scientific societies, industry, and various patients' and citizens' associations. As proponents, they may recommend the assessment of technologies at any stage. Proposals are made through a notification system set up by Agenas and must include all the documents needed for subsequent prioritization processes. For each assessment proposal, recent technical reports and ongoing projects carried out by international agencies (such as the EUnetHTA POP Database) must be verified. Details are confirmed via a checklist (Table 23.1).

The application of this checklist allows Agenas to prepare

- A list of technologies to be prioritized by the Cabina di Regia per l'HTA
- A list of excluded technologies and reasons for exclusion
- A list of available technical reports

The following criteria are considered during the prioritization process:

- The potential impact of the technology on the specific clinical pathway to which it is addressed
- The ethical and social implications of the technology, with a particular focus on quality of life and sustainability of care
- The potential organizational impact of the technology (including the risk that distribution and use of the technology will not be equitable)

TABLE 23.1

Checklist for the Assessment Proposal

Checklist	Description	Exclusion Criteria	Inclusion Criteria
Comprehensiveness of information	The proposal reports comprehensive and relevant information in all the form domains	Lack of information in one or more domains in the form	Presence of clear or relevant information in every form domain
Availability of HTA report published in the last 3 years and ongoing assessment projects in the European Network	Presence of recent HTA reports or ongoing assessment projects	Availability of recent HTA reports	Absence of HTA report or presence of partial assessment or ongoing assessment projects

Source: Ministry of Health [Italy], National HTA programme for medical devices. Strategic document. Internal document. Rome: Ministry of Health, 2016.

- The potential economic and financial impact of the technology
- The technical relevance of the technology in the clinical pathway
- The uncertainty about the effectiveness of the technology
- The epidemiological relevance of the referred conditions (frequency and severity)

The Cabina di Regia per l'HTA will be responsible for assigning the assessment to the different collaborating centers according to the development of the program. Agenas is recognized as the operational arm of the Cabina di Regia per l'HTA, being responsible for the coordination of assessment activities. The specific assessment responsibilities of each of the collaborating centers will depend on the characteristics of the technology to be assessed, the procedures to be followed, timing and expected remuneration, and the expected assessment process. The steering committee will be responsible for the final decision, which will be made according to preliminary recommendations, that is, use, no use, experimental use, and conditional use (Ministry of Health [Italy], 2016).

The implementation of activities scheduled in the new HTA national program is more likely to occur if the involvement of healthcare organizations and stakeholders can be guaranteed. In order to facilitate systematic, timely, and responsible involvement, the HTA body has set up three working groups committed to appraisal, methodological guidance, education and communication, and monitoring. In order to ensure long-term sustainability of these processes, capacity-building activities will be established (Ministry of Health [Italy], 2016).

Implementation and Transferability

The activity of the three working groups began in April 2017. The national appraisal network group will participate in and contribute to the establishment of national and regional guidelines for the appropriate use of MDs. The working team will not be fixed; instead, it will be defined on a case-by-case basis subject to the specific technological area of interest. Members will be chosen by the Cabina di Regia per l'HTA, and stakeholder representatives will be involved in the process.

The methods, education, and communication working group will be responsible for producing drafts and proposals on relevant issues, such as the use of assessments produced by the EUnetHTA network, the recognition of technical methods for HTA, and the identification of cost-saving technologies. An educational program involving members of the working group will attempt to align knowledge and experience. The dissemination of evaluation results will help to promote their use at both the regional and the local level to inform decisions regarding device adoption, introduction, and discontinuation. Finally, the monitoring group will monitor the whole process and provide feedback.

The HTA program is expected to promote greater awareness of HTA methodologies and benefits, and to encourage uniform decisions on the uptake of new MDs. Furthermore, it is expected that the ready availability of such authoritative evaluations will also encourage the use of HTA reports to inform the decision-making process in those jurisdictions where HTA is not used.

Prospects

While successful pilot trials of MDs are not required, the systematic collection of effectiveness data could further improve the decision-making process. To this end, registries may represent a strategic tool that could be used to establish the longevity of the technology, to gather comparative effectiveness data, and to fully understand their ideal organizational settings and the clinical procedures they are used for. Moreover, registries should collect data to compare the quality of MDs and to identify any possible complications of long-term use. Such evidence should be integrated with disease-specific national registries. Registries should be activated at the national level, while pilot experiences could be developed at a regional level. This approach should be applied to all expensive and highly sought-after technologies, or those for which strong evidence on clinical outcomes is still lacking.

Information collected from users could provide evidence-based data on cost-effectiveness for research relevant to the regional or local decision maker. Finally, any funds saved could be reinvested in the system.

Conclusion

The national HTA plan will provide a structured process for MD adoption and procurement throughout Italy. To begin this process, three working groups, representing a wide variety of stakeholders, have been established within the Cabina di Regia per l'HTA. This effort will lead to the coordination and rationalization of research efforts, and the alignment of decisions concerning the uptake and discontinuation of MDs, and should eventually encourage widespread commitment to HTA culture.

24

Malta

The National Cancer Plan: Strengthening the System

Sandra C. Buttigieg, Kenneth Grech, and Natasha Azzopardi-Muscat

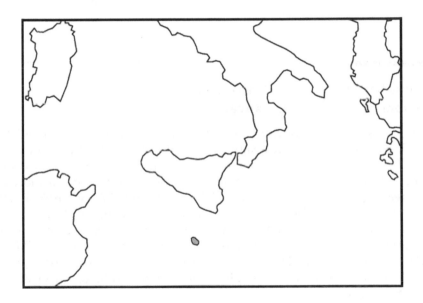

CONTENTS

Maltese Data

- Population: 436,947
- GDP per capita, PPP: $37,899.2
- Life expectancy at birth (both sexes): 81.7 years
- Expenditure on health as proportion of GDP: 9.75%
- Estimated inequity, Gini coefficient: 27.9%

Source: All data are from the World Health Organization and World Bank. Latest available data used as at October 2017.
GDP = Gross Domestic Product
PPP = Purchasing Power Parity

Strengthening Malta's Health System Response to Deal with Cancer

Malta, the smallest European Union (EU) member state, with a population of 425,384, has a universal health system where care is mainly provided freely by the state, alongside a small private sector (National Statistics Office [Malta], 2014). In Malta, cancer mortality as a proportion of all mortality is increasing—rising from 24% in 1997 to 26.6% in 2007 and 26.9% in 2013 (England, 2013). The Maltese cancer strategy, launched in 2011, requires strong public health leadership and transdisciplinary competence and collaboration (Ministry for Health, the Elderly and Community Care [Malta], 2011). By adopting a One Health approach in the prevention of cancers, Malta aims to achieve earlier diagnosis, lower mortality, and better survival rates for treatable cancers.

Background

The ministry has undertaken several strategic health initiatives. The 2011 cancer plan listed specific targets, in an effort to prevent and reduce incidence, prolong survival, provide high-quality and equitable services, and ensure optimal quality of life to cancer patients in Malta (Ministry for Health, the Elderly and Community Care [Malta], 2011). This is in line with the 2014 National Health System Strategy, which aims to respond to demographic changes and epidemiological trends; increase equitable access, availability, and timeliness; improve quality of care; and ensure sustainability. As a result of these initiatives, the Maltese health system has registered progress in its efforts to reduce mortality across all age groups; indeed, the Maltese population currently enjoys the highest healthy life expectancy (estimated to be 71 years at birth) in the EU, with amenable mortality declining sharply (WHO, 2015e). There is low unmet need, and access to health services is generally good (Eurostat, 2016). In 2015, the National Health Systems Performance Assessment (HSPA) found that, compared with other countries in the EU, Malta's health system was relatively responsive, with reasonably good financing, quality, access, and health status (Grech et al., 2015). While three public hospitals were recently granted to a private-sector provider in the form of a 30-year concession, the Maltese government is committed to continuing to provide free high-quality care, including cancer treatment (Ministry for Energy and Health [Malta], 2014).

Following the EU Council recommendation on cancer screening (European Commission, 2003), a national cancer screening program was commissioned in late 2008, and began administering breast cancer screening in October 2009 and colorectal screening in 2012 (Ministry for Health [Malta], 2017). There are also plans to introduce a cervical screening program in due course. The strategic plan included commissioning a 113-bed oncology center, using EU structural funds, equipped with state-of-the art diagnostic and therapeutic equipment. This center received its first patients in December 2015. Capacity was also strengthened in the medical consultant, oncology, and hematology sectors. Health policy development and implementation resides at the national level, while the recent publication of a Patient's Charter seeks to empower patients (Ministry for Energy and Health [Malta], 2016). Waiting times for diagnostic investigations, surgical interventions, and the treatment of cancer patients are improving. Staff levels, however, need to increase, as there is a lack of trained healthcare professionals across disciplines in cancer care, most particularly in the areas of therapeutic radiography and medical physics (OECD, 2013). Cancer patients also require free access to up-to-date oncology formulary. This concerted approach to address cancer from every angle, from preventative measures to diagnosis, from curative through to palliative care, has resulted in improved outcomes as registered in the Eurocare-4 study (Verdecchia et al., 2007).

Other critical success factors in the fight against cancer include

- Malta's accession to the EU
- Improved availability of data relating to cancer mortality
- Benchmarking with EU countries
- Increasingly active non-governmental organizations (NGOs), particularly for breast cancer, through the lobbying capacity of European networks, such as Europa Donna (http://www.europadonnamalta.org.mt).

Success Story

Cancers are the second most common cause of death in Malta (Department of Health Information and Research [Malta], 2017). In 2000, cancer survival in Malta lagged behind developed Organisation for Economic Co-operation and Development (OECD) and EU countries (Azzopardi-Muscat, 1999). For example, the Eurocare-4 study revealed that Malta's 76% 5-year survival of patients diagnosed with breast cancer in 2000–2002 was the seventh lowest out of 31 countries (Stevanovic and Fujisawa, 2011). Five-year cancer survival has since improved, even though Malta's statistics remain below those of other developed OECD and EU countries (Azzopardi-Muscat et al., 2014). In particular, cancer mortality before the age of 75 years has increased from 33.6% in 1997 to 39.5% in 2007.

Malta registers approximately 1900 people annually with newly diagnosed cancers (excluding non-malignant skin cancer), and with an age-standardized incidence rate of 242.9/100,000 per year. The risk of getting cancer before the age of 75 years is 23.7%, while about 800 people (27% of annual mortality) die from cancer each year (CancerIndex, 2012).

The 2011 cancer plan is based on interventions that target the personal motivation of at-risk members of the population. In line with the social cognitive theories of planned behavior and protection motivation (Conner and Norman, 2005), patients' educational level, behavioral control, and subjective norms are considered important predictors in understanding their behavior, likelihood of participating in screening, and motivation to lead a healthy lifestyle. The heightened public knowledge about preventable cancers, often due to the efforts of NGOs, put pressure on policy-makers to adopt a multi-sectorial approach and to emphasize public health actions related to the environment and lifestyle habits in the 2011 plan. These actions include efforts to

- Reduce tobacco consumption and passive exposure to tobacco smoke
- Control alcohol consumption
- Further promote the adoption of healthy eating and maintenance of healthy body weight
- Prevent exposures to carcinogens in the environment and at work (Ministry for Health, the Elderly and Community Care [Malta], 2011: 7)

Indeed, the future of cancer management in Malta will be determined by several factors, namely, the extent to which

- Public health is given due importance by the European Commission
- The EU's cross-border directive (European Union, 2011) provides sharing of resources across member states
- The privatization of three public hospitals impacts equity in healthcare

The global cancer burden doubled between 1970 and 2000, and it is estimated that it will double again between 2000 and 2020, and nearly triple by 2030 (Boyle & Levin, 2008). Figures released by Cancer Research UK (2017) claim that cancer rates will climb nearly six times faster in women than in men over the next 20 years, mainly due to widespread smoking in women and obesity-related cancer types that only affect women—factors that are highly relevant in the Maltese context. The average number of annual smoking-attributable deaths during the period 1999–2013 was 396 in males and 111 in females; 42% of the male and 37% of the female smoking-attributable deaths were due to cancer (Department of Health Information and Research [Malta], 2014). In 2015, Malta registered the highest obesity rate in the EU—reducing obesity is a priority action in the 2017 EU presidency, which is to be hosted by Malta (Cuschieri et al., 2016; Harris, 2016). Such factors will compel policy-makers to continue investing in cancer prevention and treatment.

Impact

Because improvements in healthcare delivery require a myriad of interventions at various levels, they can take time to translate into tangible outcomes. Fortunately, the 2011 cancer strategy was concomitant with a firm political commitment to dedicate financial, human, and capital resources, a factor that has helped ensure that cancer detection and care have become more effective. More work still needs to be done. Indeed, a second cancer plan was launched in October 2017, setting out an integrated and comprehensive plan of action for the next five years (2017–2021) (Ministry of Health, the Elderly and Community Care [Malta], 2017). This plan focuses on developing bespoke cancer pathways, screening for new cancers, introducing personalized medication, and improving postsurvival and palliative care services.

Specialized personnel, who have been trained as part of the first plan, are now expected to train new staff, thus ensuring that numbers of trained staff reach critical mass. Furthermore, adopting integrated One Health approaches creates synergies between preventive, curative, and palliative care services, as well as between hospital and primary care. The two cancer strategies will also serve as models for developing other programs targeting non-communicable diseases.

Prospects

Next steps include the introduction of integrated cancer care pathways and innovative procurement methods for cancer drugs and services through managed-entry agreements. A directorate for Cancer Care Pathways within the Ministry for Health was created to offer support and guidance, and to help increase access to care, care coordination, and personalized care for cancer patients in Malta. The development of innovative treatment modalities, such as the intelligent surgical knife (the iKnife), genetic therapy, immunotherapy, and the molecular targeting of cancer cells through techniques such as DNA cages and mRNA enhanced therapies, are shaping future management directions. However, early detection remains central to the fight against cancer, and necessitates more sensitive screening techniques. DNA sequencing will also become a reality as computer and artificial intelligence algorithms become increasingly possible. None of this is possible, however, without the concerted collaboration of cancer centers, as no one country or center can go it alone. In Europe, specifically, the development of European reference networks is likely to affect the way in which cancer patients will be treated in the not-too-distant future, with cancer pathways possibly involving a cross-border element.

Conclusion

While more work remains to be done, particularly when it comes to cancer prevention, providing access to innovative treatments and medicines, and the development of palliative care services, strategic, system-level investment has begun to yield the desired results in Malta. Remarkable improvements in survival rates have been demonstrated, notably for those with malignant melanoma, breast, testicular, thyroid, and prostate cancers. Breast cancer mortality has fallen significantly and 5-year survival rates have increased to 85.9%, which is higher than the EU average (83.5% in 2008–2013). For other cancers that are less amenable to early diagnosis, improvements are yet to be registered. In conclusion, while Malta, as a small state, is not expected to establish itself as a major center of reference and will continue capitalizing on the benefits of EU membership, it has nonetheless made significant achievements in fighting cancer.

25

The Netherlands

Reform of Long-Term Care

Madelon Kroneman, Cordula Wagner, and Roland Bal

CONTENTS

Dutch Data

- Population: 17,018,408
- GDP per capita, PPP: $50,898.1
- Life expectancy at birth (both sexes): 81.9 years
- Expenditure on health as proportion of GDP: 10.90%
- Estimated inequity, Gini coefficient: 28.0%

Source: All data are from the World Health Organization and World Bank. Latest available data
used as at October 2017.
GDP = Gross Domestic Product
PPP = Purchasing Power Parity

Background

Dutch healthcare is divided into two systems: curative care (including hospital, primary, and mental care) and long-term care. Since 2006, curative care has been financed through income-related premiums as well as community-rated premiums that citizens pay to preferred health insurers for a defined and compulsory universal benefits package. Citizens are listed with a general practitioner (GP) of their choice. GPs have a gatekeeping function: for access to medical specialist care, a referral is needed. The system is also based on managed competition in three markets, with healthcare providers competing for contracts with private insurers—and for patients. This competition has led to greater transparency and improved quality.

Care trends are currently undergoing a shift from hospital to primary care, and as a result, GPs are increasingly responsible for coordinating care for the chronically ill. In addition, small community teams of nurses and social workers support frail elderly to continue to stay at home.

Until 2015, long-term care was centrally organized and financed mainly via payroll contributions under the *Algemene Wet Bijzondere Ziektekosten* (Exceptional Medical Expenses Act). This act, which was passed in 1967, was originally intended to protect citizens from catastrophic healthcare expenditure—as might be experienced due to long-term admission into a nursing home, for instance. Over the years, the act expanded to cover all kinds of long-term care, including residential care, home care, and care for patients with psychiatric disorders and with learning and sensory conditions, resulting in complex and costly regulations. Eligibility criteria for residential care became very broad, and large numbers of patients were moved into institutional care unnecessarily; many of these patients, given adequate support, should have been able to stay in their own homes (Kroneman et al., 2016). In 2014, almost one in 20 residents in the Netherlands were recipients of some form of long-term care services (Van Ginneken and Kroneman, 2015). In the light of increasing healthcare expenditure and the aging of the population, policy-makers feared that long-term care would become unsustainable. The economic crisis of recent years made it clear that further growth in expenditure must be curbed (Batenburg et al., 2015).

Reform of Long-Term Care

In 2015, the *Algemene Wet Bijzondere Ziektekosten* was replaced by the Long-Term Care Act, and the Health Insurance Act, Social Insurance Act, and Youth Act were reformed. These reforms aimed to keep people living in their own homes as long as possible, and to encourage the utilization of social network support whenever possible. To ensure that patients can access care in close proximity to home, and that individualized treatment and effective coordination can be provided, social support, home help, and youth care (mental healthcare for children and parenting support) are now the responsibility of local municipalities. Home nursing care, however, which is regarded as curative (primary) care, remains a national-level responsibility (Kroneman and Maarse, 2015).

Access to residential care in residential homes and nursing homes is now limited to those who are in need of supervision 24 hours per day (mentally, physically, or medically). Again, this type of care is still organized at the national level, financed through health insurers. An overview of the changes in long-term care legislation is summarized in Figure 25.1.

While the 2006 reforms included transitional measures, the 2015 reforms were implemented almost overnight, creating significant financial and organizational difficulties (Kroneman et al., 2016; Van Ginneken and

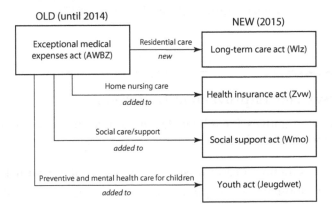

FIGURE 25.1
Changes in long-term care in the Netherlands. (From Van Ginneken, E., and Kroneman, M., *Eurohealth*, 21(3), 47–50, 2015.)

Kroneman, 2015). Municipalities, in particular, were unprepared for their increased care responsibilities. The central government decreased funding for non-residential care because it was assumed that municipalities could organize care more efficiently. Clients, at the receiving end of the abrupt changes, were understandably anxious about receiving less care (Verver et al., 2017).

The reforms created a number of other challenges:

- Although there has always been a large informal care sector, Dutch citizens were nevertheless accustomed to being able to turn to professional care rather quickly. Municipalities now have the responsibility to promote self-reliance and informal care, but do not have legal options to reinforce this.

- Since decentralization, many municipalities (rather than a few central insurers) require contracts and remuneration to provide services. Care providers complain that as a result, the administrative burden has increased.

- Recipients complain that the waiting periods for assessments and care decisions are too long. The assessment procedures are unclear, and some clients feel that the municipality uses standard care options instead of tailor-made solutions.

- GPs need extra time to attend to the care and management of elderly who remain living at home; in addition, there is a growing demand for emergency care, leading to capacity problems.

- Nursing homes must deal with increased care complexity because only those elderly who have high care needs are now admitted to nursing homes.

- As a result of decentralization, variations in both care and costs have developed between municipalities (Ieder[in], 2017).

Current Systems Improvement Initiatives

A number of initiatives have been taken as a response to the challenges described above. These include

- The strengthening of administrative capability at the municipal level
- The development of a support structure for the assessment of care needs
- The improvement of quality and efficiency in nursing homes
- The development of elderly-friendly hospital environments

Efforts have also been made to smooth transitions between care providers to make it possible for the elderly to remain at home longer. These last initiatives have focused on the development of eHealth "smart home" applications, and have worked to strengthen social support networks and develop age-friendly neighborhoods. The *Even Buurten* (Come by for a While) initiative, for example, develops neighborhood networks to care for frail elderly (van Dijk et al., 2013).

The National Program for Elderly Care, an extensive program to improve long-term (social) care, was launched in 2008. The program, which ran for 8 years, focused on developing better integrated care for the elderly, as well as improving elderly participation in care. The regionally based program used university hospitals as the main drivers initially, and then built up a regional network of health and social care providers, insurers, municipalities, and seniors' organizations (Wehrens et al., 2017).

Other programs focused on developing tools for professionals and carers in home care settings, as well as more institutionalized long-term care. New types of organizing and financing care, and new governance arrangements, were developed. Underlying values were put under the spotlight, and a broader and more holistic definition of health, with an emphasis on self-management and self-reliance, was developed (Huber et al., 2011). This shift aligns with the current political focus on creating a "participation society," in which citizens are called on to become more self-reliant, and societal resilience is encouraged. In line with this, there have been a number of regional initiatives to change financial and organizational structures, aiming to develop population health management practices, rather than relying on traditional fee-for-service provisions.

At the same time, however, the public has become concerned about levels of provision, quality of care, and equal access. Populist discussions from both the left and right sides of the political spectrum have increased. Given the pressure on the system to provide long-term care and the many uncertainties surrounding the best way forward, further experimentation, innovation, and evaluation are needed.

Toward Population Health Management

Community care teams and networked forms of healthcare and social care that use innovative financial and organizational arrangements are playing an increasingly significant role when it comes to the provision of long-term care in the Netherlands. While networking is not a completely new phenomenon, the current scale of network formation, as well as its reach—spanning social and healthcare, education, and housing—is unprecedented. At the same time, there are still many uncertainties, and there has been much experimentation in different regions of the Netherlands.

Many of these new arrangements can be seen as forms of population health management—the development and implementation of an integrated health system working to improve the health of a whole population. Following the much-cited *Kinzigtal* program in Germany (Hildebrandt et al., 2012), as well as discussions in the United States concerning accountable care organizations and the triple aim (Berwick et al., 2008), many regions in the Netherlands have begun to collaborate with health insurers to come to arrangements that focus on population health. Such programs can have either a regional focus or a focus on specific target groups (like frail elderly). Organizationally, they can take different forms, for example, working as loosely coupled networks or through lead organizations (Provan and Kenis, 2007). Financial arrangements vary, with some networks using pay-for-performance (P4P) schemes and others working on a fee-for-service basis (Lemmens et al., 2017).

Enhanced population health management, which creates the organizational conditions for collaboration across organizational and sectoral boundaries, holds the promise of further integration of health and social care, as well as a stronger focus on prevention and self-management. Success stories on a smaller scale already exist, such as the creation of dementia networks, ParkinsonNet (http://parkinsonnet.info/) and integrated diabetes care. However, there are many challenges ahead if these initiatives are to be expanded:

- Collaboration, especially across sectoral boundaries, is difficult to establish given the variations in legal and financial arrangements, professional practices, and diagnostic definitions.
- Data infrastructures to follow patients or clients across organizational domains remain an issue. While some regions have been active in building patient portals, their functionality and reach are as yet rather limited, and integration of information remains a huge challenge (Aspria et al., 2016)
- Financial arrangements that stimulate collaboration across organizations and sectors are still hotly debated. Some initiatives are now experimenting with shared savings models or use P4P schemes, but outcomes are not yet clear (Eijkenaar, 2013).

Prospects

Given the many uncertainties, the current "experimental phase" is likely to last for a number of years. Careful monitoring of the effects of different initiatives, as well as the conditions for successful implementation and expansion, is required. New forms of governance might provide greater flexibility than the current top-down models (Sabel and Zeitlin, 2012).

Conclusion

While recent reforms have encouraged many regional and municipal developments in healthcare provision, they have also created unprecedented problems. Current initiatives to overcome these difficulties include the implementation of population health management programs, which hold the promise of establishing better integrated and networked (social) care arrangements. A number of issues still need to be addressed, and experimental forms of governance will be necessary to ensure further development.

26

Northern Ireland

Developing a Framework to Support Building Improvement Capacity across a System

Gavin Lavery, Cathy McCusker, and Charlotte McArdle

CONTENTS

United Kingdom Data

- Population: 65,637,239
- GDP per capita, PPP: $42,608.9
- Life expectancy at birth (both sexes): 81.2 years
- Expenditure on health as proportion of GDP: 9.1%
- Estimated inequity, Gini coefficient: 32.6%

Source: All data are from the World Health Organization and World Bank. Latest available data used as at October 2017.

Note: Data from England, Scotland, Wales, and Northern Ireland.

GDP = Gross Domestic Product

PPP = Purchasing Power Parity

Background

Changing demographics, in combination with heightened public expectations and the increasing role of advanced technology, mean that many care systems are encountering unprecedented staffing, logistical, and resource challenges. Around the world, growth in healthcare funding has flattened and is no longer keeping pace with the growth in demand. Many countries have accepted that the current model of care for their population is neither affordable nor sustainable. The picture in Northern Ireland does not differ from that described above (Figures 26.1 and 26.2).

Over the last 15 years, the avoidable harm that occurs in healthcare has been increasingly recognized. According to some studies, between 3% and 16% (or more) of patients experience an adverse event, and there are few

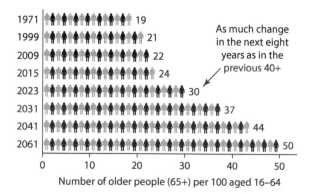

FIGURE 26.1

Northern Ireland population projections (2015–2061). (From Bengoa, R., Systems, not structures: Changing health & social care, Expert Panel Report, 2016. Retrieved from https://www. health-ni.gov.uk/sites/default/files/publications/health/expert-panel-full-report.pdf.)

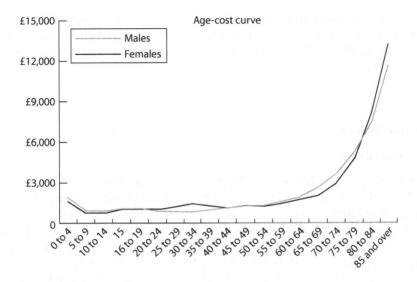

FIGURE 26.2
Age–cost curve for health and social care in Northern Ireland. (From Bengoa, R., Systems, not structures: Changing health & social care, Expert Panel Report, 2016. Retrieved from https://www.health-ni.gov.uk/sites/default/files/publications/health/expert-panel-full-report.pdf.)

signs that the rate is decreasing (Shojania and Thomas, 2013). In Northern Ireland, the ability to learn from such adverse events is a key strand of work under the DoH strategy Health and Wellbeing 2026 Delivering Together, and a culture in which (non-harmful) near-miss events are reliably reported and reviewed (Donaldson et al., 2014) is growing. However, anxiety surrounding the cost and sustainability of medical care, along with recent recognition of the concept of avoidable harm, has generated interest in promoting continuous improvement of services using structured, evidenced-based, collaborative approaches derived from other industries and contexts. Many healthcare leaders believe that an improvement science or quality improvement (QI) approach can lead to care that is safer and more effective, compassionate, and sustainable (Berwick, 2013).

The 1.8 million citizens of Northern Ireland have access to health and social care (HSC) via a single integrated system. QI may provide a way to engage staff in achieving widespread system change. Small numbers of staff have received, or are receiving, advanced-level training in QI using an approach based on the Institute for Healthcare Improvement's (IHI) breakthrough collaborative model (Institute for Healthcare Improvement, 2003). However, as in other systems, there are many challenges. How can it improve at scale? How can large numbers of staff be engaged in shaping new models of care—especially when staff surveys show that some already feel overwhelmed, undervalued, and unable to influence decision makers? If such engagement is achieved, how will it be possible to ensure that staff have the capability and capacity to drive improvement? How can a workforce who are prepared and equipped to take lead roles in a future HSC service be developed? How can a system that reflects on its own performance, continually adjusting to deliver better care, be created?

Success Story

Our chosen exemplar is the development of the Attributes Framework (AF) for Health and Social Care, a program that aims to support leadership in QI and safety (Department of Health, Social Care and Public Safety [Northern Ireland], 2014). The AF is a multiprofessional approach, accepted across all areas of HSC, that addresses the problem of developing a workforce that is both engaged and equipped to improve care across a complex system. Prior to 2015, a number of courses and programs focused on leadership, while others claimed to provide training in improvement. These programs, however, were profession specific and varied significantly in terms of content, approach, and terminology. None produced an ethos of systematic improvement within their core audience or across the system. Those running such programs did not think yet another training program focused on QI would be successful—identifying that time and opportunity to train staff was itself a challenge, and there was growing recognition that training, as a sole intervention, did little to change behavior.

In 2014–2015, a multiprofessional group representing HSC front-line staff, in conjunction with relevant professional educators and academics, worked to develop a framework of attributes (knowledge, skills, and attitudes) that we might expect of individuals who would be successful QI leaders in HSC. The aim was to develop an agreed framework across all professions that would describe the attributes of an improver at multiple levels, from front-line workers to members of the board. The developmental process itself used an improvement approach by providing many opportunities for the workforce and their employing organizations to shape the framework as it evolved. A specific effort was made to engage those undergraduates who would be the future users of such a framework.

The final version of the AF has 39 statements, written in simple, unambiguous language that describes the attributes related to safety and improvement expected of those working at four levels within the system. Figure 26.3 shows the design of the framework, and Figure 26.4 shows some of the attributes at Levels 1–4. Staff can self-assess using a simple tool (Northern Ireland Practice & Education Council for Nursing and Midwifery, 2014).

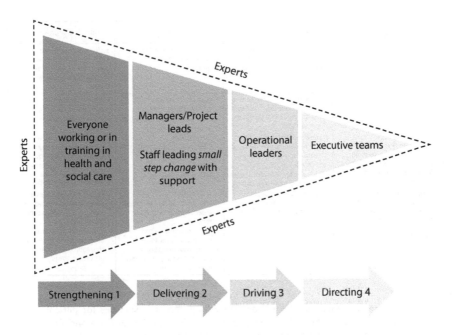

FIGURE 26.3
Levels of the attributes framework. (From Department of Health, Social Care and Public Safety [Northern Ireland], Supporting leadership for quality improvement and safety: An Attributes Framework for Health and Social Care, 2014. Retrieved from https://www.health-ni.gov.uk/sites/default/files/publications/dhssps/Q2020%20Attributes%20Framework.pdf.)

What can be achieved with the AF in the next 5–10 years? The framework was designed to

- Help individuals assess the attributes they currently possess (and those they require) to lead QI. This allows individuals to identify learning and development needs for current (or future) roles.
- Help organizations build workforce capability and capacity, thus ensuring that staff can participate in, and lead, initiatives that develop quality care and services.
- Assist in the training and education of QI and patient safety providers, in order to develop cross-disciplinary programs that are appropriate, synergistic, and comprehensive.

Today, the AF is accepted across Northern Ireland's system as the road map for future QI training and is being used by both organizations and individuals. Existing training is being mapped to the framework and then modified or augmented to ensure that it fulfills the requirements of the appropriate level. From 2017 onward, all medical, nursing, and social work students in Northern Ireland will already fulfill AF Level 1 requirements at the point of graduation from university.

Strengthening foundations for improvement	Delivering improvement	Driving improvement	Directing improvement
Everyone working or in training in health and social care.	Staff and those in training, who can lead small-step-change(s), with support, in their service.	Staff who lead team(s) or service(s) within their organization.	Staff charged with leading quality improvements across their organization and/or across the health and social care system.
• I understand why and how we put patients/service users at the center of everything we do • I understand what contributes to the safety of patients/service users and work with my colleagues to identify problems and help reduce risks • I understand what is meant by quality improvement and collect information in my area to aid improvement in patient/service user care and services	• I understand how the culture in my workplace influences the quality and safety of care and services • I recognise my responsibility to question the way we work in order to improve care and services • I am able to work with a team to achieve small-step-change • I can explain and use PDSA cycles to make small-step-change to care and services	• I communicate effectively with diverse audiences • I mentor and teach others about improvement methodology • I understand, use and present data to improve care and services • I influence, negotiate and lead improvements in care and services • I strive to motivate and lead improvements in care and services • I strive to motivate and energise my colleagues • I demonstrate resilience in order to lead improvements in care and services	• I lead improvements in care and services, aligning priorities and removing barriers • I encourage, promote and support a learning culture in and/or across organizations, learning from engagement with patients/service users and their carers/families • I direct the implementation and spread of improvement methodology across services boundaries • I use evidence-based tools, or accepted guidance, to ensure that appropriate resources are used in the organization

FIGURE 26.4

Attributes framework (Section 1) for those providing care and services for patients, service users, and their carers. PDSA. (From Department of Health, Social Care and Public Safety [Northern Ireland], Supporting leadership for quality improvement and safety: An Attributes Framework for Health and Social Care, 2014. Retrieved from https://www.health-ni.gov.uk/sites/default/files/publications/dhssps/Q2020%20Attributes%20Framework.pdf.)

Plan Do Study Act (PDSA) cycles are often used with the Model for Improvement (Langley et al., 2009). What are we trying to achieve? How will we know that change is an improvement? What changes can we make (to bring about an improvement)?

The future of our healthcare system requires a move from care being centered around hospitals to one in which citizens receive the majority of care and support at, or close to, home. Hospitals will provide specialist care that cannot be delivered elsewhere (Bengoa, 2016). This will require widespread reform of processes and systems and a workforce that is ready and able to change and improve. In the 2016 policy document "Health and Wellbeing 2026: Delivering Together," the health minister stated, "It is clear to me that, in order to achieve our ambition for health and social care, we need to establish an infrastructure capable of supporting, enabling and driving the improvements we seek" (Department of Health [Northern Ireland], 2016: 21). This political recognition has provided a fresh impetus to continue to develop and embed in practice the AF.

Impact

The AF was launched by the Department of Health and has become part of the overarching strategy for the future of HSC (Department of Health, Social Care and Public Safety [Northern Ireland], 2011). Since its genesis originated in the desire for uniformity and synergy for QI across all the relevant professions, academics, and policy-makers, few obstacles to acceptance by specific professions or organizations have emerged. The process of mapping and modifying existing courses means that training and education bodies are keen to support and engage with the AF. HSC organizations are using the framework to plan and track the growth of their QI infrastructure.

An online training module for AF Level 1 is available to all HSC staff, and the in-house training programs run by HSC organizations are being aligned to AF Level 2. AF level badges are presented to "graduating" senior leaders and front-line staff at special awards ceremonies, thus helping to increase the visibility and validity of the programs. The AF assessment tool is often used pre- and post-training to assess the effectiveness of any QI-related training, and staff are encouraged to use it for self-reflection prior to annual reviews, appraisal, and revalidation.

Implementation and Transferability

There are more than 65,000 HSC staff across Northern Ireland. Each employing organization has agreed targets for QI training. In the first 6 months, 1000 staff were trained to AF Level 1; by April 2017, 10% of all HSC staff will be trained to AF Level 1, and that is set to increase to 30% by April 2018. The aim is to have all HSC staff trained to AF Level 1 by 2020 and to have appropriate numbers of staff trained at higher levels. Discussions are currently underway to determine whether consultants (medical) and ward sisters or charge nurses should have progressed to AF Level 2 on taking up their posts, and whether clinical directors and associate medical directors and service managers should have AF Level 3 capability.

Beyond 2020, a system should have been established in which there are sufficient numbers of staff trained to lead continuous improvement at the front-line within or across services (AF Levels 2–3) supported by a small number of QI experts (Level 4). At this point, all HSC staff, having achieved AF Level 1, will be familiar with the ideas, concepts, and terminology of QI, making such progress easier.

Prospects

QI experts are expected to satisfy all the attributes from levels 1 to 4. Currently, the few individuals with such training in Northern Ireland have achieved this level of QI expertise by experience and training gained elsewhere in the world, at least in part. A decision must be made as to how we train our Level 4 "experts" at scale. Senior leaders and board members should also have Level 4 (but not necessarily Level 1–3) attributes. Some early prototypes of Level 4 training for senior leaders have begun in some organizations. Further work on agreed content and format is required, and a sustainable approach should be developed over the next 1–2 years.

Conclusion

The AF approach brings Northern Ireland closer to a place in which all HSC staff are equipped to commit to the idea that "everyone in healthcare really has two jobs when they come to work every day: to do their work and to improve it" (Batalden and Davidoff, 2007). By aligning the ideas and language of policy-makers, executives, managers, educators, and clinical and front-line staff, the AF will help make QI the universal language and approach for health system improvement.

27

Norway

Bridging the Gap: Opportunities for Hospital Clinical Ethics Committees in National Priority Setting

Ånen Ringard and Ellen Tveter Deilkås

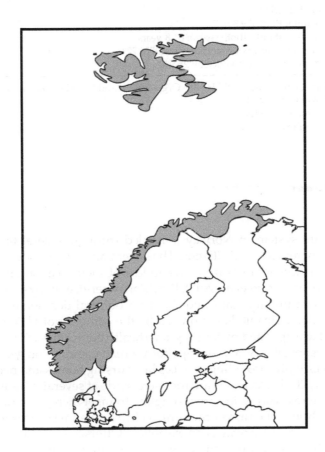

CONTENTS

Norwegian Data

- Population: 5,232,929
- GDP per capita, PPP: $59,301.7
- Life expectancy at birth (both sexes): 81.8 years
- Expenditure on health as proportion of GDP: 9.72%
- Estimated inequity, Gini coefficient: 25.9%

Source: All data are from the World Health Organization and World Bank. Latest available data
used as at October 2017.
GDP = Gross Domestic Product
PPP = Purchasing Power Parity

Background

The healthcare system in Norway is funded through general taxation, and
coverage is universal for all citizens. The system aims to provide high-quality
healthcare services to all citizens, regardless of their location and socioeconomic position (Ringard et al., 2013). Provision of services is, by and large, public. Norway is a unitary state, and the main political decision-making body
is the parliament. Hospitals are owned by the Ministry of Health and Care
Services, but are operated by four regional health authorities (RHAs). Primary
care is the responsibility of municipalities, which commission general practitioner services while providing services like nursing homes and neonatal care
(Deilkås et al., 2015). Decisions are therefore made at several different levels.

Health systems worldwide are struggling to meet increasing demands for
healthcare while at the same time confronting limited resources for meeting
those demands, and Norway is no exception (Ottersen et al., 2016). The question of how best to meet this challenge has been high on the political agenda
in Norway over the last four decades. The recent decline in oil prices and the
resultant decrease in state revenue have also prompted a renewed interest
in the issue of priority setting. Traditionally, much attention has been given

to determining principles for the prioritization and allocation of resources (Norheim, 2003). In 2016, three essential criteria for priority setting were defined by the third Norwegian Committee on Priority Setting in the Health Sector; these comprise the health benefit criterion, the resource criterion, and the severity criterion (Ministry of Health and Care Services [Norway], 2016; Prioriteringsutvalget, 2014).

In addition, there has been a discussion of how best to manage the decision-making or priority setting process (Daniels and Sabin, 2008; Ringard et al., 2010). The focus has been on whether priorities should be set at the individual level (i.e., in the meeting between doctors and patients) or at the national level (e.g., by creating high-level national decision-making bodies) (Norheim, 2008; Mørland et al., 2010; Ministry of Health and Care Services [Norway], 2015). Less attention has been devoted to the question of how to handle discussions and decisions at the intermediate or hospital level, and how these can be linked to discussions on the other two levels.

While clinical ethics committees (CECs) do provide already established forums for such discussion at a hospital level, the role of the CEC as an arena for connecting individual and national interests has so far received less attention (Førde and Pedersen, 2011). This chapter argues that linking all three decision-making levels within the hospital sector may have a positive impact on the coordination of messages, and could help provide solutions to the problem of managing scarce healthcare resources. CECs have an increasingly significant role to play in both the identification and resolution of problems, and can help to bring important ethical questions to national prominence.

Role of CECs

The first Norwegian CECs were established in 1996. Today there are 38 committees, and all hospital trusts have at least one. CECs are cross-disciplinary; members are drawn from diverse clinical settings within the hospital, and include external members who contribute legal or ethics competence. Some also have patient or lay representatives as members (Pedersen and Førde, 2005; Førde and Pedersen, 2011). The committees provide ethical guidance and advice in the hospital context, identifying, analyzing, and, when possible, solving ethical dilemmas related to patient care. Currently, the CEC's main objective is to handle specific ethical problems in individual cases, and to provide a forum where healthcare staff can discuss problems related to patient treatment both prospectively and retrospectively.

According to their mandate, however, the committees have also been given the task of contributing to an increased understanding of what values are at stake in questions of resource allocation and priority setting in the hospital. Hospitals are currently responsible for the evaluation and decision-making in relation to new medical devices and other technologies (e.g., new procedures). Thus, CECs will be expected to provide broader advice on more general questions regarding specific resource allocation in relation to the established principles (i.e., health benefit, resources, and severity) in the future. A recent survey of CEC protocols shows that such a development is slowly taking place. Over the period 2007–2013, about 30% of all cases considered by one CEC concerned matters of principle. Some originated in a dilemma related to a particular patient, while others, when initiated, had a more general scope. Most of the discussions led to adjustments within the hospital organization or in clinical practice through development and updating of guidelines (Førde and Ruud Hansen, 2014).

Impact

In the survey, some cases discussed in individual CECs were found to have influenced national healthcare policy (Førde and Ruud Hansen, 2014). The following case serves as the starting point for our study on how CECs may play an even more significant role in national policy development.

The case in question originated in a hospital department, and was initially brought to the attention of the local CEC in the western part of Norway by a doctor treating patients with lung cancer. The local CEC was asked, in October 2015, to consider whether a cancer medicine that a patient had bought privately could be administered in the public hospital. The cancer drug in question was at that time not offered by the public healthcare system, but was being considered by the national system for health technology assessment (HTA). The dilemma for the doctor was whether it was ethical to administer this drug in a publicly funded system built on the philosophy of universal and equal access for all.

The local CEC considered the question and came to the conclusion that administration of the drug would constitute a violation of equal-access principles and advised against it. In addition, they decided in December 2015 to petition the National Council for Priority Setting for guidance on the topic (National Council for Priority Setting, 2015). What makes this case of particular interest is that it also came to the notice of the Ministry of Health and Care Services, which at the time was in the process of preparing a white paper on priority setting. The case was therefore included as a specific topic for discussion in the white paper (Ministry of Health and Care Services [Norway], 2016).

Interestingly, the ministry came to the same conclusion as the local CEC, basing its decision on the same issues of equity as the committee. The white paper was discussed in parliament during the fall of 2016; however, this discussion did not alter the conclusion.

Implementation and Transferability

A case originally identified by an individual clinician and discussed by a local CEC traveled all the way to the highest national decision-making levels. The fact that this particular case was discussed in parliament might be a bit out of the ordinary, but the role played by the local CEC serves to highlight the importance of establishing competent advisory bodies at the meso-level or hospital level.

Several factors are key to solving priority setting dilemmas: substantive goals and criteria need to be established, legitimate processes must be observed, and coherent policy instruments must be in place (Daniels, 2000; Ottersen et al., 2016). These coherent policy instruments must exist at both the individual and the national level. CECs or similar organizations should not only provide advice at the individual level, but also act to ensure that these singular problems are considered at the national level, thereby increasing coherence of the system.

CECs or similar bodies have been established in many countries around the world. They are most common in the United States, but they can also be found in several countries in Europe (Pedersen and Førde, 2005). All countries have to decide how best to allocate scarce healthcare resources, and those with already established CECs should be able to make use of these forums to help inform such advice or decision-making. How questions are generated, how they are dealt with on the hospital level, and whether there are national or higher-level bodies to extend and reinforce such decisions will, of course, vary substantially.

Prospects

The ramifications of this one case are instructive—CECs could fill an important role in the creation of a comprehensive and coherent system for advising and deciding on important topics related to the use and allocation of scarce resources.

Ultimately, however, success depends on a number of factors:

- CECs need a clear mandate, which they currently have in Norway, to engage with topics of more general interest.
- CECs should ideally include members whose competencies encompass legal, economic, and organizational aspects of priority setting, as well as clinical and ethical aspects.
- A higher-level body, similar to the Norwegian National Council on Priority Setting, is needed to provide a broader arena for discussion of overarching principles than is possible at the local hospital level.

- There should be mechanisms in place for converting advice from the council to formal policy. In Norway, a national priority decision body, or *Beslutningsforum*, has recently been established within the hospital sector for just this purpose. Comprising the CEOs of the four RHAs, the forum ensures that final deliberations and decisions can find their way from the national body all the way back to the individual clinicians and patients.

Conclusion

In this chapter, we argue that multi-level healthcare systems need to establish organizational instruments at all levels to identify, discuss, and elevate priority setting problems. As the Norwegian experience shows, already established hospital CECs can help to coordinate decisions relating to priority setting at the clinical and national levels. We expect that this CEC function will become even more important in the future.

28

Portugal

Prevention of Antimicrobial Resistance through Antimicrobial Stewardship: A Nationwide Approach

José-Artur Paiva, Paulo André Fernandes, and Paulo Sousa

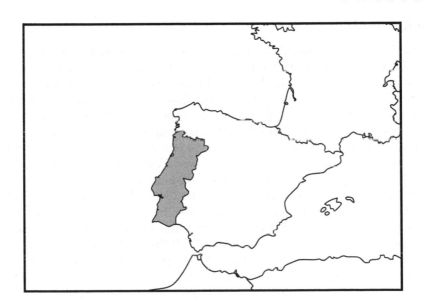

CONTENTS

Portuguese Data

- Population: 10,324,611
- GDP per capita, PPP: $30,624.2
- Life expectancy at birth (both sexes): 81.1 years
- Expenditure on health as proportion of GDP: 9.50%
- Estimated inequity, Gini coefficient: 36.0%

Source: All data are from the World Health Organization and World Bank. Latest available data
used as at October 2017.
GDP = Gross Domestic Product
PPP = Purchasing Power Parity

Background

In the 1940s, the availability of antibiotics transformed medicine, permitting the use of innovative techniques and procedures that save more lives and improve quality of life. However, by the 1950s, penicillin resistance had become an issue, and since then antimicrobial resistance (AMR) has become a substantial and increasingly intractable problem. By the beginning of the new millennium, more than 70% of the pathogens causing healthcare-associated infections were resistant to at least one of the antibiotics commonly used in their treatment (Centers for Disease Control and Prevention, 2001). In 2013, more than 2 million people were diagnosed with antibiotic-resistant infections in the United States, with at least 23,000 dying as a result (Centers for Disease Control and Prevention, 2013a). In the European Union (EU), AMR is responsible for 25,000 deaths and more than €1.5 billion of healthcare costs and productivity losses annually. If this trend continues, mortality attributable to AMR in 2050 will be 390,000 in Europe and 10 million globally (O'Neill, 2016). This could result in a reduction of the European GDP by between 1% and 4.5% by 2050 (Klynveld Peat Marwick Goerdeler, 2014).

Carbapenems are last-line antibiotics, used for hospital treatment of severe infections. Very common bacteria, such as Enterobacteriaceae, are becoming resistant to these antibiotics, due to the production of carbapenemases. The spread of these highly resistant bacteria is at very different stages in European countries. Outbreaks of carbapenemase-producing Enterobacteriaceae (CPE) have been reported in several states: 13 out of 38 countries reported interregional spread, and in 2015, four of these reported widespread CPE. Only three out of 38 countries replied that they had no single case of CPE (Albiger et al., 2015; European Centre for Disease Prevention and Control, 2016). In some countries, more than half of *Klebsiella pneumoniae*–causing infections are resistant to carbapenem antibiotics. Because of this, prescriptions of the antibiotic colistin almost doubled in Europe between 2010 and 2014, particularly in those countries that report high levels of carbapenem resistance. Resistance to colistin is now spreading (Falagas et al., 2014). Even frequent and previously easy-to-treat infections, such as urinary tract infections, may now pose problems (Redgrave et al., 2014).

The world is facing an "antibiotic crisis" and antibiotics, being on the verge of losing their efficacy, must be viewed as scarce and time-limited resources. It is something of a catch-22 situation: due to the high incidence of AMR, more potent and broad-spectrum antibiotics are being used, and as antibiotic use increases, AMR increases. In fact, there is ample evidence of a direct relationship between antibiotic consumption and antibiotic resistance, both at the individual and the population level: indeed, countries with higher antibiotic consumption have higher levels of AMR (Luepke et al., 2017; Redgrave et al., 2014).

Given its severity, AMR is a high priority politically within the EU and for the World Health Organization (WHO).

The agenda to decrease antibiotic use includes

- Preventing infections and their spread
- Improving monitoring and surveillance
- Avoiding unnecessary use of antibiotics: using them only when needed, for the least time possible, and with the least possible spectrum
- Developing research and innovation toward new effective antibiotics
- Developing new ways to tackle infection without harming bacteria (European Council, 2016)

Portugal, in common with all Southern and Eastern European countries, has a high rate of AMR. In 2011, it had one of the highest prescription rates of quinolone, a synthetic broad-spectrum antibiotic, and the highest hospital consumption of carbapenems in Europe (European Centre for Disease Prevention and Control, 2014).

Prevention of Antimicrobial Resistance

The National Program for the Prevention of Antimicrobial Resistance was created by the Portuguese government in 2008. This became part of the National Program for the Prevention of Infection and of Antimicrobial Resistance in 2013 (Ministério da Saúde [Portugal], 2013a).

Three vectors of intervention were defined to promote the judicious use of antibiotics.

Joint Awareness

First, to emulate the spirit of WHO's One Health initiative, which was developed to encourage equal collaborations between physicians, veterinarians, dentists, nurses, and other health-related disciplines, a number of associations representing different stakeholders came together to form the Portuguese Alliance for the Preservation of the Antibiotic. This alliance agreed on a 10-point memorandum that was signed on November 18, 2011 (Table 28.1).

National Publicity Campaign

There is good evidence that public campaigns promoting responsible antibiotic use may be associated with reductions in overall antibiotic use (Filippini et al., 2013; Saam et al., 2017). In light of this, a national antibiotic awareness campaign was developed in late 2011 and early 2012. The Portuguese antibiotic awareness campaign was implemented under the slogan "more antibiotics, less health." Very simple key messages were promoted, namely, "do not self-medicate with antibiotics," "unused antibiotics should be returned to the pharmacy," and "do not treat colds with antibiotics." The campaign involved primary care centers, hospitals, schools, and big-event arenas. Five million flyers and 10,000 posters were distributed, 200 billboards and public

TABLE 28.1

Portuguese Alliance for the Preservation of the Antibiotic 2011 Memorandum

---

Recognition that the antibiotic is endangered and needs to be protected, as it is a unique medication producing potential individual benefit but individual and collective damage

Elimination of self-medication, promotion of compliance to medical prescription and of the pharmacist role as a pedagogue in antibiotic use

Consolidation of the infection control activity and prevention of multidrug-resistant microorganisms cross transmission, both in the community and in the hospital, namely promoting biosafety and vaccination in human and veterinary medicine

Promotion of research on epidemiology and antimicrobial resistance, in human and veterinary sectors; and of easy access of clinicians to data on antibiotic consumption and prevalence and susceptibility of microorganisms

Release and implementation of guidelines on the use of antibiotics in human and veterinary medicine

Implementation of structures and processes of antibiotic stewardship in all health units, including primary and continued care, as stated in General Directorate of Health 028/2011 of 15/07/2011

Development of the use of rapid microbiological tests, that will allow reduction of duration of empirical antibiotic therapy and of avoidable antibiotics use

Implementation of rewarding policy for health units that attain good results, decreasing antibiotic consumption and antimicrobial resistance

Promotion of a swift investigation, development and marketing of new, innovative and useful antibiotics

Promotion of the practices put forward by the European Platform for the Responsible Use of Medication in Animals (EPRUMA)

Source: Portuguese Alliance for the Preservation of the Antibiotic, Memorandum, 2011, November 18. Retrieved from https://www2.arsalgarve.min-saude.pt/portal/sites/default/files//images/centrodocs/dia_europeu_antibioticos_2013/memorando_APAPA_dia_antibiotico_2013.pdf.

transport signs were placarded in the three main Portuguese cities, and a daily TV spot was presented for 30 days. All digital and traditional media avenues were exploited. The campaign was repeated, at a much smaller scale, in the winter of 2013–2014.

National Antimicrobial Stewardship Program

In 2013, law 15423/2013 (Ministério da Saúde [Portugal], 2013a) was issued, stating that all hospitals should implement an antimicrobial stewardship (AMS) program by the end of 2014. The aims of AMS programs are to increase the appropriateness of antibiotic treatment, improve clinical outcomes, reduce resistance, and control costs (Davey et al., 2017). Interventions to help physicians prescribe antibiotics properly fell broadly into two categories:

1. Restrictive techniques, which apply rules to make physicians prescribe properly
2. Educational or persuasive techniques, which provide advice or feedback to help physicians prescribe properly

National guidelines that recommend restricting the use of antibiotics to not more than 24 hours postsurgery in surgical prophylaxis and to not more than 8 days (except for rare exceptions) in infectious disease therapy were published by the General Directorate of Health. To help practitioners understand the issues and develop strategies to reduce quinolone and carbapenem use, short courses were held in each of the seven health regions, and prescriptions of these two classes of antibiotics, although allowed to be started, should be discussed with the bedside physicians and eventually validated by the AMS team in the first 96 hours after prescription.

It is estimated that the implementation of this three-vector plan should lead to a minimum 5% decrease in global community and global hospital consumption of antibiotics, and to a minimum 10% decrease in community quinolone and hospital carbapenem consumption.

Prospects

Since the implementation of the 5-year plan, there has been further high-profile media discussion about AMR. By the end of 2015, 78% of Portuguese hospitals and 11% of primary care centers had implemented AMS programs (Fernandes et al., 2016). In addition, the educational course on reducing antibiotic use was replicated inside each of the health regions. Available data show a largely positive trend, with some areas achieving better than expected results (Table 28.2).

TABLE 28.2

Consumption of Antibiotics in Portugal in Defined Daily Doses per 1000 Patients per Day

Year	Global Community Consumption of Antibiotics	Community Consumption of Quinolones	Global Hospital Consumption of Antibiotics	Hospital Consumption of Carbapenems	Hospital Consumption of Quinolones
2011	23.72	2.91	1.73	0.14	0.22
2012	23.04	2.61	1.67	0.14	0.18
2013	19.04	2.18	1.64	0.15	0.18
2014	20.32	2.12	1.55	0.14	0.17
2015	21.25	2.05	1.57	0.13	0.15

Source: European Centre for Disease Prevention and Control, Summary of the latest data on antibiotic consumption in the European Union, 2017, Retrieved from https://ecdc.europa.eu/sites/portal/files/documents/Final_2017_EAAD_ESAC-Net_Summary-edited%20-%20FINALwith% 20erratum.pdf.

The program needs to be sustained and even expanded. Community quinolone and hospital carbapenem consumptions are still higher than the European mean, respectively, and unnecessary use of these and other antibiotics still exists.

The data feedback process is being further developed, with improved communication, interactivity, timeliness, and reach. A hospital data package linking antibiotic consumption and AMR variables and indicators is being prepared for hospital physicians.

In 2016, a law was enacted that links hospital reimbursement to an assessment of structures, processes, and results in the field of infection control and adequate use of antibiotics (Ministério da Saúde [Portugal], 2016). A system of external auditing, assessing structure and processes in the field, is due to be implemented, and this will also be linked to hospital funding. These measures are essential to boost administration boards' support for the cause and to ensure that antibiotic use is a priority in terms of hospital management goals.

Conclusion

The nationwide three-vector plan designed by the Portuguese National Program for the Prevention of Infection and Antimicrobial Resistance is designed to reduce and hopefully eliminate the unnecessary use of antibiotics. The effort has thus far succeeded in reducing the consumption of antibiotics, mainly of quinolones and carbapenems, two antibiotic classes that induce significant collateral damage in terms of the emergence of AMR. A special focus on further reduction of carbapenem consumption is currently a priority in terms of hospital management goals.

29

Russia

The Future of Physicians' Specialization

Vasiliy V. Vlassov and Mark Swaim

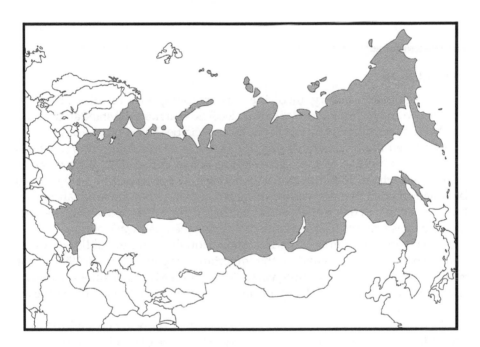

CONTENTS

Russian Data

- Population: 144,342,396
- GDP per capita, PPP: $23,162.6
- Life expectancy at birth (both sexes): 70.5 years
- Expenditure on health as proportion of GDP: 7.07%
- Estimated inequity, Gini coefficient: 41.6%

Source: All data are from the World Health Organization and World Bank. Latest available data used as at October 2017.
GDP = Gross Domestic Product
PPP = Purchasing Power Parity

Background

By the eighteenth century, medicine was grappling with rapidly expanding knowledge and technologies (Williams, 2000). So that multiple standards of care for the same condition were not in conflict, specialties based on physician affinity developed out of generalist practice. By the late twentieth century, nations diverged in a number of specialties earning official recognition, but a singular trend of growth in specialization was unchanged. Specialization may seem to be ornamentation that denotes technical prowess and knowledge advancement, but it is essential in that it alters workforce self-organization and delivery of care, and channels patients into more finely distinguished pathways of care.*

Specialization legitimates medical professionalism in the public eye, especially when physicians validate the importance of generalists. In 1999, major medical organizations promulgated a "Charter on Medical Professionalism" that espoused professionals, including specialists, as activists in healthcare reform (Haynes et al., 1986).

When does a new (sub)specialty germinate? When does it crystallize? New specialties emerge when physicians recognize a scope of services employing a set of technologies dictated by a set of problems to be treated. A specialty or its subspecialty may form around an organ, a set of related organs, or a particular anatomical area. However, while this is a necessary precondition, a specialty will only develop if its objectives align with those of the society in question, along with its healthcare system. Thus, the hepatitis C viral epidemic led to a

---

* For this review we searched MEDLINE using next strategy: (specialization[mh] OR specialty OR specialized) AND (training OR education) AND (doctors OR physician* AND 2000:2017[dp] AND (trend* OR future OR progress* OR tendency OR perspective). Excluded were issues of race and gender pertaining to specialization, physician (e/im)migration, international cooperation, aging and reimbursement. A reference list was bolstered by deploying the similar search strategy at scholar.google.com and snowballing. The Russian perspective is enriched from author VVV's collection and the database "Rossiiskaia Meditsina".

new American specialty, transplant hepatology. Every specialty must be recognized by the system's power brokers and represented in the healthcare structure. A hospitalist proctologist designation, for instance, lacks standing where all in-hospital care (such as in post-Soviet Russia) is provided by hospitalists.

Early-career physicians inexorably seek more (sub)specialty training because specialties are perceived to be more prestigious, providing a better lifestyle, improved work conditions, higher incomes, and scope of services consistent with personal interests. The Western model perceives a difference in income between the (sub)specialist and the generalist; this difference may be substantial in other nations, and may have been greater in other eras (Sloan, 1970).

The trend to seek subspecialization is widespread. In all affluent nations and across all specialties, moving into subspecialties is the norm: it is ubiquitous (Brotherton et al., 2005). Most developed nations face an increasing oversupply of subspecialists, but this abundance is not equally distributed. Canada, for example, has had at least 60% of ophthalmology residents pursue fellowships (e.g., in vitreoretinal surgery) over the past 25 years (Fisher, 2016). Trainees compete for more rarefied credentials and access to better teachers. For poorly understood reasons, subspecialists within pediatrics are less common (Mayer and Skinner, 2004).

A tendency to seek a narrower scope of practice is prevalent also in both middle- and low-income nations. In fact, the orientation among all doctors in all countries in the twentieth century trends away from general practice (Newton and Grayson, 2003). However, low purchasing power reduces the growth of subspecialist cadres in poorer nations. In many post-Soviet countries, primary care physicians are in short supply, but specialists, paradoxically, are even less accessible: they emigrate from the public health sector to the private clinics and "cherry-pick" patients.

Because the general practice workforce has collapsed, along with public esteem for general practice, a number of corrective initiatives have emerged, some durable and showing signs of success (Martin et al., 2004). For example, some U.S. medical schools secure students biased toward primary care, and the number of pediatrics residents seeking subspecialty training in the United States has decreased significantly (Mayer and Skinner, 2004). The reasons for such changes are unclear, however; it may be that certain educational drives are retaining residents in family medicine and in proximity to where such care is needed, or it may be a consequence of less quantifiable sociological factors (Fagan et al., 2015).

Future Reflections

In most post-Soviet countries, getting subspecialty training is simple. Residencies are short (1–2 years) and the specialization is merely 6 months of additional training. Doctors in these programs are treated as students, and are paid shamefully low stipends. The need to extend residency training is widely accepted, but resources are scarce and legislative will to change regulations relating to doctors-in-training is absent. To fill primary care positions, Russia recently took a strange turn by changing federal healthcare law (law 323): residency is currently not required for new graduates who choose to practice primary care. Russia has thus forfeited the main achievement of medical education in the twentieth century, at great cost to physician expertise. Sooner or later, the Russian system will need to establish a reasonable period of residency training, despite the fact that this runs counter to the pressing need for a greater number of primary care doctors.

While specialization was organ-oriented until the early twentieth century, later specializations cropped up around new pathophysiological concepts, for example, infectious diseases and endocrinology. Expanding knowledge and technological advances led to arcane, asymptotically narrow subspecialties, such as vascular neurology (Adams and Biller, 2014) and neurohospitalist (Chang and Pratt, 2012). The pendulum has swung too far.

The proliferation of specialties leads to problems of who best manages a patient: which specialty should deal with the eye problems of diabetes, for example? The eye was once the province of a singular (sub)specialist, but now many speak for the organ. Prostate cancer is reckoned with differently by generalists, urologists, oncologists, and radiation oncologists. Bewilderingly, in some quarters this has provoked calls for the creation of neosubspecialties, such as vulvology (Micheletti et al., 2002), to overcome the issues incurred by treatment by too many diverse subspecialties of one organ, tissue, or site.

The transformations of both medicine and society alter the disposition of medical graduates toward specialty, and always will. The fact that we now live longer lives, with a related increase in the likelihood of suffering, has sparked the emergence of palliative care medicine, just as decades earlier similar factors led to a geriatrics specialty. The United Kingdom now has more palliative care specialists than the combined number of oncologists and neurologists (Doyle, 2007).

What governs and regulates subspecialization? The most acute demonstration of the management of subspecialization may be the United States, where market forces provide dominant but incomplete control (Stoddard et al., 2000). Market forces are an inadequate mechanism when it comes to timing: training and its duration dictate deficit and supply. In other countries, shortages in some specialties and surfeits in others have led to efforts to implement centrally regulated training volume. Regulation of this kind may be guided by such methods as the U.S. FutureDocs Forecasting Tool, which estimates the supply of physicians and capacity to meet population needs up until 2030 (Ricketts et al., 2017).

The American system perceives its specialist supply issues as having been exhaustively studied, but this is inaccurate (Mayer and Skinner, 2004). The need for either specialists or subspecialists varies wildly by region in the

United States, and the reasons for this are unclear. Some analyses of physician graduate training statistics suggest that the American physician workforce will fall short and that the nation needs more graduates and residency positions (Jolly et al., 2013). Short-term interventions never work, however, because physician training is lengthy and cumbersome; indeed, this fact alone may make policy interventions deleterious. Population demographics do suggest that many countries, from the United States to post-Soviet countries, would be best served by considering overall physician numbers rather than responding to the community demand for specialists.

Impact

The increasing proportion of specialists comes at a cost to the proportion of generalists, and subspecialists grow in prevalence at a cost to the pool of specialists. Will there be a ceiling on the numbers of (sub)specialists? Likely, each society will contrive its own solution. Technologies will always advance, and societies will change, and these will limit or drive specialization, but more specialization is likely to be the prevailing trend for the next decades. The human brain arguably copes well with what can be mastered in about 3 years of training, and the expansion of medical knowledge disturbs an equilibrium that tends to move away from generalism. In budget-conscious systems, a cap on the subspecialist workforce is inevitable. An obvious "natural" ceiling for subspecialization is that the very narrow specialty may not provide enough patients to keep a physician engaged (Ricketts et al., 2016). As yet, however, this mechanism does not appear to act on a large scale anywhere.

Prospects

Can we remedy the side effects of growing specialization? One remedy is the development of the more collegial, team-based approach. This has long been promoted, but the system gravitates to the fragmentation of care. Time is a barrier. Another, more problematic approach is to give generalist training to subspecialist physicians (Levi, 2017). In essence, this approach is not new— all systems of subspecialist training have grown out of generalist training. However, such an approach is unlikely to be popular: subspecialists may ignore non-relevant expertise and filter out cases that disturb their homogenous practice.

An alternative remedy has been proposed: the expert-generalist, that is, the general practitioner who has been trained in a subspecialty area

(Fins, 2015). This approach is not radically new either, and is commonplace in some countries, such as the United Kingdom and Canada.

Conclusion

It is expected that we will continue to see an increase in the subspecialization of physicians in successful healthcare systems, along with initiatives to reverse this trend. Physicians' areas of expertise are likely to become increasingly idiosyncratic, which may have an adverse effect on healthcare quality and lead to greater inequality when it comes to accessing specialist care.

30

Scotland

Deliberative Engagement: Giving Citizen Involvement Meaning and Impact

Richard Norris, Andrew Thompson, and David R. Steel

CONTENTS

United Kingdom Data

- Population: 65,637,239
- GDP per capita, PPP: $42,608.9
- Life expectancy at birth (both sexes): 81.2 years
- Expenditure on health as proportion of GDP: 9.1%
- Estimated inequity, Gini coefficient: 32.6%

Source: All data are from the World Health Organization and World Bank. Latest available data used as at October 2017.
Note: Data from England, Scotland, Wales, and Northern Ireland.
GDP = Gross Domestic Product
PPP = Purchasing Power Parity

Background

The National Health Service (NHS) in Scotland is undergoing major change. Its broad aims have been stable for many years: promotion of quality, safe, patient-centered, and effective care; transformation of the way services are delivered to shift the balance of care from hospital to community settings; and integration of health and social care with a greater focus on prevention, anticipation, and self-management. Structurally, however, there has been little change, with the only major development being the creation in 2016 of new joint bodies to integrate NHS health and local authority social care services.

As in most developed countries, economic pressures on health service budgets emanating from the global financial crisis of 2008 and its after effects have driven this agenda—resulting in policies of austerity. Likewise, the aging population and advances in medical technologies are putting greater strains on the system's ability to provide care at sustainable levels of volume and quality.

Where the Scottish health system distinguishes itself from other comparable systems is in the Scottish political environment, which reflects its particular identity, history, culture, and small population. Eschewing conventional neoliberal policies fostered elsewhere, Scotland has chosen an alternative path toward providing welfare and social solidarity based on a partnership approach, benefiting from a stable quality infrastructure, strong clinical leadership, and tight networks (Thompson and Steel, 2015, 2017). Citizens, whether through civil society organizations, or as patients, carers, or active individuals, have a stake in these policies, but it requires more than a formal place at the table to ensure meaningful engagement. Current developments are attempting to open up the potential of deliberation to give a voice to citizens, thus enabling a meaningful conversation that can deliver health gains and reduce inequalities.

Success Story

The four principle means of supporting and driving improvement in health and social care systems and services in Scotland are

1. Improvement methodologies
2. Quality assurance
3. Implementation of evidence-based guidance and recommendations
4. User, carer, and citizen participation

The first three are widely accepted, implemented, and evaluated across the NHS and social care services in Scotland (McDermott et al., 2015). Policy injunctions to establish or increase participation are often a response to rising citizenship expectations, political and democratic pressures, and rhetoric around human and consumer rights. There is also, particularly across the United Kingdom, a rising tide of case law on legal challenges to public authorities for failure to engage or consult adequately. However, explicit linkages to identifying and measuring improvement in outcomes are rarely made in practice in relation to user, carer, and citizen participation.

The Our Voice initiative (https://www.ourvoice.scot/) is a national program being conducted in partnership with citizens by government, statutory, and third-sector agencies, and impacting three different levels:

1. Individual interaction with service providers
2. Local communities and local services
3. User, carer, and citizen voices to influence national policy

This is a whole systems approach that builds on existing engagement frameworks, tests new approaches, and links the different levels together. While designed specifically to deliver improvement to the Scottish health and social care services, Our Voice has reviewed participation approaches from across the United Kingdom and other parts of the world, as well as engaging with stakeholders in Scotland. It has sought to learn from best practice and apply this to develop a flexible and evolving framework in Scotland.

Our Voice began as a series of public engagement activities undertaken in the autumn of 2014, followed by a collaborative design event in May 2015 to present the Our Voice framework to stakeholders, with a follow-up event in November 2015. The statutory Scottish Health Council and the third-sector intermediary Health and Social Care Alliance Scotland have been commissioned to deliver key aspects of the Our Voice framework, and the Scottish government has led on developing proposals for a leadership coalition and communications. The framework is now emerging with the establishment of a national Our Voice Citizens' Panel in 2016, local community projects working with a diverse range of groups, and a program of engagement with users and carers of maternity and neonatal services to influence a new national strategy.

The Scottish government (2016) has recently published a Health and Social Care Delivery Plan that aims to develop a system that

- Is integrated
- Focuses on prevention, anticipation, and supported self-management
- Will make day-case treatment the norm, where hospital treatment is required and cannot be provided in a community setting
- Focuses on care being provided to the highest standards of quality and safety, whatever the setting, with the person at the center of all decisions
- Ensures that people get back into their home or community environment as soon as appropriate, with minimal risk of readmission

It sets out specific targets for a range of actions to be achieved between 2017 and 2021. The plan confirms that the Our Voice framework will be used to support engagement on the work of the delivery plan. Our Voice sets out a number of objectives concerning user, carer, and citizen involvement:

- At the individual level, people should be fully involved in decisions about their treatment and care, and they should be empowered and supported to give feedback on the care and services they receive. Their feedback should be used to drive and inform continuous improvement to services. Integrated stakeholders will work together to develop systems for responding to feedback that are accessible, manageable for staff, and capable of being transferred across settings.
- At the local level, a peer network will support people to engage purposefully in local planning processes. Guidance, tools, and techniques will build people's capacity to get involved in and lead local conversations. Particular support will be given to those whose voices are not always heard and to develop local networks of people who are willing to get involved.
- At the national level, a citizen voice "hub" will tap into existing structures and networks, gathering intelligence on issues of concern and involving as wide a range of people as possible in improving services. Strategic gathering and analysis of individual stories on topics of national interest will provide policy-makers and health and care providers with powerful evidence for improvement. Citizens' panels will create opportunities for people to engage in national policy debate.

- A leadership coalition of health and social care service users, carers, and leaders from the NHS, local authorities, and the third sector will guide the development of the framework, work to maintain the momentum, and act as champions for a stronger citizen voice within their organizations. It will be independently chaired by a member of the public.

Impact

The collaboration has sought and gained commitment from across the health and social sectors, including users, carers, the public, third-sector organizations, the Scottish government and Parliament, NHS, local authorities, and health and social care integration authorities. All have an investment in ensuring that, as services develop, they continue to maintain the confidence of citizens, with new developments tested and reviewed, and learning shared across the framework. Examples of this include

- Conducting public discussion groups on a wide range of topics across Scotland
- Commissioning an Our Voice Citizens' Panel of 1300 people from all parts of Scotland broadly representative of the Scottish population
- Supporting and testing a variety of approaches at the local level, in particular to support newly integrated health and social care partnerships
- Establishing citizens' juries
- Implementing thematic projects (a current example is a joint project by the Scottish Health Council and the Health and Social Care Alliance exploring user and carer experiences of social isolation in an island community)

Local "focus group" discussions conducted across Scotland on identified themes through Scottish Health Council local offices have been particularly valuable in enabling and supporting the input of user, carer, and public views on a wide variety of topics, for example, cosmetic surgery, after-hours care, a register of interests for healthcare professionals, and organ transplants. The local discussion group approach is a relatively informal and low-cost method of conducting deliberative engagement with a variety of stakeholder groups on a potentially wide variety of topics. While the value of the discussion group is apparent, a better view of the particular strengths and weaknesses of this model can be obtained by developing other more structured deliberative mechanisms (e.g., the citizens' panel and citizens' juries).

Discussion groups can also be combined with other approaches. A mixture of approaches was used to gather views to influence policy on maternity and neonatal services; these included engaging with local groups, one-on-one discussions by telephone, face-to-face conversations, and questionnaires (Scottish Health Council, 2017).

Implementation and Transferability

This is an evolving framework. The current objective is to test and refine approaches, and measure outcomes in terms of how the approaches have led to improvements, for which an outcomes framework is being developed. The leadership coalition will review the operation of the Our Voice framework and make recommendations for its further development based on outcomes. The transformational changes that will need to be made in the next 10–15 years in health and social care services will only succeed if they are supported by the public, with user and carer views and preferences foregrounded prominently. These changes will either reshape existing services to meet the needs of future generations or fail to achieve the necessary improvement and undermine the political consensus that currently exists in Scotland for health and social care. Success will be measured by the continuing level of public confidence and political consensus in the Scottish health and social care system. It is also important that user, carer, and public involvement gets beyond tokenism and what has been referred to as a "technology of legitimation," which means that a necessary part of this process will involve challenge and discomfort for service providers and policy-makers (Harrison and Mort, 1998).

Prospects

The first citizens' jury is about to be commissioned (looking at shared decision-making), and the Scotland-wide Our Voice Citizens' Panel just conducted its first member survey (November 2016), which included questions on social care, pharmacy, and oral health. These initial steps need to be evaluated and refined, in keeping with both the quality of the engagement as judged by the participants, and the outcomes and influence achieved. The leadership coalition, once constituted, will review the progress of the various Our Voice components, identify how further progress can best be made, and give support and advice.

Participation as an activity is dependent on context. There is no one "true" or "best" approach to engaging with the general population or even particular communities. There is also no one public, citizen, or user view. Different approaches and methods are required if a balance is to be struck between the views of users and carers, and the views of the wider public. Knowledge relating to how particular models perform in different settings will be shared. Much of the knowledge gained by testing these different models will be applicable not only across Scotland, but also in other parts of the world. Testing and evaluation is an integral approach in Our Voice, so that the whole system can apply lessons learned from different parts of the framework.

Conclusion

The Our Voice framework is an ambitious and collaborative attempt to develop and harvest a range of approaches in order to determine which models are most beneficial. The aim of health and social care improvement will also require an understanding of how citizen involvement links to the other drivers mentioned earlier. While these have an internal focus, there is broader interest in any analytical generalizations that might be of value beyond Scotland and its particular system and culture.

31

Spain

*How Can Patient Involvement and a Person-Centered
Approach Improve Quality in Healthcare? The
Patients' University and Other Lessons from Spain*

Laura Fernández-Maldonado, Sergi Blancafort Alias, Marta Ballester
Santiago, Lilisbeth Perestelo-Pérez, and Antoni Salvà Casanovas

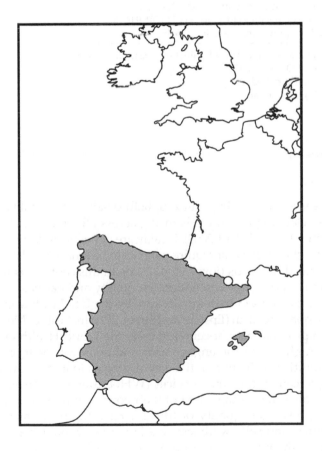

CONTENTS

Spanish Data

- Population: 46,443,959
- GDP per capita, PPP: $36,309.8
- Life expectancy at birth (both sexes): 82.8 years
- Expenditure on health as proportion of GDP: 9.03%
- Estimated inequity, Gini coefficient: 35.9%

Source: All data are from the World Health Organization and World Bank. Latest available data used as at October 2017.
GDP = Gross Domestic Product
PPP = Purchasing Power Parity

Background

Chronic diseases are a leading cause of both death and disability worldwide, with rates accelerating globally, advancing across all regions and affecting all socioeconomic classes (WHO, 2002). Healthcare systems are facing major social and economic challenges that require patients to be empowered and capable of making informed decisions (Jovell et al., 2006). In this context, patient-centered care has come to preoccupy discussions on quality and has been enshrined as one of six key elements in high-quality care by the U.S. Institute of Medicine's report on the quality chasm (Epstein and Street, 2011; Institute of Medicine, 2001).

In a bid to promote patient-centered care, the World Health Organization (WHO) has highlighted the engagement and empowerment of people and their communities as the first of five key strategies in its framework for integrated people-centered health services (WHO, 2016e). Empowered patients have control over the management of their condition in daily life. They take action to improve their quality of life and have the necessary knowledge, skills, attitudes, and self-awareness to adjust their behavior and to work in partnership with others, where necessary, to achieve optimal well-being (EMPATHiE Consortium, 2014).

Empowerment interventions aim to equip patients (and their informal care-givers whenever appropriate) with the capacity to participate in decisions related to their condition to the extent that they wish to do so, to become "co-managers" of their condition in partnership with health professionals, and to develop self-confidence, self-esteem, and coping skills to manage the physical, emotional, and social impacts of illness in everyday life (EMPATHiE Consortium, 2014).

Empowerment interventions most frequently address one or more of three dimensions: patient education and health literacy (HL), support for self-man-agement (SM), and shared decision-making. Since the 1990s, there have been several definitions of HL (Suñer and Santiñà, 2016). In short, HL refers to hav-ing the knowledge, motivation, and skills to access, understand, appraise, and apply health information; form judgments; and make decisions regarding healthcare, disease prevention, and health promotion (Sørensen et al., 2012).

Despite institutional backing for patient empowerment, the concept is still not embraced by all healthcare professionals and healthcare managers. As a key reference point, the Eurobarometer Qualitative Study on Patient Involvement concluded that the concept of patient involvement was not clearly understood by either patients or practitioners, and that the percep-tion of improved cooperation as an effective means to achieve better health outcomes was rarely mentioned (European Commission, 2012).

Worldwide efforts to promote and implement empowerment interventions need to be recognized, however. Illustrative of the interest generated world-wide, two recent European projects (both carried out by the Avedis Donabedian Research Institute) found considerable scientific evidence of patient empower-ment interventions in their literature reviews: in the EMPATHiE study, 101 sys-tematic reviews were identified for the period 2000–2013, and in the ProSTEP study, 257 systematic reviews in 18 chronic diseases were identified in the period 2010–2016 (EMPATHiE Consortium, 2014; European Commission, 2015). These figures demonstrate the relevance of patient empowerment pro-grams that are succeeding despite the lack of official recognition.

The gap between general practice (as illustrated by the Eurobarometer study) and existing evidence of positive impacts (as illustrated by EMPATHiE and ProSTEP) makes it particularly important to take note of successfully implemented patient empowerment interventions.

In this chapter, we describe one such successful initiative, the Patients' University (UP) project. The UP not only provides patients with the neces-sary skills and tools to enhance their involvement and improve their health outcomes, but also redefines the meaning of the partnership between patients and practitioners.

The Patients' University Project

In order to respond to the current burden of chronic diseases, a number of Spanish regional health administrations have launched initiatives based on different methodologies, formats, and assessment systems aimed at promoting patient involvement in decision-making affecting their health (Nuño-Solinis et al., 2013). In 2006, the Fundació Salut i Envelliment (Health and Aging Foundation) UAB (http://salut-envelliment.uab.cat/lafundacio/en/) launched an initiative called the Patients' University (http://www.universidadpacientes.org/). Its mission is to meet patient information needs and promote HL and SM to both patients and health professionals through educational programs and research projects based on both quantitative and qualitative approaches. The UP approach, which is aligned with a patient-centered perspective in health systems, not only responds to the complexities of modern healthcare systems, but also reflects a future investment in fostering the patient's voice in decision-making forums.

The Spanish Patients' Views

Traditionally, patients have rarely been considered active agents in health research, so their influence on the establishment of health priorities and strategies has been minimal. In recent decades, however, healthcare systems have increasingly encouraged patients and citizens to view themselves as strategic stakeholders in improving healthcare quality and to participate in health research in various ways. Their participation leads to a deeper understanding of patients' views in relation to healthcare.

According to studies undertaken by the UP—including the "Patient of the Future" study (2003), which was designed to examine patients' views of arthritis disease and how quality of care could be improved, along with the quality of cancer care in Spain—patients are typically emotionally and intellectually overwhelmed during their interactions with healthcare services (Arrighi et al., 2015; Coulter and Magee, 2003; Gabriele et al., 2012). For instance, they frequently receive unintelligible information, or have to make lifestyle adjustments that are not aligned with their personal goals, often having to do so without the appropriate support. When patients fail, for any of several reasons, to communicate important information to health professionals, the resulting isolation and lack of participation in key decisions may ultimately pose a serious threat to their health. The adherence study demonstrated that communication is fundamental in helping patients cope with disease and in mediating in the recovery process (Pagès-Puigdemont et al., 2016). Patients who have a good relationship with their health professional, or who receive specific information and training, not only understand their condition and treatment more fully but also are more likely to adhere to treatment regimes.

Addressing the Health Literacy Needs of Patients

Current research into HL reveals that many long-term patients do not fully understand their condition and feel incapable of cooperating with health professionals in any significant way (Fernández-Maldonado et al., 2013). The fact that patients with chronic illnesses are also often overcome by feelings of uselessness, lack of autonomy, and isolation can have negative repercussions on overall health.

In 2012, Fundació Salut i Envelliment conducted a nationwide survey to measure social HL perception and values for the first time in Spain (Fernández-Maldonado et al., 2013). The original data reported in this DataWatch came from a survey conducted in 2011. The findings were based on interviews with a random sample of 3000 Spaniards aged 18 and over, whose responses were weighted to represent the adult population of the country. This study identified specific drawbacks of the healthcare system and pinpointed necessary improvements to strengthen population skills in health management, but especially in vulnerable groups affected by disabilities and adverse social determinants. Other initiatives, such as the appointment tool kit for patients (http://www.universidadpacientes.org/kitdevisitamedica/) developed by Fundació Salut i Envelliment and Esteve, aim to strengthen doctor–patient relationships, improving communication and providing better information. This multimedia and multilingual tool kit has been adapted and applied by primary care services operated by several regional administrations in Spain.

Chronic Disease, Patient Education, and Self-Management

There is growing interest in the development of SM support programs in Spain, where one particular initiative—the Chronic Disease Self-Management Program (CDSMP)—has had a particular impact. This program specifically targets patients with chronic conditions (Lorig et al., 1999).

The CDSMP, in the form of a training model culturally adapted to the Spanish setting, has been widely implemented by the UP and has made a significant contribution to the development of strategic plans for chronic conditions in Spain (Blancafort et al., 2013). It has guided the development and implementation of training programs targeting patients, health professionals, and volunteers in 10 Spanish regions.

Patients' organizations play a key stakeholder role in healthcare systems, but their representatives require advocacy skills that will enable them to manage relationships and the complexity of changing environments more effectively. To respond to this need, the UP launched an online training program in Latin America for organizations interested in pursuing accreditation in patient advocacy. This kind of hands-on training is an effective way of encouraging patient representatives to develop their advocacy skills and to share best practices.

The Patients' University as a Trend

In 2012, the Spanish Ministry of Health created a network of health schools for citizens to integrate initiatives by regional governments and to foster information exchange and support between patients and institutions such as the UP (Ministerio de Sanidad, Servicios Sociales e Igualdad [Spain], 2012). The UP is building partnerships and strategic alliances with Spanish and international stakeholders, with a view to bringing significant benefits to patients and their families. The establishment of public forums, where views and starting points can be shared, helps to build bridges and create spaces for dialogue so as to establish common ground and understanding. These meetings also help to garner support for recommendations to policy-makers. They also facilitate the monitoring of information, ensuring that programs can be updated and improved in response to the active patients' needs.

Conclusion

The UP project to improve health promotion is based on core initiatives aimed at engaging patients, healthcare professionals, management, and policy-makers through research, training, and the provision of information. The initiatives fostered by the UP ultimately result in the compilation of solid evidence that forms the basis for recommendations made to healthcare professional associations, public administrators, and policy-makers, with the purpose of raising their awareness of, and engagement with, patient involvement and person-centered care.

32

Sweden

The Learning Health System

John Øvretveit and Camilla Björk

CONTENTS

Swedish Data

- Population: 9,903,122
- GDP per capita, PPP: $49,174.9
- Life expectancy at birth (both sexes): 82.4 years
- Expenditure on health as proportion of GDP: 11.93%
- Estimated inequity, Gini coefficient: 27.3%

Source: All data are from the World Health Organization and World Bank. Latest available data used as at October 2017.
GDP = Gross Domestic Product
PPP = Purchasing Power Parity

Background

Sweden, like many other countries, is finding it challenging to provide healthcare that not only meets citizens' expectations but also is affordable. Citizens and employers are resisting increases to taxes to pay for healthcare, yet expectations are higher than ever, fueled by comparisons to customer service standards in other industries, along with constant publicity about new medical discoveries. Demand is also increasing due to an aging population, and there are personnel shortages in a number of healthcare specialties and services.

In other industries, digital technologies have been harnessed to increase productivity (Evans and Wurster, 1999). However, the digitalization of healthcare has been relatively slow. This delay may be due, in part, to concerns about privacy and security. The fact that effective healthcare relies on a culture of personal relations, where trust and personal contact are important, may also explain the delay. Replacing physical with virtual visits is one way to increase productivity; however, these may not be favored by all patients and practitioners. More analysis is needed to identify how best to use these technologies.

The Swedish government built a nationwide digital infrastructure in the 1990s, providing a framework for the development of health information technology. As yet, however, there is no nationwide sharing of electronic health records (EHRs) or patient data between health and social care. The aim is to establish interoperability between record systems in the coming years.

One area of digital health that has received government support is the establishment and maintenance of the 106 national quality registers that form a key part of the Swedish learning health system (LHS). These long-running and highly developed national clinical databases cover a number of health conditions and procedures. They are recognized as being of strategic significance for the country's competitive position in global health industries, and have made an impact in European and global healthcare markets (Larsson et al., 2012).

The Learning Health System Concept

This chapter discusses the Swedish healthcare system's efforts to integrate digital data and technologies into healthcare provision and research via the LHS, and examines the resulting benefits.

We define an LHS as a health system that collects, stores, analyzes, and presents clinical and other data at the time and place that users require the data, and in a way that enables better informed everyday decision-making. Such systems also support longer-term learning and improvement through research and quality improvement projects (Øvretveit and Keel, 2014).

Digital data can be used for multiple purposes in an LHS. The benefits of being given the correct data in the correct format at the right time in the right place are numerous, and have ramifications for all stakeholders, including patients, clinicians, and service delivery managers. The advantages of having timely access to such information will also extend to researchers and will shape future strategic planning and population health management.

Patients

The Swedish national rheumatoid arthritis quality register is an example of a highly successful LHS. Patients are able to learn from their data by completing a series of a patient-reported outcome measurement surveys online or at the clinic. The compiled data can be used to build a comprehensive picture of how an individual's pain scores vary over time, relative to their current treatments.

Clinicians

The national rheumatoid arthritis quality register also provides rheumatologists with additional data relating to their patient, as well as to other patients, suitably de-identified (Øvretveit et al., 2013).

Service Delivery Managers

A number of health systems and hospitals provide hospital managers with "performance dashboards" that present key data about costs, patient satisfaction, and clinical outcomes on a visual display.

Strategic Planning and Population Health Management

In order to target community-based population health, "hot spots" and other mapping systems use data from the census and the healthcare system to identify small geographic areas with a large number of patients using emergency departments (Diamond et al., 2009).

Researchers

Part of the concept of an LHS is that data generated in the course of everyday healthcare can be used for more timely and lower-cost research, as well as for research that would not otherwise be possible. For example, data from the orthopedic register has been used to identify those procedures and prostheses that have high reoperation rates. This has led to Sweden currently having among the lowest international reoperation rates and costs (Larsson et al., 2012). More of these national registers are extracting data from the patient's EHRs and are also incorporating patient-reported outcome measures into the EHRs.

Transferability

Sweden has only been able to achieve these benefits and further develop its national and local LHS because of a number of complementary conditions and initiatives. Other countries and health systems would need to assess whether they have these conditions or what they might do to create similar conditions in order to build an LHS to collect and make effective use of the data.

In 2015, a national organization for governing all Swedish healthcare quality registers, Nationella Kvalitetsregister, was created. This centralized body supports the use of data for research and quality improvement in hospitals and other services, and ensures that both time and resources are dedicated to developing nationwide registers. Data from all registers can be accessed from a central website: http://kvalitetsregister.se/englishpages/findaregistry/allswedishqualityregistries.2028.html. Individual hospitals and county health systems are able to use the register's systems and link data to their own EHR data to allow targeted analyses and presentation of information for patients and the other users described above.

A key lesson for others is that LHSs need to be designed for, and with, users of the system. Users and technical experts need to identify those decisions that can be better informed by data analyses, and design programs that allow these analyses to be utilized on a practical level. This requires investment in design teams and in software at the national, county, and hospital levels.

To build an LHS, leaders will need to know how digital data can enable more fact-based decision-making in routine operations and in strategic planning. They will need to identify priorities for designing dashboards and other analytic systems to present the data in an actionable form, at the time and place it is needed. A strategic approach is needed. Managing data is becoming as important to the operations of future healthcare services as managing personnel. Effective use of data will decide which services survive in competitive markets and which services are able to improve their care to patients and reduce costs.

Analysis of how health systems are building an LHS reveals that a distinct transformational infrastructure is needed to ensure that consistent effort and investment is continued over a period of time before the LHS delivers benefits to the users noted above (Øvretveit et al., 2016). Such an infrastructure is comprised of three parts:

1. A structure of accountability for building an LHS that involves leaders and users at each level of the organization, designers, and information technology (IT) staff
2. A strategy of actions over time carried out by leaders and project teams to prioritize subjects and build the systems
3. A system of supports and resources to help build the LHS over time

In Sweden, this transformational infrastructure was made up of the national governance organization for the quality registers, and government and regional groups for building the data exchange systems, and developing the legal and regulatory policies. In addition, distinct structures within counties and hospitals also support the development of systems to collect data from EHRs, and link to other data to perform some of the functions described earlier.

Conclusions

Sweden is advanced in its information technology infrastructure; the population's computer literacy is also high. The "digitalization of health" has been slow but is now gaining speed due to a number of influences, not least the imperative to reduce costs, improve quality, and meet expectations that consumers bring to healthcare from their experience of other services. This chapter considered how the country is building an LHS that uses digital data collected in the process of everyday healthcare to inform a variety of decisions and to learn how to more effectively provide services and enable self-care.

Sweden's system can be translated to other countries and health systems, but a key consideration is to build an infrastructure that allows data exchange and database linking. Realizing the benefits also requires technicians to partner with users to develop ways to present the information in the best possible form to support medical decision-making. Such systems make more timely and lower-cost research possible, and provide additional research opportunities. Researchers have an important role in designing these systems, and in helping to ensure that unwarranted causal conclusions are not drawn from the correlational analyses that are increasingly used in "big data" analytics assessments.

33

Switzerland

Teamwork and Simulation

Anthony Staines and Adriana Degiorgi

CONTENTS

Swiss Data

- Population: 8,372,098
- GDP per capita, PPP: $62,881.5
- Life expectancy at birth (both sexes): 83.4 years
- Expenditure on health as proportion of GDP: 11.66%
- Estimated inequity, Gini coefficient: 31.6%

Source: All data are from the World Health Organization and World Bank. Latest available data
used as at October 2017.
GDP = Gross Domestic Product
PPP = Purchasing Power Parity

Quality Improvement in Swiss Health and Healthcare

Switzerland is a confederation of 26 states (cantons), and its official languages define the country's four regions: French, German, Italian, and Romansh. Health, specifically healthcare quality, is mainly the responsibility of the cantons. This has led to a diversity of regulations, associations, and initiatives in health in general—particularly when it comes to health improvement approaches. Synergies at the national level are often a challenge because of language barriers. A debate is currently underway about the confederation taking a stronger role in this regard. Patient safety, however, is one area that receives the highest degree of leadership at the national (system-wide) level, through the auspices of the Swiss Patient Safety Foundation. National breakthrough collaboratives have been organized on such topics as surgical safety, medication reconciliation, and bladder catheterization (Staines et al., 2017), and a voluntary incident reporting system is also coordinated by the foundation.

At the national level, quality indicators have been developed by the National Association for the Development of Quality in Healthcare, and are now publicly reported. Their number is still modest, but they are slowly gaining acceptance.

At the meso-level, there have also been some regional breakthrough collaboratives (Staines et al., 2017). Efforts to reduce treatments that are not evidence-based are being made under the umbrella of the Choosing Wisely initiative.

At both the meso- and the micro-level, improvement initiatives have included tools and strategies such as care maps, Lean process improvement, the model of the European Foundation for Quality Management, voluntary accreditation, and peer review. To facilitate procedures where improvements in quality demand scale, highly specialized medicine has been coordinated through agreements among all the cantons.

Switzerland is among the countries devoting the highest proportion of gross domestic product to healthcare. It is therefore under pressure to contain costs and increase efficiency. However, the Swiss population holds its health system in rather high regard and believes its healthcare to be high quality. Social forces therefore oblige politicians to maintain the status quo and uphold the current healthcare infrastructure and system. While healthcare professionals have become used to the presence of quality and safety initiatives, such professionals frequently regard these as constraints to their professional autonomy or as attempts to have them focus on compliance to rules and guidelines. The digitalization of healthcare will influence improvement efforts, although Switzerland and its healthcare providers are not at the forefront in this area.

Team Training and Simulation

Team training and non-technical skills training through simulation are likely to be major contributors to patient safety and quality of care in the next decade. Salas et al. define team training as "a set of tools and methods that, in combination with required [team-based] competencies and training objectives, form an instructional strategy" (Salas et al., 2008: 906). The authors describe typical content as including "mutual performance monitoring, feedback, leadership, management, coordination and decision-making" (Salas et al., 2008: 909). The field of aviation has provided models of team training through its concept of crew resource management. When applied to healthcare, that approach has led to the goal of using all available resources (information, equipment, and people) to obtain quality and safety of care. Simulation, which has additional benefits, has been defined as "a technique that creates a situation or environment to allow persons to experience a representation of a real event for the purpose of practice, learning, evaluation, testing, or to gain understanding of systems or human actions" (Lopreiato et al., 2016: 33).

Teamwork training has been shown to have a positive impact on team performance (Salas et al., 2008). Recently, studies have also shown the impacts on outcomes, such as reduction in mortality (Forse et al., 2011; Neily et al., 2009). Compared with no intervention (Cook et al., 2011), simulation training of health professionals has been shown to be associated with substantial effects on knowledge, behavior, and skills, and in addition has moderate effects on patient outcomes.

Team training pilot projects have been carried out in some Swiss hospitals, and the Team Strategies and Tools to Enhance Performance and Patient Safety (TeamSTEPPS) concept is spreading (Agency for Healthcare Research and Quality, 2017). These projects rely less on compliance and more on developing resilience than previous patient safety initiatives, an advance that is promising for both complementarity with other initiatives and acceptance by professionals. The concept fits with Hollnagel's urge to move from Safety I, the reactive approach of focusing on counting adverse events and learning from them, to Safety II, the resilient approach of focusing on developing the ability to succeed under varying conditions (Hollnagel, 2014).

Simulation was initially oriented toward practicing technical skills on models of patients or body parts. It has grown to include non-technical skills training, such as teamwork and communication skills. Simulation is complementary to teamwork training.

Impact

Over the coming years, the Swiss health system is likely to continue to centralize specialized medicine. This will lead to the emergence of new subspecialties and the concentration of technology. To maintain job attractiveness for health professionals, shorter work hours will become the norm, leading to greater staff numbers and many more handovers. The average length of hospital stay will continue to decrease, leading to higher patient turnover. A projected lack of trained professionals is likely to lead to the delegation of new tasks to new professions and to hiring professionals trained in other cultures and languages. The pace of work will probably increase, with greater reliance on technology. All this will add to the high complexity inherent to healthcare, leading to archetypal complexity. Success is likely to come from the mastering of such complexity. We believe that teamwork and simulation have the potential to contribute positively to this situation.

Both team training and simulation promote team building, preparation, rehearsal, and reflection on processes. Teams practice in a protected, controlled environment until the procedures become natural. They are encouraged to identify pitfalls and are trained to avoid them. Team training and simulation are generally well received by clinicians, who consider them appealing activities.

In Switzerland, health professionals are perceived to be well trained. However, there is no reason to believe that Swiss healthcare generates fewer errors or adverse events than that of other countries. This suggests that the potential for improvement probably relies more on teamwork skills and attitudes than on clinical knowledge.

Patients' expectations regarding care coordination are likely to rise over the coming decade, as will the expectations of health professionals about multidisciplinary care. Retention challenges will incentivize providers to meet such expectations. The idea of "never the first time on a patient" is also increasingly influential among professionals as well as the general population, encouraging the use of dummies and simulated environments.

In the TeamSTEPPS model (Agency for Healthcare Research and Quality, 2017), the patient is part of the team. Most simulation scenarios are built around patients and their needs. At a time when healthcare is developing strategies to increase patient involvement and patient-centeredness, these concepts are likely to be well received.

An important factor in the widespread application of team training and simulation in the coming decade will be the capacity of research to produce evidence of its positive impact on patient outcomes. Such evidence would encourage leaders to invest in this area and increase the willingness of professional associations and scholarly societies to recommend it.

Spreading Team Training and Simulation in Switzerland

We believe that the first stage in spreading team training and simulation throughout Switzerland will involve the launch of pilot projects, mainly in highly complex environments, in an increasing number of healthcare organizations. Next, major organizations will establish their own teamwork training program and set up simulation training facilities. The numbers of such programs and facilities are then likely to expand. It is anticipated that smaller organizations will pool their training resources. The final stage will involve the integration of teamwork training and simulation into the undergraduate curriculum, taking the opportunity to mix different streams of professions in common multidisciplinary training. This type of integration is already underway in some programs.

The demand for team training and simulation is likely to increase, while budgets are likely to stagnate. Fortunately, it is possible to start small; a simulation center is an asset, but not an absolute prerequisite. The key is to appoint well-trained instructors, coaches, and change agents. The training time of healthcare professionals will be the most significant expense.

We believe that endorsement by the Swiss Patient Safety Foundation, as well as by scholarly societies, would be an impetus to the dissemination of team training and simulation. It is probable that the requirements for continuous education and skills updating will increase, and as Swiss culture values professional education, teamwork training and simulation may well become an accepted training component.

If teamwork training and simulation become embedded in undergraduate education (including multiprofessional streams), and if most healthcare organizations include these elements in their continuing education, they are likely to become part of the Swiss healthcare culture.

Once all professionals have been educated in teamwork and with continuing education in place, there is the potential to move to new levels of care. Healthcare institutions could use simulation scenarios based on their own critical incidents and identified risks. Undergraduate education and regionally pooled simulation centers could develop standard teamwork skills, and organizations could develop internal training customized to address their own risks.

Prospects

Further success may be linked to the integration of team training and simulation in the design and implementation of a global quality and safety improvement system. It will also require the fostering of a learning culture, research and measurement, and leadership support. Research coordination may also help accelerate the production of robust evidence. Cooperation between academic institutions and healthcare organizations will be necessary and could benefit from national support.

The spread will require the development of an organization or a network able to respond to emerging requirements, such as developing pedagogical material in all official Swiss languages, as well as providing centers to train the trainers and promote the exchange of experience through conferences. Such steps will also enhance visibility. Branding, similar to that of TeamSTEPPS (Agency for Healthcare Research and Quality, 2017), may help in that respect.

The success of team training and simulation will also be determined by the promotion of a just culture (Frankel et al., 2006). In such a culture, professionals are not afraid to report errors, and organizations strive to learn through analyzing errors and risks in a simulated environment. Such organizations rehearse strategies to prevent errors and mitigate adverse events through early detection and mutual support.

Conclusion

In Switzerland, team training and simulation are likely to spread and develop over the next decade. They will not be a substitute for other quality improvement approaches, which will still be required. They may, however, become the basis on which other initiatives will be founded. Team training and simulation are reliant on a just culture, but they will also contribute to the spread and implementation of such a culture.

Patient safety is evolving. It began with an emphasis on detecting harm and on complying with rules that reflected best safety practices, while avoiding naming, blaming, and shaming. Teamwork and simulation emphasize all that is positive: developing resilience to errors through team detection; mutual support within the team; reflection on safe practices, processes, and pathways through simulation; and becoming a learning organization.

Teamwork and simulation combine two fundamental approaches in patient safety: multiprofessional cooperation (teamwork) and individual accountability (acquiring and using teamwork skills). They therefore have the potential to boost patient safety and its acceptance by healthcare professionals.

34

Turkey

Moving Quality in Healthcare Beyond Hospitals: The Turkish Accreditation Model

Mustafa Berktaş and İbrahim H. Kayral

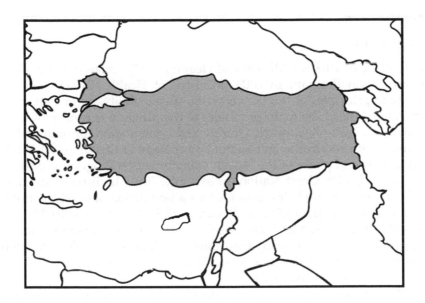

CONTENTS

Turkish Data

- Population: 79,512,426
- GDP per capita, PPP: $24,243.9
- Life expectancy at birth (both sexes): 75.8 years
- Expenditure on health as proportion of GDP: 5.41%
- Estimated inequity, Gini coefficient: 40.2%

Source: All data are from the World Health Organization and World Bank. Latest available data
used as at October 2017.
GDP = Gross Domestic Product
PPP = Purchasing Power Parity

Background

The Republic of Turkey's Ministry of Health (MoH) declared a new Health Transformation Program in 2003 (Ministry of Health [Turkey], 2003). In addition to principles of human centrism, sustainability, participation, and competition, one of the main principles of the program is continuous quality improvement in healthcare. Quality and accreditation for qualified and effective health services is an essential component of this program (Akdağ, 2008). The second phase of the Health Transformation Program is currently underway (Turkish Great National Assembly, 2016). In Turkey, diversified healthcare services are being provided for a population of nearly 80 million people within more than 1,500 healthcare facilities by around 800,000 healthcare staff. With limited and diminishing resources, it is obvious that programs will need to be critically transformed to increase the efficiency of the administration of such large-scale healthcare services (General Directorate of Health Research [Turkey], 2016).

The MoH has already enacted multiple legislative regulations to improve patient safety and satisfaction, helping to ensure health quality and patient safety on a national basis. Efforts have also been made to manage a number of risks at the micro-level, such as those associated with severe malpractice problems, surgical errors, and so on. However, more needs to be done, and continued government commitment is necessary if the quality of healthcare services is to be maintained.

Although maintaining a high standard of healthcare service also depends on many other factors, including those of a sociocultural, political, economic, and technological nature, one of the most challenging issues is the need for an overall transformation of healthcare service quality and patient safety. Such a major cultural transformation will only succeed if the necessary reforms continue to be made, and essential health sector infrastructure continues to be strengthened.

One way in which the health system can be significantly strengthened is through the implementation of a reliable and comprehensive accreditation system. The pressure of competition, along with an increase in patient expectations, encourages the implementation of such programs. The increase in patient expectations also requires a simultaneous consideration of both the quality of healthcare services and related cost issues (Kayral, 2014).

Future Reflections

Turkey has made great progress in improving the provision of quality healthcare, particularly when it comes to acute care in hospitals. In the course of this process of improvement, important knowledge has been gained. But this knowledge alone is not enough, and successful reform outcomes should be extended to other areas where healthcare is provided.

The standardization and accreditation of health services outside the hospital system is necessary to ensure the delivery of high-quality healthcare. For instance, improved and accredited school health services could effectively target and promote the health and well-being of students and school personnel. Standardization and accreditation is also essential where medical care is delivered in establishments such as nursing homes and child protection institutions, hotels and motels, boarding houses, and shopping centers, all of which cater to a great volume of people.

There are nearly 16 million students in 61,000 schools in Turkey. Twenty-six thousand of these are primary schools, and the total number of students studying in these schools is in excess of 5 million. Nearly 5 million students attend Turkey's 17,000 middle schools. Eleven thousand high schools cater to more than 4 million students (Ministry of National Education [Turkey], 2016). Schools can become the site of both health instruction and practice: health education can be taught, along with sports and physical education; bathrooms and the general school environment can be made more sanitary; and first aid services can be provided for staff and students.

Shopping centers, too, can more actively promote health. There are currently 346 shopping centers in Turkey, and millions of people take advantage of the services they provide (Council of Shopping Centers, 2016). The creation of healthier environments in these centers, including better ventilation, stricter building codes, the upkeep of grounds, improved sanitation, and the provision of efficient first aid care for emergencies, will further ensure shoppers' well-being.

Healthcare reform in Turkey began with the institution of quality standards in 2005. These efforts were designed to produce a high level of quality within healthcare institutions. As a result of these improvements, a new obligation has arisen to design better healthcare in accordance with the national accreditation system, which has been built to conform to international principles. These principles were first endorsed in 2012, and the process of accreditation by the International Society for Quality in Health Care (ISQua) began in 2013. Since this date, a series of workshops and meetings have been held to decide on the nature of standards set for the accreditation program. These meetings culminated in the issuing of standards of accreditation in health by the MoH, to be accredited by ISQua, in 2013. A number of other accreditation initiatives have been launched since; these include setting standards for dental health, dialysis, and medical laboratories (Department of Quality and Accreditation in Health, 2015). These new standards will encourage health institutes to produce their own solutions for a variety of health challenges, including patient safety, efficient management, and promoting a healthy work–life balance.

Following the establishment of an accreditation system for health institutions, accreditation programs will be developed for those institutions that provide health services, but whose primary purpose is not necessarily medical. Initially, the types of institutions that are most in need of such programs must be selected and standards for these institutions developed. Pilot studies should then be initiated in these organizations. As a result, accreditation programs can be finalized and the standards implemented.

Another important issue that will be considered in these reforms is quality culture. Supportive programs should also be developed for the promotion of a quality culture in health institutions, to help ensure that the regulation process is as straightforward as possible. Risk management strategies and continuous improvement tools can be implemented to help integrate quality culture.

Impact

Health services need to be standardized, accredited, and of the highest quality, whether care is provided in hospitals or in other community settings, such as schools and shopping centers. It is also important to take precautions to ensure that the incidence of accidents is reduced. Healthcare centers themselves need to adhere to health and safety regulations, as patients are potentially at risk from falls, cross-infection, and so forth. If standardization of medical services and quality culture is properly established in institutions such as hospitals, other institutions will ultimately benefit. In order to facilitate the spread of these improvements, a number of awareness-building efforts, such as short films, project awards, and good practice awards, will be instituted.

Implementation and Transferability

The standards set in health institutions will also be sought wherever health services are provided. Community awareness in this matter will ensure that accreditation programs are demanded.

Raising public awareness through advertising campaigns, good practice awards, and so forth will strengthen these services and ensure that service provision adheres to the highest possible standards, thus promoting longevity.

Prospects

Accreditation programs for health institutions will need to be widespread if targeted programs are to be developed. As a first step, standardization in health institutions should become a permanent expectation and part of the health culture. If accreditation programs are to be successful in other non-health institutions, it will be necessary to develop legal arrangements to implement accreditation programs for these institutions and to cooperate with the MoH and health institutes, in addition to other ministries and agencies. There should be a constant flow of information relating to accreditation— updates, changes to regulations and definitions, news of improved outcomes, and so forth—to all stakeholders via websites and public announcements.

Good practices introduced in non-health organizations can be presented and demonstrated in forums such as congresses, fairs, and conferences, to provide information to other countries. Thus, sharing of experiences will allow different countries and cultures to embrace these programs and develop similar programs using different examples.

Conclusion

Turkey has come a long way in improving the quality of healthcare, and the process of reform has added to the knowledge across the field. Improvement in healthcare in Turkey was initiated by the introduction of quality standards in 2005. These efforts were designed to produce a certain quality of care within healthcare institutions; however, these efforts alone are not enough. In order to improve the health of the population further, the knowledge that has been acquired during the period of reform, particularly in hospitals, should be extended to other areas where healthcare is provided.

Schools, for instance, should protect and promote the health of students and school personnel. This will involve a comprehensive regulatory program that ensures the provision of a safe school environment and health education, along with first aid services. Nursing homes, accommodation facilities such as hotels and motels, child protection institutions, and shopping centers should also seek accreditation. The provision of safe, high-quality healthcare is not just the responsibility of hospitals; it should be developed and supported in all public domains.

35

Wales

Realizing a Data-Driven Healthcare Improvement Agenda: A Manifesto for World-Class Patient Safety

Andrew Carson-Stevens, Jamie Hayes, Andrew
Evans, and Sir Liam Donaldson

CONTENTS

United Kingdom Data

- Population: 65,637,239
- GDP per capita, PPP: $42,608.9
- Life expectancy at birth (both sexes): 81.2 years
- Expenditure on health as proportion of GDP: 9.1%
- Estimated inequity, Gini coefficient: 32.6%

Source: All data are from the World Health Organization and World Bank. Latest available data used as at October 2017.

Note: Data from England, Scotland, Wales, and Northern Ireland.

GDP = Gross Domestic Product

PPP = Purchasing Power Parity

Background

Wales has a population of 3 million people. Its government provides a publicly-funded healthcare system within the framework of the United Kingdom's National Health Service (NHS). The Organisation for Economic Co-operation and Development's (OECD) review of healthcare quality in the United Kingdom (2016a) concludes that NHS Wales is set up to achieve quality improvement. Responsibility for assessing health needs, planning services, and organizing care rests with health boards covering Wales. In addition to planning services, health boards are responsible for the delivery of the majority of secondary care services; however, they also work in partnership with NHS trusts, primary healthcare providers, local authorities, other public bodies, and private-sector organizations to secure and deliver the range of services needed for their populations. Wales has an integrated healthcare model informed by a planned approach aligned to policy, delivery, and improvement. Since 2015, health policy in Wales has emphasized the importance of "prudent healthcare," an approach through which healthcare is coproduced by providers and users with the aim of simplifying care and reducing harm and waste.

Safe High-Quality Primary Care in an Era of Universal Coverage

More than 90% of healthcare encounters in Wales take place in a primary care setting, yet little is known about the safety of such encounters. In 2012, the World Health Organization (WHO) convened a Safer Primary Care Expert Group, which concluded that there was a paucity of research on the nature and burden of healthcare-associated harm in primary care around the world (Sheikh et al., 2013). The priority given to universal health coverage by the WHO and other global agencies, together with the continuing emphasis on primary care as the bedrock of successful health systems, is prompting debate about how to design high-quality and safe, yet cost-effective care models (Evans et al., 2013). Wales has the potential to delineate systemic risks to patients receiving primary healthcare services, and to share this knowledge with other countries. It is also accumulating considerable experience in driving system-wide quality improvement in healthcare.

Fundamental Importance of a Culture of Learning

The 1000 Lives Campaign in Wales has achieved substantial reductions in harm and mortality through interventions, including care bundles and checklists. This work, which has yielded a decade's experience and learning, has so far focused mainly on hospital care. Wales now has the opportunity to extend its improvement programs to establish data-driven primary care patient safety and improvement.

This approach would align well with systems scientist Peter Senge's concept of the learning organization (Senge, 1990). Senge sees this as "an organization that exhibits adaptability, learns from mistakes, explores situations for development, and optimizes the contribution of its personnel" (Wilkinson et al., 2004: 105–106). Such organizations require infrastructure to facilitate a range of activities and processes that will establish a culture of learning. Major investments have been made in countries and health systems to recognize that a key part of this infrastructure is effective patient safety incident reporting systems. They provide a conduit for learning and the prevention of harm.

Existing incident reporting systems continue to disappoint. For example, the National Reporting and Learning System in England and Wales, which has collected 15 million incident reports over more than a decade, cannot show reductions in risk, harm, serious error, and death commensurate with the effort and costs of the system (Carson-Stevens et al., 2015). A 2013 review of safety and quality of NHS care in England by Professor Don Berwick, the preeminent global leader in this field, *A Promise to Learn—A Commitment to Act: Improving the Safety of Patients in England*, reaffirmed that organizational learning is the key to improving patient safety (Berwick, 2013). The chief medical officer for England's report "An Organisation with a Memory," more than a decade earlier, had said the same thing (Department of Health [United Kingdom], 2000). Few informed observers would dissent from the view that the NHS has been slow to understand how to generate and act on learning from healthcare-associated harm.

This message of missed opportunities is not unique to the United Kingdom—there are few places where incident report data are being more effectively utilized to inform improvement, or as Macrae (2016: 74) puts it, "We collect too much and do too little."

Only one in three patient safety reports from primary care in England and Wales include contributory factors that describe the circumstances, actions, or influences on the development or risk of an incident (Runciman et al., 2009). In the absence of this detail, the ability to understand and prevent harm is limited. One patient safety incident occurs every 26 seconds in the NHS in England and Wales. This chilling statistic calls for a transformation in the effectiveness of learning from unsafe care. The Primary Care Patient Safety Research Group at Cardiff University has developed a data-driven improvement model that is informed by analyses of patient safety incident reports (Figure 35.1).

This data-driven learning system asks the following:

- How can we maximize the usefulness of data provided by staff?
- How can analysis of these data regularly inform our improvement agendas?
- How can we engage staff with data they have provided?
- How can we demonstrate that we have acted on their concerns? (Williams et al., 2016)

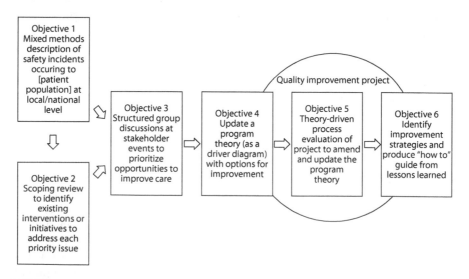

FIGURE 35.1
A theory-driven approach for evaluating improvement developed by the Primary Care Patient Safety Research Group at Cardiff University. (From Carson-Stevens, A. P., Generating learning from patient safety incident reports from general practice, Doctoral dissertation, School of Medicine, Cardiff University, 2017.)

Finding answers to these fundamental questions will require a functioning patient safety incident reporting and learning system employing state-of-the-art data science methods to analyze and visualize data. It will also mean that staff will have to be proficient in quality improvement methods to examine system structures, processes, culture, and performance. This will ensure that there is no duplication of effort, that opportunities are taken to design new ways of working, and the means to give every patient high-quality care is identified (Leatherman et al., 2010).

A data-driven learning organization must have robust mechanisms for ensuring that patient safety incidents in care are reported, systematically interrogated in near real-time, and used to redesign care processes. Automated classification of incident reports will be essential to free up time for risk managers to investigate and thoroughly understand underlying contributory factors, and to work with leaders on improvement initiatives. Staff commitment will not be sustained unless they have confidence that they are being listened to and believe that the organization is committed to becoming a safer place to work and deliver services. The organization will be seeking to corroborate its insights from existing research studies, as well as developing novel and innovative solutions for practice.

Prospects

Fear of blame is a major barrier to promoting an underlying safety culture in health systems. As a small nation, with devolved responsibility for health and social care, Wales has started to acknowledge the shift from rewards, punishments, and pay-for-performance, to supporting healthcare professionals who are, above all, motivated by patient-centered values and learning (Berwick, 2016). This approach has been formalized through learning networks and communities of practice involving multiple stakeholders and national agencies. Wales now needs to create an environment that has strong teamwork, communication, and situational awareness.

At present, there are numerous channels to report patient safety incidents. These systems need to interact to achieve a more complete national picture of preventable harm. Healthcare leaders and managers must demonstrate commitment to minimizing blame-laden reporting to achieve a culture of open reporting among healthcare professionals and staff in primary care. This should extend to patients and carers.

National-level analyses of patient safety incidents can identify priorities for quality improvement in primary care, particularly where there is no guidance from existing patient safety organizations (Cooper et al., 2017, Williams et al., 2015). Rigorous methodological work on an existing national database has shown how these data can be improved, with richer narratives from front-line staff yielding learning for improvement purposes. In the future, the health professional workforce must receive patient safety education and training about the rationale for reporting and awareness of human and system factors contributing to harm.

Conclusion

The value and utility of a patient safety incident reporting system as a key informant for an evolving data-driven healthcare improvement agenda in Wales is now being recognized. NHS leaders, their organizations, professionals, and staff must now seek to maximize what can be learned from patient safety incidents to prevent healthcare-associated harm and make their services safer.

36

Central and Eastern Europe

Strengthening Community-Based Family Care and Improving Health Equities

Jeffrey Braithwaite, Wendy James, Kristiana Ludlow, and Russell Mannion

CONTENTS

Central and Eastern European Data

- Population: 448,969,224
- GDP per capita, PPP: $22,727.2
- Life expectancy at birth (both sexes): 78.4 years[a]
- Expenditure on health as proportion of GDP: 7.4%[a]
- Estimated inequity, Gini coefficient: 31.3%

Source: All data are from the World Health Organization and World Bank. Latest available data
used as at October 2017.

[a] Data unavailable for Macedonia and Kosovo.

GDP = Gross Domestic Product

PPP = Purchasing Power Parity

Background

This chapter focuses on Central and Eastern Europe (CEE)—the group of countries that make up the most easterly part of Europe. While definitions vary, CEE covers up to 27 countries (Table 36.1), from Germany and Austria in the west, to the Baltic Sea and Finland in the north, across to the border of Greece in the south, and extending to the east all the way through to Russia. Elsewhere in the book, we have provided individual chapters from Germany, as well as Austria, Estonia, and Russia, and we cover Kazakhstan in our chapter on Central Asia.

Although the CEE countries are grouped geographically and are frequently regarded as a bloc for political purposes, they are not homogenous. The group is economically diverse, ranging from lower-middle-income countries, such as Moldova, Armenia, Georgia, and Kosovo, to upper-middle-income countries, such as Croatia, Serbia, Bulgaria, and Bosnia-Herzegovina, and higher-income countries, such as Poland, Slovakia, and Slovenia. However,

TABLE 36.1

Classification of CEE Countries and Territories, with Population Data (Millions)

Countries Most Commonly Classified as CEE		Countries Sometimes Classified as CEE
- Poland, 37.9	- Albania, 2.9	- Russia, 144.4
- Romania, 19.7	- Lithuania, 2.9	- Germany, 82.7
- Czech Republic, 10.6	- Slovenia, 2.1	- Ukraine, 45.0
- Hungary, 9.8	- Macedonia	- Kazakhstan, 17.8
- Bulgaria, 7.1	(FYROM), 2.0	- Azerbaijan, 9.8
- Serbia, 7.0	- Latvia, 1.9	- Belarus, 9.5
- Slovakia, 5.4	- Kosovo, 1.8	- Austria, 8.7
- Croatia, 4.1	- Estonia, 1.3	- Georgia, 3.7
- Bosnia and	- Montenegro, 0.6	- Moldova, 3.5
Herzegovina, 3.5		- Armenia, 2.9

as former Soviet bloc or Soviet-influenced nations now finding their place in Europe, they share common ground in their efforts to make economic, intellectual, and social progress individually, and as a region, for mutual and bilateral benefit. While many CEE countries are European Union (EU) member states, others, such as Bosnia-Herzegovina, Albania, and Montenegro, are still in the process of negotiating their accession.

Challenges

As we show throughout this book, while all health systems exhibit common reform features (Braithwaite et al., 2017a,b), and every reform process in every health system is a variation on similar themes, there are distinctive challenges in CEE countries. In this region, old Soviet-inspired politics frequently hold sway when dealing with health system matters, often trumping evidence-based change. Evaluation of reform is inconsistent and infrequently conducted (Braithwaite et al., 2015, 2016); however, where sound evaluations of progress have been conducted (see, e.g., Jakovljevic et al., 2017; Rechel and McKee, 2009), considerable differences emerge between the various systems' structure, financing, performance on the human development index, life expectancy, and other measures, such as the proportion of out-of-pocket expenses relative to GPD allocations for health. Where the CEE countries differ most, however, is in their approach to health system inequities and in their ongoing efforts to strengthen primary healthcare and family medicine, which the World Health Organization (WHO) has argued are cornerstones of an effective, universal health system.

Natural Health Policy Laboratory

This part of Europe has been described as "a remarkable but inadequately exploited natural laboratory for studies of the effects of health policy" (Mackenbach et al., 2013: 1125). Like most of the countries presented in this book, the recent history of CEE health systems is one of change and reform under challenging conditions; many CEE countries have had to completely transform their traditional monolithic, centralized, state-funded systems to fit the requirements of their new EU status over a relatively short time.

In terms of inequities, there is clearly an east–west divide in Europe, with poorer life expectancy, more all-causes of death, worse mortality trends, and insufficient allocations of national GDP to healthcare most prominent across CEE countries. Whereas the average life expectancy for Europe as a whole is over 80 years (and in the mid-80s for women in some wealthy Western European countries), in the CEE region it can be as low as 60 years (WHO, 2016a). Thus, although there are differences in the patterns of care and delivery by country, as a bloc, CEE citizens clearly do not enjoy the same benefits as those in wealthier, more effective health systems (Greer et al., 2013; Mackenbach et al., 2013; Rechel et al., 2013).

There have been some successes, albeit qualified ones. At the fall of the Soviet Union, there was little by way of tobacco control policies, the management of alcohol at national levels was sporadic, and there were few initiatives to promote decentralization of systems to empower regions and local communities. The proportion of the population eating a balanced diet predicated on healthy choices, including fruits and vegetables, was low. Over time, in response, in Moldova, for instance, health resources have been allocated more equitably, and Ukraine, Georgia, Armenia, and Uzbekistan have slowly decentralized their systems of governance and some fiscal allocations (Rechel and McKee, 2009; Rechel et al., 2013). Rates of infectious diseases have decreased, and road safety has improved in most CEE countries due to targeted EU investment and infrastructure programs. Most CEE countries have introduced private health insurance, with some using out-of-pocket payments to complement taxpayer-financed care. And some CEE countries have made efforts to strengthen primary care via general practice and family medicine (Rechel and McKee, 2009).

If health systems in the CEE region are successful in putting effective primary care and family medicine at the heart of the system (Groenewegen et al., 2013; Oleszczyk et al., 2012) and narrowing health inequities (Mackenbach et al., 2013; World Bank, 2014), their populations would benefit immeasurably, with services being delivered closer to patients, and fewer lives lost (Shi, 2012; Starfield et al., 2005).

Progress

What has been the progress with these twin initiatives—family, community-based care and tackling inequities—and the impact of success? Not as much as we would like is the short answer. Ten years ago, Seifert et al. (2008) called for improvements in primary care based on more effective family medicine across the CEE region, adding to earlier recommendations from Svab et al. (2004). They noted that even though family medicine was more likely than in the past to be embedded in university curricula, postgraduate training had been established, recertification of clinicians was more likely, and guidelines had been introduced, progress was uneven and slow. To build momentum and accelerate progress, they argued for better-coordinated exchange of information and collaboration within the CEE region and beyond, publications of successful case study exemplars, and further support for the family medicine discipline and its contributions to improved care.

Greater political, policy-maker, and health systems commitment is required if inequities within countries and across the CEE region, are to be addressed. Following a recent health reform round table in Vilnius, Lithuania, UK policy analyst Tessa Richards summarized the problems: weak primary healthcare, insufficient involvement of patients, too many inpatient beds (consuming more resources than the acute sector should in a balanced health system), and corruption. She also pointed to where the solutions lie: in collaborating for common purpose, benchmarking against best (or even, just better) performance, and changing the public's prevailing perspective that health is the responsibility of the government rather than the individual (Richards, 2017).

Implementation and Transferability

There are pockets of the CEE where more effective primary care and family medicine have been provided. There is evidence that Croatia, Lithuania, Macedonia, and Moldova have attempted to strengthen their community-based care systems. Latvia (Tragakes et al., 2008) and Estonia (Koppel et al., 2003; Lember, 2002) report increasing public satisfaction with their primary care services. In regard to equity of care, access, and resource allocations, there have also been some improvements. Slovenia, which spends more per capita on healthcare than its CEE peers (OECD, 2015b), has enjoyed a rise in life expectancy, as has Croatia (WHO, 2016b). Maternal mortality has been reduced in Bulgaria and Macedonia (World Bank, 2016c). CEE countries can learn from each other, and benefit from the experiences of those countries that have made progress, when things are demonstrably improving.

A key success factor when changes have been made appears to be the extent to which a CEE country is oriented toward EU policies and has access to EU knowledge and funding. Such access helps strengthen knowledge and can stimulate greater levels of political willingness to improve the health status of the population. Indeed, should the CEE region as a bloc be able to approximate the health systems of Western Europe over the next 5–15 years, or even make encouraging progress toward this, the gains would be immense.

Prospects

Theoretically, the CEE region could achieve their hoped-for gains; however, there is not as clear a trajectory as many would like to see. The development and implementation of policies designed to strengthen family medicine and narrow disparities, we argue, would be the logical place to start. Reforms in these areas would inevitably benefit the 362 million people in the region, help build momentum for other gains, and raise standards and morale across the countries of the bloc.

Conclusion

The opportunities for improvement in the CEE region are immense. Human capital can be built, and countries can learn from each other as progress, or even modest change, is made. Further models for change and case studies are available to the west of the CEE, in the rest of Europe. Overall, there are signs of progress, but more is needed. The peoples of the CEE deserve nothing less.

37

Central Asia

From Russia with Love: Health Reform in the Stans of Central Asia

Jeffrey Braithwaite, Wendy James, Kristiana
Ludlow, and Yukihiro Matsuyama

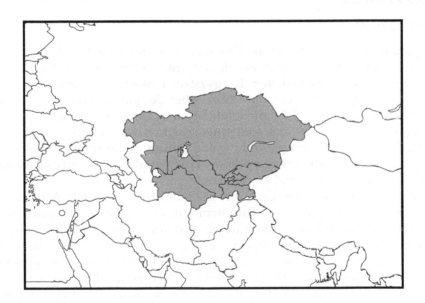

CONTENTS

Central Asian Data

- Population: 70,125,427
- GDP per capita, PPP: $11,038
- Life expectancy at birth (both sexes): 69 years
- Expenditure on health as proportion of GDP: 5.1%
- Estimated inequity, Gini coefficient: 40.2%

Source: All data are from the World Health Organization and World Bank. Latest available data
used as at October 2017.
GDP = Gross Domestic Product
PPP = Purchasing Power Parity

Background

Triggered by the collapse of the Union of Soviet Socialist Republics (USSR) in 1991, a cascading series of events occurred that meant that the Central Asian countries of Kazakhstan, Kyrgyzstan, Tajikistan, Turkmenistan, and Uzbekistan separated from the USSR. After decades of being part of the Russian Empire (1721–1917), and later the Soviet Union, the shift to independence was a significant and sometimes challenging undertaking. Each of these countries has a complex ethnic and cultural composition: while many citizens have Russian heritage, others in the population identify as Eastern Turkic, Eastern Iranian, and Mongolian.

Once independence was declared, the four original Central Asian countries (Kyrgyzstan, Tajikistan, Turkmenistan, and Uzbekistan) joined with neighboring Kazakhstan, recognizing that they had much in common, and have since worked to support each other in their respective economic, political, and social transitions. In terms of healthcare, their Soviet-imposed Semashko-modeled systems, characterized by centralized decision-making, were seen as needing to shift to more Bismarck- or Beveridge-oriented models,* or even to embrace mixed-finance models (Antoun et al., 2011).

Over the quarter century that has passed since the breakup, "the Stans"† have all implemented significant health systems reforms. This chapter focuses on both analysis of these reforms and their projected future.

---

* The Semashko, Bismark, and Beveridge models all provide a version of universal healthcare. The Semashko model, developed in the USSR, is centrally organized and integrated, with all health personnel employed by the state. In the Beveridge model, healthcare is provided and funded by the government; while there are many public hospitals and state-employed personnel, private practice is possible. In the Bismark model, a non-profit insurance system covers healthcare costs, and while still universal and highly regulated, the majority of doctors and hospitals are private.
† The suffix *stan* comes from the Persian language and means "land of."

Geographically, the Stans occupy the land west of the Caspian Sea, bordering China to the east. The total population exceeds 70 million: Uzbekistan has a population of more than 31 million, Kazakhstan more than 17 million, Tajikistan more than 8 million, Kyrgyzstan more than 6 million, and Turkmenistan more than 5 million.

It is never easy to induce change in health systems, but modernizing healthcare in a region where traditional values—and enmities—still hold sway, and whose underpinnings are deeply rooted in tribalism, has special challenges. Traditionally inhabited by Nomadic tribes, the region was a crossroads for the famous trading route, the Silk Road. Some of the population still live a nomadic or semi-nomadic lifestyle in a physical environment that is far from domesticated: much of the terrain of these countries consists of high mountains and passes, deserts, treeless grassy steppes, and dry rugged, non-arable land. The GDP per capita is on the low end of the spectrum: from Tajikistan (US$818), Kyrgyzstan (US$1139), and Uzbekistan (US$2128), through to Turkmenistan (US$7522) and Kazakhstan (US$8585). By way of comparison, Russia's GDP per capita is US$10,885, the United States' is US$59,609, and the world average is $15,880 (International Monetary Fund, 2017).

Post-Soviet Health System Reforms

While there is currently a wide variation between the five countries in terms of total health expenditure per capita, the GDP has increased across the region since the 1990s, and all member states are making efforts to improve. Life expectancy has risen significantly in all Stan nations. The socialist Semashko model of healthcare, the model that operated for many years in the USSR, was originally designed to provide a comprehensive, one-size-fits-all care system, free at the point of delivery, paid for through taxation, and available to all citizens. Decision-making was centralized and the system was accountable to the Communist Party. Historically, this highly centralized, deeply hierarchical system lagged behind Western standards in the acquisition and development of tests, equipment, and technologies.

While the story of the reform of the Stans' health systems is by no means a linear one* (European Observatory on Health Care Systems, 2002), generally it has been a tale of modernizing healthcare and introducing market-based reforms (Nemec and Kolisnichenko, 2006). As we have seen elsewhere in this book, changes to health systems do not come easily, and it is not a trivial task to alter the foundations of healthcare, including the key principles on which the system is based; nevertheless, marked development has occurred. The

---

* In Kazakhstan, for instance, the number of village "hospitals" fell from 684 in 1994 to 208 in 1997 (European Observatory on Health Care Systems, 2002).

major transformations of the Stans' systems have been fiscal: the replacement of a taxpayer-funded model, the introduction of health insurance financing, the introduction of private players into the market, and the acceptance of private payments or copayments. The goals of these reforms were to offset the burdens on the state for healthcare costs, to ensure there were price signals in the marketplace, and to inject more resources into the system.

Challenges and Achievements

An ideal health system in the Stans' region will have a safety net for those with less capacity to pay, serve vulnerable groups and minorities, and have a platform of a decent, universal primary care system on which other services are based. Services to patients would also be provided via good-quality secondary and tertiary care, again with universal coverage, grounded in equity. It would also be a system where those with the capacity to pay do so, within means.

Some policy analysts keeping a watchful eye on health reform over the last two decades in these Central Asian countries have noted ongoing inequities, including problems of access, insufficient funding available to pay-for-care, ineffective delivery systems, and questionable quality of care (Rechel et al., 2012). Nevertheless, there have been various signs of success. For example, a new family medicine model has been established in Uzbekistan, along with human resources training, and in Kyrgyzstan, practitioners now receive direct financial payments, and formal user fees-for-service have been authorized (Fidler et al., 2009). In Kazakhstan, the number of people utilizing the health system has increased significantly: in 2012, almost 60% of the population had consulted a health professional in the past 4 weeks if they thought they needed it; simultaneously, the number of informal cash payments for health services decreased (Balabanova et al., 2012).

Rechel et al. (2012) have suggested that successful reforms across the region share a number of common factors. These include improved organization and governance of the system, more effective interactions with international aid donors, and external advisory expertise. Other success factors include reorganizing healthcare in line with better funding models, and placing emphasis on pilot improvement projects before moving to the take-up and spread of initiatives at the national level. An important success factor, supported by the international literature, is the action of gathering sufficient support across stakeholder groups for proposed changes and diffusion strategies (Braithwaite et al., 2017a,b; Rechel et al., 2012).

So, while there has been some progress, there are many more obstacles to overcome and gains to be made. It is easy to come up with a blueprint for what should happen in the Stans' health systems: decentralizing decision-making, bolstering private-sector involvement, investing in infrastructure, acquiring new technology and pharmaceuticals, and strengthening primary care. These reforms must, of course, be part of an overarching strategy of fiscal reform, with a view to providing opportunities to bolster health insurance, taking a population health approach, and strengthening the legal and regulatory landscape (Fidler et al., 2009).

These are ideals and not achievements, however. The Stans inherited inefficient health systems, and face the challenge of traditions and tribal inertia; in addition, their politicians do not always seem to have the best interests of the population at heart. If GDP and health sector financial allocations are stagnant or falling at the same time as reforms are attempted, then progress in healthcare will be derailed by more immediate problems of daily living and surviving.

Having said that, as financial and improvement measures begin to be put in place, even countries in a region that has the most modest of GDPs can begin to think about increasing access to care, acquiring relevant technology, and creating centers of relative excellence. Another hard-to-achieve but worthwhile future goal is the creation of an evidence-based or -oriented system that accesses international research, but also provides local data to continuously improve systems of care and provide evidence for the efficacy of new systems of care.

What Is Next?

Given that some of these goals are not yet attainable, a number of authors have analyzed the progress made thus far and have suggested ways to hasten the improvement process. One policy analysis group suggests prioritizing monitoring and evaluation, building in feedback loops along the way, so that momentum is built, with progress measured and strengthened over time (Fidler et al., 2009). Another group of policy analysts argues that cooperation between the Stans, encouraging more open and accountable relationships, and developing international aid donors to secure more external resources and expertise are important in the next-stage strategy (Ulikpan et al., 2014). A third group of analysts argue that the influence of the Stans' emerging research institutes, medical schools, and universities will create generational change in the workforce through education and research (Adambekov et al., 2016).

Recent international research (e.g., Ulikpan et al., 2014) argues that the progress made in Kyrgyzstan and Tajikistan, which can be measured by more effective development, cooperation, and better health outcomes, is eminently replicable across other cultures. This study attributed progress to shifts toward more open management cultures and transparent processes, and new and better relationships with international aid donors and investors. As measured by health status and socioeconomic data, Uzbekistan and Turkmenistan have made less pronounced progress than the other states, while Kazakhstan's indicators show the most comprehensive improvements (World Bank, 2017a; WHO, 2017a).

Conclusion

Progress may be difficult, but the Stans have shown, despite a degree of haphazardness and slow progress, that identifiable gains can be made. Increased political stability and an improving economic situation, in combination with a willingness to embrace new models and approaches, and greater cooperation between the Stan nations, should see further advances.

Part IV

Eastern Mediterranean

Samir Al-Adawi

The World Health Organization (WHO) classifies 21 countries as Eastern Mediterranean. These 21 countries span the two largest continents, Asia and Africa. Some of these countries are part of West Asia, while others are located in the Mediterranean basin stretching from North Africa to Asia Minor. The Gulf states and those lying adjacent to the Arabian Sea and those in the Horn of Africa also belong to this region. These countries, regarded by many as the cradle of civilization, were the birthplace of the Abrahamic religions. The landscape is mostly arid and semiarid with pockets of mountainous terrain.

The demographics, economies, and healthcare systems of the region vary. Some countries in the Eastern Mediterranean are the most affluent in the world, while others are considered to be developing, and others are somewhere in between. These differences necessarily determine standards of healthcare. The region is home to approximately 583 million people, and many countries are currently in transition: experiencing a "youth bulge," undergoing rapid urbanization, and dealing with the influx and efflux of refugees. While environment-related and infectious diseases predominate, the prevalence of lifestyle-related diseases is also increasing.

Part IV contains contributions from eight Eastern Mediterranean countries and one region. Three core concepts run through the nine chapters: information technology as the way forward (Lebanon and the United Arab Emirates [UAE]); the improvement possibilities of new approaches to education, manpower, and intersectoral collaboration (Pakistan, Iran, Yemen, and Jordan); and potential solutions to the challenges created by lifestyle diseases and increased longevity (Oman and Qatar). This section concludes with an overarching chapter on the Middle East and North Africa (MENA) region, providing some coverage to those countries for whom a full chapter was not available.

Ali Mohammad Mosadeghrad addresses what he sees as the Iranian hospital system's lack of robust quality and safety "checks and balances." Exploring recent efforts to improve the undertakings of Iran's Office for Accreditation of Healthcare Institutions, Mosadeghrad looks at the value of introducing independent bodies to further enhance the health accreditation system, and discusses reforming the standards themselves.

The Jordanian healthcare system is among the top five in the world, and Jordan is a favored destination for medical tourism. Reem Al-Ajlouni and Edward Chappy propose strengthening the workforce through continued professional development and embracing lifelong learning to ensure that the health system maintains its success and shields itself from the challenges of the future.

After many years of civil war, Lebanon has only recently become more stable, and is currently (along with Jordan) a regional "safe haven" for refugees. Nasser Yassin, Rawya Khodor, and Maysa Baroud discuss how the introduction of eHealth and m-health into the health system could heighten the efficiency of healthcare delivery, particularly among the underserved and frequently isolated refugee population.

The Sultanate of Oman has one of the most equitable and efficient healthcare systems, but future challenges, including emerging refractory diseases of affluence, must be still be addressed. Ahmed Al-Mandhari, Huda Alsiyabi, Samia Al Rabhi, Sara S. H. Al-Adawi, and Samir Al-Adawi assess the Healthy Villages initiative and the changes required to equip the public with knowledge of how to make everyday health-conscious decisions.

Syed Shahabuddin and Usman Iqbal tackle ways to overcome existing healthcare inequity in the Islamic Republic of Pakistan. They suggest two mandates for the future. One is reform of health worker training and education. A second mandate involves embracing an evidence-based approach tied to local needs.

The state of Qatar is one of the most affluent in the world. One of the consequences of growing affluence is increased longevity; however, urbanization and acculturation have begun to erode Qatar's traditional family-based safety networks. Within such a background, Yousuf Al Maslamani, Noora Alkaabi, and Nagah Abdelaziz Selim grapple with the necessary development of hospice palliative care, with an emphasis on intersectoral coordination.

While the UAE has one of the best healthcare systems in the world, epidemiological studies indicate that the country is increasingly experiencing chronic lifestyle diseases. Subashnie Devkaran looks at the UAE's efforts to develop a national unified medical record system to ensure continuity of care and help develop evidence-based prevention strategies.

Yemen is currently besieged by civil war, and its health system is on the brink of collapse. Like other crisis-ridden regions, both environment-related and infectious diseases are rife, as are lifestyle diseases. Despite this predicament, Khaled Al-Surimi looks at the potential of intersectoral collaboration between primary and public health to work for the benefit of the country.

Finally, health systems in the countries in the MENA region are widely diverse and include some of the wealthiest, as well as the most impoverished. All, however, are systems in transition. Jeffrey Braithwaite, Wendy James, Kristiana Ludlow, and Subashnie Devkaran consider some of the recent organizational reforms in the region, their impact, and their potential to transform population health into the future.

38

Iran

Hospital Accreditation: Future Directions

Ali Mohammad Mosadeghrad

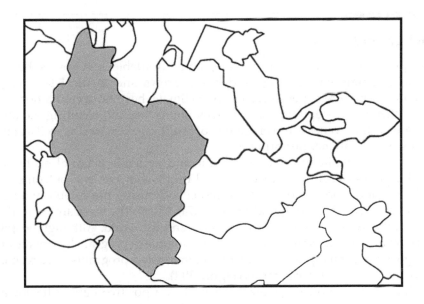

CONTENTS

Iranian Data

- Population: 80,277,428
- GDP per capita, PPP: $17,046.4
- Life expectancy at birth (both sexes): 75.5 years
- Expenditure on health as proportion of GDP: 6.9%
- Estimated inequity, Gini coefficient: 37.4%

Source: All data are from the World Health Organization and World Bank. Latest available data
used as at October 2017.
GDP = Gross Domestic Product
PPP = Purchasing Power Parity

Background

The provision of high-quality, safe, and affordable services is becoming increasingly challenging as demand grows and medical technology becomes ever more expensive. In such a climate, hospital accreditation may be an effective strategy for continuously improving and ensuring the quality, safety, effectiveness, and efficiency of healthcare services, as well as measuring organizational performance.

Hospital accreditation is a systematic and systemic external evaluation of a hospital's structures, processes, and results (outputs and outcomes) by an independent professional accreditation body using pre-established optimum standards (Mosadeghrad et al., 2017a). Currently, there are more than 120 healthcare accreditation programs around the world. Although hospital accreditation was initially introduced as a voluntary program by private and non-governmental agencies, government-mandated programs have recently been initiated in many countries (Figure 38.1).

While accreditation is generally viewed in a positive light, a healthcare accreditation literature review reveals controversial and inconclusive findings and mixed results. Some research suggests that accreditation is an effective strategy for enhancing health systems, improving the quality of healthcare services, and promoting informed decision-making and accountability (Braithwaite et al., 2010; Nandraj et al., 2001). However, other research questions the efficacy of the healthcare accreditation program in improving quality of care (Sack et al., 2011) and patient satisfaction (Heuer, 2004). Within such a context, the aim of this chapter is to discuss the status of Iran's hospital accreditation system and to chart some relevant future directions.

Hospital Accreditation in Iran

Iran has a population of about 80 million and an estimated GDP (purchasing power parity) per capita of US$18,135. In 2016, there were 925 hospitals

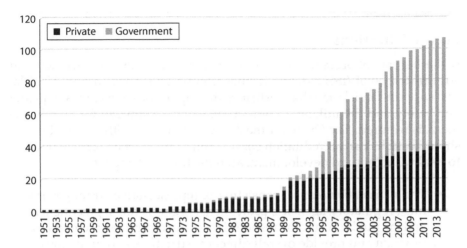

FIGURE 38.1
The trend of healthcare accreditation worldwide.

in the country: 660 public, 75 semi-public, and the remainder (190) private. All hospitals must go through a statutory licensing process to be eligible to provide services. Since 1962, all hospitals have been evaluated annually using mainly infrastructural standards. The results of these evaluations are used to determine the reimbursement of the hospitals by insurance companies.

Iran's current hospital accreditation program, known as Accreditation Standards for Hospitals (ASH), replaced the old hospital evaluation program in 2012 (Mosadeghrad, 2016). The Office for Accreditation of Healthcare Institutions, administered by the Ministry of Health, is responsible for the annual accreditation of all Iranian hospitals—developing and encouraging the use of hospital standards, conducting accreditation surveys, issuing accreditation certificates, and dealing with hospital appeals.

Since its introduction, ASH has had a positive impact on hospital services. It has, however, been less effective in improving medical services and has not translated into enhanced quality and safety of services and patient satisfaction (Shahebrahimi, 2016; Zoleikani, 2015). Too many or unclear standards, less emphasis on performance and outcome standards, poor hospital management and leadership commitment, increased staff workload, lack of physicians' involvement, lack of resources (particularly personnel) to implement standards, and financial burden were the main barriers to the implementation of accreditation standards. Furthermore, the surveyors themselves were not well trained, had little motivation for their work, and were inconsistent (Mosadeghrad et al., 2017a,b; Yarmohammadian et al., 2013).

Future Directions

If the benefits of accreditation are to be maximized, the organization and administration of ASH requires further reforms. ASH needs national legislative support. Legislation that enshrines the importance of healthcare quality and safety, and the utilization of accreditation as a strategy for its achievement, is required. The Office for the Accreditation of Healthcare Institutions should establish an accreditation council and three independent committees to deal with standard development, accreditation, and appeals:

- The *accreditation council*, consisting of representatives from government regulatory agencies, professional organizations, practitioners, and the public, would be responsible for governing the accreditation program and provide overall direction, structure, and guidelines.
- The *technical committee* would be made up of representatives from health-related professional scientific associations and academic organizations, and would be responsible for the drafting, periodic review, and updating of accreditation standards, as well as producing training and guidance documents for surveyors and hospitals.
- The *accreditation committee* would manage the process of hospital accreditation and provide the necessary documents for the accreditation council to decide the outcome of a hospital's accreditation status.
- The *appeals committee* would deal with hospital appeals against their accreditation results.

The credibility of an accreditation program depends on its methods, standards, and surveyors. The processes and methods of hospital accreditation also need reforming. Hospital accreditation currently relies on annual scheduled on-site surveys in Iran. This may result in hospitals' compliance with the standards only in the months prior to the accreditation survey. It is strongly suggested that self-assessment, unannounced surveys, local or national patient satisfaction surveys, and review of hospital key performance indicators are more dependable and would greatly enhance the credibility of ASH.

The ASH model, standards, and criteria also need to be substantially changed. The health system could learn a great deal from industrial management models and strategies, which could be modified for use in healthcare institutions to enhance productivity and provide better service to consumers (Mosadeghrad and Ferlie, 2016). A quality management model was developed and tested for Iranian healthcare organizations (Mosadeghrad, 2012a,b, 2013). Accordingly, an accreditation framework is proposed consisting of 11 constructs, of which seven are enablers and four are results (Figure 38.2).

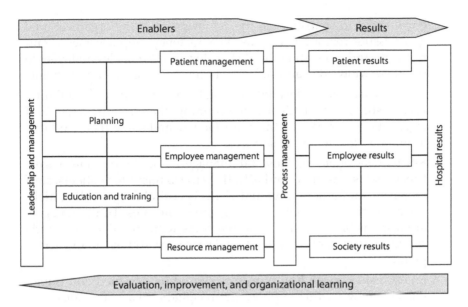

FIGURE 38.2
A model of hospital accreditation.

The enablers and the results constitute 65% and 35% of the accreditation scores, respectively. Each construct includes several subconstructs, standards, and assessment criteria:

1. *Leadership and management*: The role of leadership and management in driving continuous quality improvement in hospitals is critical. Hospital managers should inspire and drive quality management change, support a culture of continuous improvement, and provide the framework for planning, directing, coordinating, providing, and improving healthcare services to respond to patient and community needs.

2. *Planning*: Hospital managers should establish a long-term vision, develop the values required for long-term success, set quality goals and objectives, formulate appropriate strategies, and develop policies and action plans to achieve predetermined goals.

3. *Education and training*: Hospital managers should recognize and nurture the development of abilities, knowledge, and skills. Adequate education and training for suppliers, employees, patients, and the public should be provided.

4. *Employee management*: Hospitals should have the right number of competent staff to meet patients' needs. It is essential to establish strategies to develop and manage employees at the level of the individual, team, and the organization as a whole. Employees should be

encouraged to contribute to the achievement of the organizational goals. Issues relating to employee competency, empowerment, commitment, and motivation should be addressed to enable employees to use their knowledge and skills in continuous quality improvement.

5. *Patient management*: Hospitals should be patient centered—building relationships with patients, determining their requirements and expectations, measuring their satisfaction, and using their feedback to improve the quality of services.

6. *Resource management*: Hospital managers should provide a safe, functional, supportive, and effective environment for patients, staff members, and other individuals. This is critical for providing high-quality patient care and achieving good outcomes. Thus, managers are responsible for creating partnerships with suppliers to provide resources to support organizational processes, strategies, and action plans.

7. *Process management*: Key processes should be designed and implemented in ways that support the hospital's goals, strategies, and action plans; improve healthcare services; satisfy patients and other key stakeholders; and achieve better performance. Such processes include appropriate initial assessment of patient needs, development of healthcare service plans, provision of healthcare services, ongoing assessment of patients' needs, and the successful discharge, referral, or transfer of patients.

8. *Employee results*: Hospital performance to a large extent depends on its employees' competencies, satisfaction, and commitment. Hospital managers should attract, develop, and retain talented employees and fully engage them in quality improvement activities. Employees should be cared for, recognized, and appreciated for their positive contribution to achieving organizational goals.

9. *Patient results*: Hospitals should attempt to identify, understand, and fulfill patients' rational needs and expectations.

10. *Societal results*: Hospitals are established to provide high-quality services to local, national, and even international communities.

11. *Hospital performance results*: Accredited hospitals are expected to have better operational, clinical, and financial performance than non-accredited hospitals.

Accreditation standards should target the hospital, the departments, and the teams, along with individual staff. A three-dimensional model comprising structures, processes, and outcomes; planning, implementation, and evaluation; and organization, team, and individual staff should be considered in developing standards and assessment criteria (measurable elements). Additionally, the assessment criteria should cover at least eight quality indicators (i.e., effectiveness, efficiency, safety, timeliness, equity, patient-centeredness, employee orientation, and continuity). The assessment criteria should be understandable, specific, measurable, achievable, and relevant. The standards should be updated using accepted best practices and current evidence. Such an approach enhances the credibility of the hospital accreditation standards and would improve hospitals' organizational and clinical performance.

There is a pressing need for consistency in the hospital accreditation surveyors' approach. Currently, teams of 20–25 surveyors are involved in the accreditation of a hospital; however, teams of 3–5 surveyors, including a doctor, a nurse, an administrator, and paramedics, should be sufficient. Surveyors should be recruited according to robust surveyor selection criteria and trained effectively. There should be a mechanism for regular assessment of surveyors' performance.

Accreditation surveyors should be encouraged to write objective detailed reports. These reports should be reviewed and validated by the accreditation committee and fed back to hospitals to develop improvement action plans. The Office for Accreditation of Healthcare Institutions should publish the names of accredited hospitals on its website and in annual reports.

The Office for the Accreditation of Healthcare Institutions, the accreditation standards, and the surveyor training programs should be accredited by an international body, such as the International Society for Quality in Health Care (ISQua), to increase the value, credibility, and reliability of its evaluations and ensure compatibility with other countries' accreditation programs.

ASH should also be integrated with other national healthcare reforms to be more effective. In 2014, the Health Transformation Plan (HTP) was launched by the Ministry of Health to promote universal health coverage (UHC); however, little was done to ensure the provision of effective, safe, and patient-centered hospital care (Mosadeghrad, 2017). Accreditation is a good strategy for improving the quality of care and could complement the HTP. If these suggestions are implemented, the healthcare system in Iran is likely to overcome some of the prevailing challenges.

Conclusion

As the results of Iran's initial hospital evaluation system were not as positive as expected, the Ministry of Health and Medical Education established a more comprehensive hospital accreditation system, ASH. This chapter has focused on reforms to the existing ASH, proposing some changes in the hospital accreditation's governance, structure, method, standards, and surveyors, to ensure improved hospital performance.

39

Jordan

Improving Quality of Care by Developing a National Human Resources for Health Strategy

Reem Al-Ajlouni and Edward Chappy

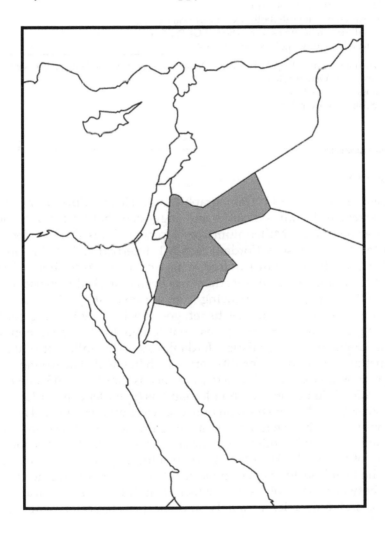

CONTENTS

Jordanian Data

- Population: 9,455,802
- GDP per capita, PPP: $9,050.1
- Life expectancy at birth (both sexes): 74.1 years
- Expenditure on health as proportion of GDP: 7.5%
- Estimated inequity, Gini coefficient: 35.4%

Source: All data are from the World Health Organization and World Bank. Latest available data used as at October 2017.

GDP = Gross Domestic Product

PPP = Purchasing Power Parity

Background

The data from the 2012 Demographic and Health Survey showed an improvement in health indicators and suggested that Jordan was moving toward achieving its Millennium Development Goals (MDGs) for health (Department of Statistics [Jordan] and ICF International, 2012). However, since 2012, the demographics in Jordan have changed. According to the 2015 population and housing census, there has been a marked increase in population, with the population growing from approximately 6.5 to 9.5 million. Jordan's role as a host country to refugees fleeing neighboring countries (including 1.3 million Syrians) has contributed to the rapid population growth (Department of Statistics [Jordan], 2016). This influx of refugees, in combination with Jordan's fertility rate, which has stalled at around 3.5 children per woman for more than a decade, means that as of 2015, one-third of Jordan's 9.5 million people are non-Jordanians (Spindler et al., 2016).

This growth is placing strain on public services and resources. These pressures on the health system are threatening previous health sector achievements, and the health landscape in Jordan has begun to shift. The Jordanian Ministry of Health (MoH) is the biggest healthcare provider and the regulator for the health sector as per public health law. However, the health sector is split between the public sector, which includes the MoH, Royal Medical Services, and university hospitals, and the private sector, which is well developed but poorly regulated. The public sector is responsible for recruiting around 48.2% of Jordan's human resources for health (HRH) (Al-Ajlouni, 2010). A major concern is the ability of Jordan's health workforce to continue to provide high-quality healthcare services. Many stakeholders believe that developing a national HRH strategy to address challenges relating to the healthcare workforce could strengthen the national health system and improve health outcomes.

The development and management of HRH is a critical component of health systems strengthening (WHO, 2000). However, HRH issues have not been comprehensively addressed by the Government of Jordan (GoJ), donors, or stakeholders. While there are many policies in place, Jordan does not have an overall strategy for HRH, nor are key systems in place to support HRH development, planning, or management. In the past few years, however, the GoJ has renewed its focus on the health sector in response to an increased burden on the health system. This renewed focus is also driven by the recognition that medical tourism has the potential to improve the economy, as well as the GoJ's commitment to achieving universal health coverage as declared in the Jordan Vision 2025 document.

The ambitious targets of the Sustainable Development Goals (SDGs), including universal health coverage, will only be achieved if dramatic improvements are made to strengthen the health workforce. Global and national efforts to achieve the health targets of the MDGs were thwarted in many countries by shortages and inequitable distribution of staff, and gaps in health worker capacity, motivation, and performance (Global Health Workforce Alliance and WHO, 2014). While Jordan met the MDGs, the momentum gained is likely to be frustrated by emerging demographic changes.

Health workers are defined as all people engaged in actions whose primary intent is to enhance health (WHO, 2006b). This includes physicians, nurses, and midwives, but also laboratory technicians, public health professionals, community health workers, pharmacists, and those other support workers whose main function relates to delivering health services. Health workers are the core of health systems: improvements to health services coverage and health outcomes are dependent on many factors, most importantly the availability of health workers (Figure 39.1).

Currently, Jordan does not have a formal HRH strategy and plan. Human resource policies and procedures do exist, but they are not coordinated or implemented. Even the public sector, which comprises the Royal Medical Services (the military), university hospitals, MoH hospitals, and primary healthcare facilities, does not have common policies and procedures. Private-sector facilities have their own policies and procedures and do not sufficiently coordinate or share HRH data. All health professionals appointed by the MoH are governed by Civil Service Bureau regulations and standards. The MoH has little or no control over these regulations and standards.

HRH leadership and governance in Jordan requires targeted attention and investment. Historically, the formulation of HRH policies has been supported

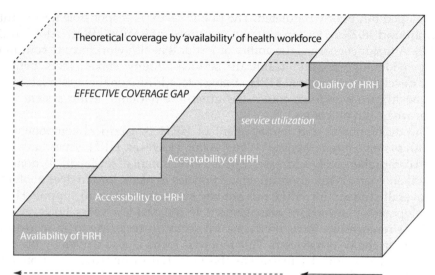

Population + health needs: Who is provided EFFECTIVE COVERAGE?

FIGURE 39.1

Human resources for health: the impact of the availability, accessibility, acceptability, utilization, and quality of health workers on healthcare coverage. (Adapted from Global Health Workforce Alliance & WHO, A Universal Truth: No health without a workforce, 2013. Retrieved from http://www.who.int/workforcealliance/knowledge/resources/hrhreport2013/en/.)

by multiple governmental and non-governmental agencies, as well as international organizations, which has led to overlap and duplication in policy and decision-making. For each policy, there is a need to develop an implementation plan with clearly defined objectives, strategies, expected results, an associated budget, and other necessary resources. The MoH 2013–2017 Strategic Plan highlighted a need to develop a comprehensive HRH policy, as well as a training plan for both technical and administrative staff at the ministry and directorate levels (Ministry of Health [Jordan], 2012). The MoH aims to work with partners to implement this plan.

Jordan has a robust education system that produces health workers for all cadres, and the numbers produced are sufficient to meet the needs of the system. However, the system suffers from a lack of formal strategies and policies relating to recruitment, development, utilization, and retention of the health workforce. Formal in-service or continuing training programs do exist, but they are not offered regularly (Al-Ajlouni, 2010). According to the MoH 2013–2017 Strategic Plan, only 28% of the workforce has received 6 or more hours of training in a 12-month period, and the MoH would like to increase this percentage.

To maintain an appropriate level of quality, the health sector must commit to a concept of long-term continuing professional development. There is a need to further review and invigorate the existing continuing medical education and continuing nursing education programs. Doctors and nurses are not the only health professionals that need ongoing training, however. There is also a need for continuing education for dentists, pharmacists, physiotherapists, nutritionists, paramedics, and radiology and laboratory professionals. The requirement for continuing professional development needs to be included within a national HRH strategy and is critical to ensure coordination and collaboration between the private and public health sectors and universities, professional associations, and councils.

A fit-for-purpose, fit-for-practice health workforce should have the competencies and quality standards required to meet the current and anticipated future population needs, and to achieve the intended policy outcomes. The stock, skills mix, distribution, productivity, and quality of the workforce must also satisfy the needs of the population.

Sustaining Healthcare Quality with a National HRH Strategy

A comprehensive national HRH strategy that ensures a fit-for-purpose, fit-for-practice health workforce will help Jordan's health sector develop a cohesive approach to health worker training and recruitment over the next 5 years. Numbers and types of workers need to be determined. An HRH strategy would assist in the distribution of the required number and types of health workers and utilize them more effectively with the goal of long-term sustainability of healthcare services and improved health outcomes.

Health workforce policies that are partial rather than comprehensive, such as those that focus on education, do not effectively address health workforce shortages or ensure equitable nationwide access to health services (Sousa et al., 2013). A national HRH strategy would address issues that are specific to Jordan, such as the provision of incentives, gender equity, strengthening of planning and information systems, training, and continuous professional development. A national HRH strategy should be aligned with global initiatives, including the World Health Organization's (WHO) Global Strategy on Human Resources for Health, which provides concrete recommendations and ideas on how to strengthen and improve the health workforce. A recent study by Van Lerberghe et al. (2014) documented the experience of low-income and middle-income countries that worked on addressing health workforce problems as one of the core constituents of their strategy to improve health outcomes. The authors suggest that health outcomes improve when health workforce challenges are addressed. It is expected that Jordan's national HRH strategy will support the achievement of the country's health outcomes and sustain any improvements to quality health services.

Transferability

While few lower- and middle-income countries have developed national HRH strategies, the WHO has recommended that comprehensive strategies should be implemented in order to avert the predicted shortage of health workers worldwide by 2030. The performance and quality of services delivered by healthcare systems depends on the knowledge, skills, and motivation of health workers (WHO, 2000). Without appropriate numbers of well-trained, effective, and committed health workers, it is unlikely that any health sector reform will be successful (Martineau and Buchan, 2000).

Prospects

Jordan began the process of developing its national HRH strategy in 2017 by identifying HRH-related policy concerns and priorities based on existing data, views, and opinions of key selected policy-makers and stakeholders. A briefing note was prepared that quickly informed policy-makers and stakeholders of the top priorities facing Jordan. A policy dialogue meeting was held to validate and revise the initial list of policy concerns, identify additional priorities, and reach a consensus on a policy agenda. Based on the priorities generated from this initiative, a draft national HRH strategy was developed to guide investments in human resources to strengthen health systems and improve health outcomes. The national HRH strategy will be approved in 2018.

Conclusion

The GoJ is concerned with the ability of the health workforce to continue to provide high-quality healthcare services due to the pressure of population growth, the influx of refugees, the commitment to achieve universal health coverage, and the desire to enhance medical tourism. Many stakeholders believe that developing national HRH could strengthen the national health system and improve health outcomes. An HRH strategy was developed in 2017 with the goal of implementing the plan over the following 5 years. It is envisioned that the HRH strategy will change the way HRH are produced, used, developed, and retained in the country.

40

Lebanon

m-Health for Healthcare Delivery Reform: Prospects for Lebanese and Refugee Communities

Nasser Yassin, Rawya Khodor, and Maysa Baroud

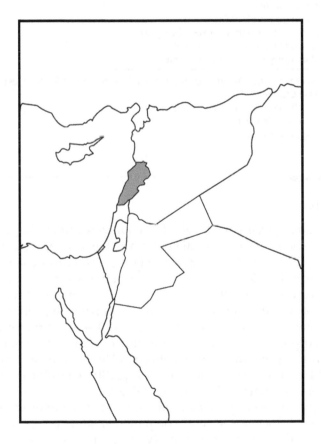

CONTENTS

Lebanese Data*

- Population: 6,006,668
- GDP per capita, PPP: $13,995.7
- Life expectancy at birth (both sexes): 74.9 years
- Expenditure on health as proportion of GDP: 6.4%
- Estimated inequity, Gini coefficient: Unavailable

Source: All data are from the World Health Organization and World Bank. Latest available data used as at October 2017.
* Includes both Lebanese and refugee populations
GDP = Gross Domestic Product
PPP = Purchasing Power Parity

Introduction

Most healthcare systems in developing countries initiate improvement or reform as a response to substantial healthcare delivery gaps, whether related to quality or financing (Van Lerberghe et al., 1997a). In Lebanon, the past decade has seen significant attempts to reform the healthcare system, mostly addressing the decentralization and autonomy of providers, the financing of healthcare systems, the mechanisms of provider payment, and the accreditation of hospitals (Van Lerberghe et al., 1997b). The most recent reform has been the incorporation of innovative health information technologies into the healthcare system through a national eHealth strategy. Its aim is to advance the health sector and improve the quality of healthcare through the use of information communication technology.

The strategy has been implemented incrementally rather than in a complete overhaul. It includes several components, namely, e-government and the standardization and interoperability of health data, e-prescription, telemedicine, e-learning for healthcare providers, and mobile health (m-health) (Ministry of Public Health [Lebanon], 2017). This chapter addresses the opportunities as well as the challenges inherent in this strategy, specifically looking at the m-health component, in the context of a healthcare system that is under considerable stress, with the influx of large numbers of Syrian refugees (1.5 million Syrian refugees have arrived in Lebanon since 2012, and now make up almost 25% of the population) adding significant pressure to an already overburdened system.

From eHealth to m-Health

The realization of the eHealth strategy, along with a commitment to apply all its components in the future, provides a compelling demonstration of recent efforts to enhance care at multiple levels. The integration of eHealth into the national health policy shows that serious efforts have been made to enhance the standards of care for communities, improve healthcare system financing, and develop intersectoral coordination between government organizations.

The adoption of the eHealth initiative will potentially provide a continuum of care at the level of healthcare organizations (mainly primary healthcare [PHC] centers and public hospitals). Current efforts are directed toward upgrading the skills of healthcare providers through the provision of training in eHealth components. Most healthcare providers at the PHC level in Lebanon are prepared for eHealth (Saleh et al., 2016), a fact that suggests that the training has been a success. eHealth will also improve provider interactions with patients and other providers, and deliver updated clinical information, helping to guide practice.

Several factors strongly influence the efforts supporting the eHealth strategy. While current and ongoing eHealth investments (infrastructure, training costs, technology upgrading, etc.) need constant economic evaluation to guarantee a cost-effective health service delivery structure, there are high hopes that this strategy will be financially advantageous over the long run. At the political level, the strong enthusiasm of policy-makers is promising; however, this enthusiasm could be diminished if the channels of communication between policy and decision makers are not well defined. However, the sociodemographic realities of Lebanon exert additional pressure on current eHealth efforts, making it difficult to ensure accessibility in underserved and remote areas where the technology could be especially helpful in improving health outcomes. Such accessibility cannot be achieved without continuous technical improvement and upscaling (technical support, follow-up, and training) of all eHealth components. Other relevant stakeholders and healthcare institutions also need to be engaged in the eHealth policy cycle to ensure sustainability.

Of the different components of the eHealth strategy, m-health—the use of mobile phone technology for health services—deserves special attention. This chapter aims to highlight the ways in which m-health will improve health access and quality in Lebanon, both for host communities and for refugees. In this context, host community refers to a community within which Syrian refugees are concentrated; 87% of Syrian refugees are hosted in the 251 most vulnerable localities across Lebanon, where 67% of the poorest Lebanese live (United Nations Development Programme, 2017c). Although there may be connectivity and other problems within refugee settlements, mobile phone penetration is high in Syrian refugee communities, where many use their phones for Short Message Service (SMS) and WhatsApp (Internews, 2013), and in Lebanese communities, where 88% of adults own a mobile phone (Poushter, 2016).

Current m-health initiatives aim at using mobile phones for health promotion, disease awareness, decision support, and treatment follow-up through an SMS-based system. This system is set to change the traditional healthcare model in Lebanon, and will eventually be linked to the patient databases (applicable to all nationalities) of the national PHC network, whereby patients in the geographic reach of all centers within the network will receive disease awareness messages and automated reminders for medical appointments. Health promotion messages and lifestyle advice will also be sent via SMS to all patients who access PHC. Such a system will be especially beneficial to Syrian refugees, who account for approximately 50% of visits to PHC centers in Lebanon, and who are limited by their finances when choosing care (Johns Hopkins University Bloomberg School of Public Health and Médicins du Monde, 2015). In such a context, m-health has the potential to improve the accessibility, delivery, and management of health services and information.

m-Health: Accessibility, Delivery, and Management of Healthcare Services

Evidence supports the positive effects of m-health on health and health services outcomes in low-resource countries such as Lebanon. In other developing countries, m-health has improved public health awareness, disease management, and patient compliance with treatment plans (Vital Wave Consulting, 2009). This technology is also shown to be cost-effective (Beratarrechea et al., 2014), a factor that improves health resources distribution and service delivery. m-Health offers communication based on wireless cellular capabilities, and is operated through small, lightweight, rechargeable devices, such as phones and tablets (Free et al., 2010). Consequently, it is particularly effective for targeting hard-to-reach communities, where the lack of clinics and healthcare workers, and limited access to health-related information, diminishes individuals' awareness about their health (Vital Wave Consulting, 2009).

An in-depth look at the Lebanese healthcare system clarifies how m-health can contribute to its efficiency and effectiveness. In Lebanon, both secondary and tertiary care are predominantly private (Ministry of Public Health [Lebanon], 2007), and 69.5% of national health expenditure is out-of-pocket (OOP), which imposes a high financial risk on household income (World Bank, 2016d). As regards public health coverage, the Ministry of Public Health (MoPH) covers 54% of the population; the rest is covered by six public health funds (Ministry of Public Health [Lebanon], 2007). For host community households and refugees, the financial costs of healthcare create discrepancies in terms of access and affordability. This has shifted the MoPH's attention toward primary care. Supporting primary care through capacity building and increased funding is necessary to the success of Lebanese healthcare. Strategically, the MoPH plans to improve its core public health functions (epidemiological surveillance and emergency preparedness) and public health programs (school health, mental health, maternal and child health, etc.) through its national PHC network (Ministry of Public Health [Lebanon], 2007), and m-health is a key component of this strategy.

Success Factors for m-Health Implementation

Potentially, m-health has a bright future in supporting primary care, and its likelihood of success is high. When it comes to access to PHC, m-health services are expected to fill what is currently a sizable gap—ensuring that the needs of all low-income communities in isolated areas are catered to. It will also address gaps in affordability, providing treatment follow-up and medical information at no cost. The SMS-based system will send reminders for treatment follow-ups, and encourage individuals to question or change certain health behaviors. This emphasis on prevention and continued care should increase referrals to PHC centers and away from specialized care. As such, m-health manages non-communicable diseases in underprivileged communities by ensuring that communities seek care in a timely fashion, irrespective of the cost to the patient.

The success of such an initiative hinges on the commitment of the MoPH to invest in m-health as part of its eHealth strategy and to integrate it into its national PHC network. Several factors at the PHC level contribute to this success, such as continuous staff training in m-health, a strong collaboration between staff and the MoPH, and robust PHC leadership and management. As for the applicability of m-health, factors include a strong partnership between the MoPH and the SMS service provider, and improvements in the formulation of SMS messages. It is important to note that high cellular mobile subscription (87%) in Lebanon (World Bank, 2016d) is a major dynamic that ensures a good m-health outreach.

The MoPH is working toward the implementation of m-health, and a number of measures have been put in place to ensure its success:

- PHC staff will be provided with technical support.
- Capacity building will be conducted on a regular basis.
- To ensure acceptance and usability, communities will be informed about this initiative at their local PHCs and through MoPH-sponsored campaigns.
- Communities will be asked their opinion regarding both content and frequency of the SMS.
- An evaluation plan will accompany m-health implementation to assess the impact of this initiative on communities.

These measures will be funded by a scheme that is expected to guarantee sustainability.

Prospects

If properly and sustainably implemented, m-health will provide significant benefits to the Lebanese healthcare system over the next 10–15 years. Through this new and innovative healthcare delivery model, healthcare services will become more efficient and of higher quality. m-Health has the potential to improve healthcare indicators at the level of the PHC; in addition, communities will be more aware of healthier lifestyles and behaviors, and will assume greater responsibility for their own care. It is possible that m-health will halt the spread of non-communicable and communicable diseases, allowing the MoPH to focus on prevention. Its positive impact on chronic disease outcomes (Beratarrechea et al., 2014) will help tackle the high burden of disease that is overwhelming Lebanon; chronic diseases currently affect at least one family member in 60% of Lebanese households in host communities, and 50% of Syrian households (Johns Hopkins University Bloomberg School of Public Health and Médicins du Monde, 2015).

Because m-health is embedded in the PHC system, it will improve MoPH community outreach efforts, providing underprivileged and rural communities with better access to care. Aside from the quality of care, future healthcare cost reductions are expected. As such, m-health aligns MoPH efforts with its overall strategic vision, and will contribute to the achievement of its specific objectives in the intermediate and long-term (Ministry of Public Health [Lebanon], 2007).

Many additional efforts can also build on this initiative. The current SMS message initiative is based on a uni-directional model of patient communication. This could be replaced by a bi-directional model where individuals are able to communicate with their PHC through a healthcare help line. It is also conceivable that the MoPH will eventually have the capacity to respond to and manage emergencies through a mobile-based mapping system. It is possible that m-health could also be used in public tertiary care in the future. Future developments will depend on initial evaluation results and on robust collaborative evidence that will be coordinated by key m-health stakeholders to diminish wasteful expenditures and promote best practice models (Heerden et al., 2012).

Conclusion

In a developing country such as Lebanon, the adoption of m-health as part of a wider eHealth strategy represents an entirely new way of care delivery, and has the potential to improve the health system, shifting the existing healthcare delivery model from a traditional to a solid and patient-centric one. Current m-health efforts offer a promising and innovative exemplar that other countries in the region can learn from, and provide robust and effective delivery of health services.

41

Oman

Paradigm Change: Healthy Villages to Meet Tomorrow's Health Needs

Ahmed Al-Mandhari, Huda Alsiyabi, Samia Al Rabhi,
Sara S. H. Al-Adawi, and Samir Al-Adawi

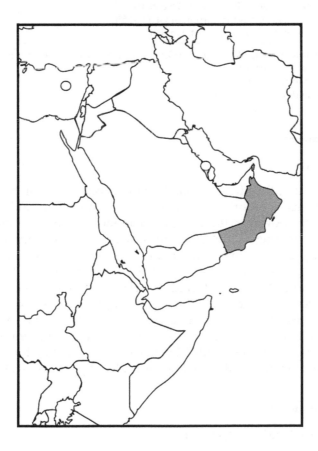

CONTENTS

Omani Data

- Population: 4,424,762
- GDP per capita, PPP: $42,737.1
- Life expectancy at birth (both sexes): 76.6 years
- Expenditure on health as proportion of GDP: 3.55%
- Estimated inequity, Gini coefficient: Unavailable

Source: All data are from the World Health Organization and World Bank. Latest available data used as at October 2017.
GDP = Gross Domestic Product
PPP = Purchasing Power Parity

Background

In recent decades, the physical health of the people of Oman has improved dramatically as a result of higher-quality nutrition and healthcare. Simultaneously, however, there has been an increase in non-communicable diseases. The existing model of health services in the country—top-down, professionally driven, and cure-oriented—is increasingly unable to deal with this new assortment of health problems. A decision has therefore been made by Oman's Ministry of Health (MoH) to upgrade the system in order to meet the new challenges (WHO, 2017f). Empirical evidence suggests that involving service users and their caregivers in planning and monitoring of the health system can simultaneously empower users and improve health outcomes (Ennis and Wykes, 2013; Semrau et al., 2016). With this in mind, the MoH has developed an initiative known as Healthy Villages (HV).

Precursor of Healthy Villages

In the 1970s, the World Health Organization's (WHO) initiative Health for All by Year 2000 was embraced across the globe (Mahler, 2016). In an effort to fulfill the WHO's time-limited objectives, Oman developed its primary healthcare by creating 116 outpatient and inpatient health centers that targeted health essentials (Oman Information Center, n.d.). Secondary care was also improved: hospital bed capacity grew from 12 beds in 1970 to 6468 in 2015 (Ministry of Health [Oman], 2015). With hospitals set up in every region of the country and primary health centers in every village or town, approximately 96% of the population now had access to healthcare services (Al-Lawati et al., 2008). The country experienced a dramatic decline in maternal and child mortality, nutritional well-being improved, and life expectancy levels rose significantly (Ministry of Health [Oman], 2015). In its year 2000 report, the WHO ranked Oman as the most "efficient" healthcare system in the world in terms of outcomes (Braithwaite et al., 2016).

Oman's first tentative step toward embracing Health for All was taken in the 1980s by initiating a program known as Experiment in Community Care (Burjorjee and Al-Adawi, 1992b). At that time, Oman's medical services were still concentrated in urban areas, as it was difficult to provide them in far-flung rural regions. Funds were limited, and there was a shortage of medical personnel (Burjorjee and Al-Adawi, 1992b). In order to alleviate the situation, a team of young Omanis with bachelor's degrees in humanities were given basic health education and taught to identify and manage common rural health problems. They were also authorized to refer some cases to hospitals. However, the experiment was discontinued in 1995 for logistical reasons.

Oman's recent demographic changes, as reflected by the increase in the composite Human Development Index (from 0.699 in 1990 to 0.781 in 2005) (United Nations Development Programme, 2017b), in combination with its improved health status, appear to substantiate the well-known McKeown hypothesis: that economic development invariably improves the health of a population (House, 2016). However, in the midst of this "miracle of health" (Al-Sinawi et al., 2012), a number of "diseases of affluence" emerged (Gostin et al., 2017) associated with new dietary habits, sedentary lifestyles, and increased longevity (Al-Sinawi et al., 2012).

These chronic non-communicable diseases are less responsive to a cure-oriented healthcare system. Rather, it is the patients' knowledge, self-awareness, and determined involvement in their own health management plans that are most likely to improve their condition. Thus, the importance of equipping the public with knowledge of how to make everyday health-conscious decisions has been recognized.

In Oman, this recognition gave rise to HV, modeled on the WHO's Healthy Cities (De Leeuw, 2001; Tsouros, 1995; WHO, 1995). The initiative began in 1986; its objectives are detailed in Table 41.1.

The concept of HV also owes something to the process of de-institutionalization that has occurred in in some countries (Johnson, 1990), as well as the now defunct "Barefoot Doctors" scheme in China (Zhang and Unschuld, 2008). In Oman, one aim of HV has been to decentralize the currently centralized and compartmentalized universal healthcare structure. In addition to

TABLE 41.1

Objective for Healthy Cities and Healthy Villages in Terms of Social and Health
Lifestyle Goals

A clean, safe physical environment of high quality (including housing quality)
An ecosystem that is stable now and sustainable in the long term
A strong, mutually supportive, and non-exploitative community
A high degree of participation and control by the public over decisions affecting their lives
The provision of basic needs (food, water, shelter, income, safety, and work) to all people
Access to a wide variety of experiences and resources, for a wide variety of interactions
A diverse, vital, and innovative city economy
The encouragement of connectedness with the traditions and heritage of village dwellers
A form that is compatible with the past and enhances the preceding characteristics
An optimum level of appropriate public health services accessible to all
High health status (high levels of positive health and low levels of disease)

Source: Adopted and modified from De Leeuw, E., *Global Change and Human Health*, 2(1), 34–45,
2001; Tsouros, A. D., *Health Promotion International*, 10(2), 133–141, 1995; WHO, *Building
a Healthy City: A Practitioner's Guide*, WHO, Geneva, 1995.

empowering users, HV also empowers local, provisional, and regional sectors, letting them chart their own paths in developing and improving health
services. Thus, all matters related to social and health lifestyles are initially
handled within different *Wilayats* (provinces and governorates within the
country).

Healthy Villages in Oman

In 1999, the MoH began to contemplate the implementation of HV (WHO,
2017c). Local communities were encouraged to collaborate with various
government agencies, including ministries of health, the interior, social
development, and education. This allowed the members of each community to play an active role in developing and improving their own
social and health lifestyles (Alhinai, 2015; WHO, 2017f). As of 2017, there
are 21 sites defined as HV in Oman (Figure 41.1), in addition to three
healthy cities and two regions piloting similarly oriented health lifestyle
projects.

At the outset, the MoH constituted regional-level *Wilayat* Health Committees
(WHCs) to engage the community and other government sectors in identifying social and health lifestyle impediments and to suggest solutions. WHCs
are headed by the *Wali* (head of the governorate) and include members from
state and public bodies relevant to the healthcare system.

FIGURE 41.1
The geographic distribution of the HV, Healthy Cities, and health lifestyle projects in the Sultanate of Oman.

As Oman has diverse ethnicities and climates, each prone to different types of endemic diseases, the scheme was flexible and allowed each community to come up with its own targets. Within each village, a comprehensive database that included individual medical records was compiled. Selected members within the community were then trained to implement the social and health lifestyle goals of their village. School pupils were taught the WHO's "eight components of health-promoting schools"—namely, health education, health services, healthy environment, school nutrition, physical education, mental health, promotion of health of school staff, and community involvement (WHO, 2017c). The pupils then shared their new knowledge. Annual reports were generated and the best-performing communities were lauded in the national media.

In 2016, a team comprising MoH and WHO officials conducted a field visit to one HV site, the village of Qumaira (Figure 41.1). The team's anecdotal and impressionistic report indicated that all the quantified parameters, such as physical activity, healthy eating, tobacco prevention, and general environmental factors, had drastically improved in Qumaira compared with in non-HV sites (WHO, 2017f). Similar outcomes were noted with respect to another village, Qualhat, which also achieved most of the tracked social and health lifestyle goals (WHO, 2017e).

Future of HV

While initial studies suggest that HV is a success, more robust methodologies are needed to quantify progress. In the interim, it is worth describing how HV supplements the presently largely geographically maldistributed and cure-oriented healthcare system:

- Oman has a large area of 312,000 km² (120,500 square miles) but a population of only 4 million people (Ministry of Health [Oman], 2015). The bulk of the population is concentrated in the north and the south, separated by a vast desert, known as the *Rub' al Khali*, or the Empty Quarter. The national capital region of Muscat in the north is the most populated region. The remainder of the northern population is distributed among a few towns, rarely with more than 60,000 people, and numerous small villages (Al-Awadhi et al., 2016). Because of the mountainous terrain, some of these villages are almost inaccessible (Burjorjee and Al-Adawi, 1992a). HV may be the right initiative to service such isolated areas.

- The country is in its crucial second phase of demographic transition, characterized by declining death rates, high birth rates, and a "youth bulge" (Ministry of Health [Oman], 2015). With a birth rate of 3.4%, it is clear that the baby boom that began in the 1970s is not yet over (Al-Balushi et al., 2016). HV may help provide the autonomy and empowerment required to satisfy an educated, youthful population.

- HV has the potential to provide Oman's rural youth with socially relevant employment in their own villages (Maben et al., 2010), thereby reducing the migration of young people to already overburdened cities (El-Khoury, 2016).

- The empowering, yet non-controversial nature of HV might spark independent reasoning and creative thinking among Omani youth, not only in healthcare but also in other areas of life.

Conclusion

Oman's twentieth-century cure-oriented healthcare system needs to be upgraded to a modern interactive one in which patients are empowered to be in charge of their own health. The WHO's concept, whereby people are allowed to choose their own health outcomes, is currently being tested in 21 HV schemes in Oman. Recent scrutiny of two of these HV suggests that the strategy of empowerment may be effective. If results from remaining test sites are also positive, this new strategy could be considered for implementation throughout Oman.

42

Pakistan

The Way Forward

Syed Shahabuddin and Usman Iqbal

CONTENTS

Pakistani Data

- Population: 193,203,476
- GDP per capita, PPP: $5,249.3
- Life expectancy at birth (both sexes): 66.4 years
- Expenditure on health as proportion of GDP: 2.61%
- Estimated inequity, Gini coefficient: 30.7%

Source: All data are from the World Health Organization and World Bank. Latest available data used as at October 2017.
GDP = Gross Domestic Product
PPP = Purchasing Power Parity

Health System Infinitives and Constraints

Over the last few decades, healthcare systems across the globe have transformed tremendously due to growth in research, education, technology, and training. These advances are more marked in developed nations, reflecting their economic growth. If Pakistan is to move forward and achieve its health goals, new initiatives and strategies must be devised to tackle the many system constraints, including a lack of resources, issues with governance, and an inadequate infrastructure.

Since the beginning of the millennium, Pakistan has adopted several new initiatives and extended existing programs in the field of preventive medicine. These initiatives include a particular focus on the healthcare elements of the United Nations' Millennium Development Goals (MDGs). Similarly, countries like Pakistan need to be working on human resources—creating better opportunities, increasing compensation, providing access to advanced technology, and offering a favorable research environment—if they are to prevent brain drain and build a skilled and expert workforce (Dodani and LaPorte, 2005).

Education and Training in Healthcare

One of the most important components of healthcare delivery is the proper training and education of healthcare professionals and all related personnel—from doctors and nurses to pharmacists and pathologists. If good health outcomes are to be achieved, all staff must undergo supervised and monitored training to do their specific job skillfully and with a high level of competence. The training of allied health professionals must also ensure that regional socioeconomic conditions, including disease prevalence, are considered. Flexible care that utilizes modern methods of management and is up-to-date in terms of technological advances is essential.

Pakistan is currently attempting to improve the quality of its medical education and training in an effort to reach the highest possible standards. However, this will be impossible unless Pakistan's unique epidemiological data, problems and threats to public health, and the social, cultural, and economic determinants of these problems are taken into account. This may require changes in undergraduate medical curricula and reforms in teaching strategies aimed at improving clinical practice and the competent management of health problems (Nasim, 2011). Some of the reforms may require broadening the role of physicians by modifying undergraduate curricula according to societal needs. Specific learning outcomes and general competencies will provide opportunities for experiences in research, policy making, and education, in a globally-oriented, interconnected, but locally grounded healthcare system (Quintero, 2014).

Ideal approaches to medical training may vary dramatically from current practice. Fundamental aspects of medical education—including such questions as whether the training programs should be time-based or competency-based, or whether the current delineation of specialties and subspecialties falls in line with the country's healthcare requirements—need to be revisited. Current teaching methods and curricula may need to be reformed to ensure that healthcare personnel can best serve the ever-changing needs of Pakistan's diverse population. Research is required in order to make informed decisions about the best clinical training locations and the best methods to utilize health information technology.

There is also a need to focus on the training of allied healthcare personnel and volunteers—whether they work in hospitals or clinics or serve the community in other ways. This should be given priority in order to facilitate smooth and flawless healthcare delivery (Azhar et al., 2009; Hafeez et al., 2011).

Some countries in the region have been implementing community-based and competency-based practice in the education of their healthcare workers and medical and health sciences personnel. However, the preservice healthcare worker syllabi in most of these countries tend to focus on the theoretical aspects of healthcare rather than providing the practical knowledge and skills relevant to the common health problems of their respective communities. Thus, even the best-educated healthcare professionals frequently find themselves ill-prepared to face the healthcare challenges presented by their local communities. Progress in this area is slow due to the lack of postgraduate training and educational opportunities (WHO-AFRO, 2015).

There is an increasing need for widely accessible information, adequately implemented training programs, and support for healthcare workers. Collaboration between government and globally-focused non-government health bodies to share resources and knowledge, and to effectively coordinate efforts in areas requiring support, would greatly benefit and improve healthcare delivery (McConnell et al., 2004).

Initially, appropriate measures of successful training will need to be specified. In order to properly assess alternative training methods, a system to routinely evaluate meaningful outcomes in terms of quality and distribution is required. An overall result-driven approach has the additional advantage of fostering innovation.

Following this, it is essential to pilot new models for financing medical training. If these programs distribute support across all levels of medical and health education, the potential to leverage public investment may also be enhanced.

Research and Technology in Healthcare

As is acknowledged worldwide, the role played by healthcare research in improving healthcare systems is crucial. Some three decades ago, a combination of clinical practice and research created a paradigm shift in the practice of medicine; as a result, modern healthcare evolved into evidence-based medicine (Evidence-Based Medicine Working Group, 1992). Successful health reforms cannot be made without the active involvement of health researchers, particularly when guidelines and rules are variable in different health systems. Studies suggest that outcomes are better in institutions that are actively involved in research, as such institutions are more likely to provide superior and efficient evidence-based care and follow treatment guidelines; however, such benefits are difficult to measure definitively (Krzyzanowska et al., 2011; Young, 2015).

Inculcating a culture of research is therefore of great importance when it comes to achieving healthcare system reform. In Pakistan, the Health Research Council, which represents both the public and private sectors, aims to promote and organize health research, ensuring its utility and relevance to the health problems of Pakistan and at the same time facilitating development of institutional capacity by training researchers (Government of Pakistan, 2016). The council oversees research projects such as surveys and journal publications, and funds projects on subjects related to human health and disease, submitted by both the public and private sectors. The key objective of the council is to harness local knowledge and generate research on health issues to guide health policy-makers.

Developments in technology are also shaping healthcare globally, with new technologies improving outcomes in safety and quality, and providing comprehensive information about internal anatomy, pathology, and physiology. This allows care providers to make informed decisions about their patients' requirements. Pakistan has recently adopted a health information system, and electronic health records are widely used. Not only do such records improve the continuity of care for individuals, but also the resulting databases will be a valuable health system resource—improving the understanding of disease presentation and treatment options, and helping to inform and redefine health policy.

Almost two-thirds of Pakistan's population (64%) live in rural areas where access to healthcare is inadequate. The use of telemedicine, which allows remote management by specialist doctors, will help to bridge the gap of time and distance. Already, the Telemedicine Project, which targets maternal and child health in remote areas, has seen positive outcomes (Oxford Policy Management, 2016).

Technology has a role to play in medical training, too. Aga Khan University's recently established Centre for Medical Innovation (CIME) uses simulation-based training in its undergraduate and graduate training programs. Simulating particular health situations helps students and professionals master certain skills in a shorter time period without compromising the safety of patients.

Conclusion

In the past 10 years or so, there has been rapid progress in the delivery of healthcare in Pakistan. Advances in technology have already had a significant impact in terms of both training and care, and further integration will no doubt lead to improved healthcare outcomes. Due consideration needs to be given to strengthening the health system itself, promoting quality assurance, and utilizing quality health research in the development of effective health policies. This in turn will ensure appropriate resource utilization. A robust and flexible medical educational system with improved specialist training is also required. A well-trained medical workforce should be supported by providing attractive career pathways, along with further education, and encouraging research.

43

Qatar

Hospice Palliative Care

Yousuf Al Maslamani, Noora Alkaabi, and Nagah Abdelaziz Selim

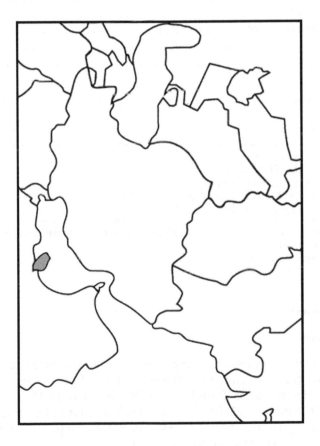

CONTENTS

Qatari Data

- Population: 2,569,804
- GDP per capita, PPP: $127,522.7
- Life expectancy at birth (both sexes): 78.2 years
- Expenditure on health as proportion of GDP: 2.19%
- Estimated inequity, Gini coefficient: 41.1%

Source: All data are from the World Health Organization and World Bank. Latest available data
    used as at October 2017.
GDP = Gross Domestic Product
PPP = Purchasing Power Parity

Background

The state of Qatar is expected to experience a rapid increase in its older population in the coming years, with the percentage of people aged 60 years or over projected to increase 10-fold: from 17,500 (3.1%) in 2000 to 172,000 (20.7%) in 2050 (United Nations, 2004). Statistics indicate that individuals above the age of 55 years currently make up less than 10% of the population. However, the numbers are expected to double over the next 10 years. In addition, the average life span in Qatar increased from 75.6 years in 2004 to 78.1 years in 2014 (Khatri, 2012; Ramadan, 2012).

As a high-socioeconomic-status country currently undergoing rapid development and urbanization, Qatar is experiencing a shift in cultural norms that may influence the values of traditional family, family ties, and welfare.

In much of the Arab world, including Qatar, the family is traditionally the main source of support and care for its older and disabled members, helping them with day-to-day activities and giving them financial assistance and emotional support (Abyad, 2001). However, changes brought about by industrialization, including the increased participation of women in the workforce, have led to the decline of the traditional extended family. Without this support network, family caregivers are more likely to face physical, psychological, cultural, and financial challenges, all of which negatively influence the quality of life of both carers and those they care for (Al-Shammari et al., 2000).

Hospice Care as a Model of Patient Care

Birth and death are the only common health events that we all experience. None of us have any control over our births, but death can often be anticipated, and we can, to some extent, help shape the quality of care that we and our loved ones will need. End-of-life care and palliative care are terms that are commonly used interchangeably. However, end-of-life care refers to those whose death is imminent, while palliative care aims at improving the quality of life of terminally ill patients and their families over the longer term. The palliative care approach involves comprehensive assessment, early identification, and treatment of pain; it also addresses the physical, social, mental, and spiritual problems that are experienced at this critical juncture in the life journey. The common diagnoses in patients accepted to hospice are cancer, dementia, renal failure, chronic liver disease, chronic respiratory disease, heart disease, and HIV/AIDS (WHO, 2015d).

Deficiencies in Existing Services

Palliative care services exist in a number of the Hamad Medical Corporation (HMC) hospitals that provide the majority of secondary- and tertiary-level care in Qatar. Currently, there are approximately 30 palliative care patients at any one time at the HMC. However, those services are highly fragmented and non-specialist, and the busy clinical environment and acute settings of hospitals are not ideal for palliative care.

Demand

A dedicated hospice with trained staff can provide the best care for dying patients, as these patients will be better served in a non-clinical environment than in an acute setting. Research from the United States indicates that hospices tend to cost less per patient than hospitals. The provision of hospice care also helps to alleviate bed pressure in acute facilities (Kelley et al., 2013).

Hospice services are aimed at improving access to care, reducing waiting times, offering patients a choice about where they wish to die, and improving the coordination of patient care, continuing care, and home care. There is a great emphasis on quality and safety improvement. The Continuing Care Strategy in Qatar, which is included under the National Health Strategy (NHS) 2011–2016, and implemented by the Ministry of Public Health as part of the National Qatar Vision 2030, supports the discharge of patients from the hospital where appropriate and assists with their continuing care needs in a more "normalized" setting, such as the patient's home or a convenient community facility. The Qatar National Cancer Plan (2011–2016), which is also included in the NHS, specifically recommended a specialist palliative care center and increased support for at-home care (Ministry of Public Health [Qatar], 2017). In support of this recommendation, the HMC is currently developing a hospice service, to be staffed by palliative care professionals, which should be operational by late 2018.

Access

Hospice care is appropriate when patients are entering the last weeks to months of life and they and their families decide to forgo disease-modifying therapies with curative or life-prolonging intent in order to focus on maximizing comfort and quality of life. This may occur when disease-modifying treatments are no longer considered beneficial or cause more harm than good. Clinicians should make a hospice referral for those patients who have been chronically ill and exhibit decreased functional status, poor quality of life, and frequent hospitalizations, and when the burden of care becomes too great for the patient's family or carers.

The planned hospice will be a specialist palliative care center with palliative specialists offering end-of-life care to all palliative patients. The hospice will have 30 beds and be located in a compound with about seven single-story villas. Patients with a terminal illness and a life expectancy of 6 months or less if the disease follows its natural course, as determined by two physicians, are eligible for hospice care. Guidelines, both disease specific and non-disease specific, will be available to assess patients for hospice eligibility.

Funding

Capital and resource requirements will be sponsored by local charitable donations, and the procurement strategy will follow the tried and tested processes developed by the HMC. A workable commercial deal has been struck with a landlord in Qatar regarding the lease costs. In addition, all necessary equipment has been ordered by the supply chain department for other projects so discounts can be arranged with suppliers.

Family and Informal Caregivers

Although hospice services will be provided at the HMC, the patient's family, friends, and informal caregivers can still support their loved ones in numerous ways—they play an essential part in ensuring the patient's social, psychological, and spiritual well-being. Family and friends may also participate in the administration of medications and injections after receiving formal training by health professionals. They may be required to act as an advocate for the patient. Family caregivers are the "hidden patients" in palliative care, and should be supported by the health team. It is important that they are able to balance their work, family, and leisure with their caregiving duties.

Public Awareness

A Hospice Palliative Care Day and an awareness campaign will be established, with the aim of increasing awareness about the availability of care under the HMC. This will help raise public awareness and understanding of the medical, social, practical, and spiritual needs of people who are living with a life-limiting illness, along with the needs of their families. Such campaigns should also raise funds to support and develop hospice and palliative care services. Social media—WhatsApp, Twitter, and so forth—will also be utilized to increase public awareness.

Training and Education

A highly skilled and experienced workforce is needed to ensure safety, along with the highest quality of care. In order to encourage the provision of holistic and continuing care, multidisciplinary teams that incorporate staff from different disciplines are being developed.

Community capacity building will ensure that clinicians, caregivers, educators, administrators, volunteers, citizens, and other partners are all equipped with the skills and knowledge they need to care for people with life-threatening conditions (Canadian Institute for Health Information, 2011). HMC's continuing medical education program will include workshops and in-service training provided by specialist palliative care physicians. Family members and friends of patients may also need some education when it comes to hospice care, as they are often uncertain as to whether a hospice is the best place for end-of-life care for their loved ones.

Research

The HMC is well placed to support any future research on the development of palliative care in the hospice setting and to build the capacity of the local workforce to develop research on hospice care effectiveness and patient experience. Financial support for such research should be available, as palliative care has been prioritized in the Qatar national research funds. Publication in peer-reviewed journals will be encouraged and local conferences organized, to ensure that any research is dispersed.

Measurable Outcomes

The following indicators will be important in measuring the success of the enterprise:

- The number of palliative care patients supported at the hospice
- The number of palliative patients who report being pain-free or relatively pain-free
- The reduction in readmission rates of palliative care patients
- The reduction in outpatient department appointments of palliative care patients
- The reduction of palliative care patients dying in the hospital

The outcomes of patient satisfaction surveys will also be instructive.

Economic Benefits versus Risks

The major financial disadvantage of hospice care is the cost of the initial infrastructure. However, this is offset by ongoing savings. It is far more cost-effective to let end-of-life patients die in a hospice than in hospital. Researchers at the Brookdale Department of Geriatrics and Palliative Medicine at the Icahn School of Medicine looked at the most common hospice enrollment periods: 1–7, 8–14, 15–30, and 53–105 days. Within all enrollment periods studied, hospice patients had significantly lower rates of hospital and intensive care use, hospital readmissions, and in-hospital death than the matched non-hospice patients (Kelley et al., 2013).

A recent observational study by Sudat et al. (2017) showed a reduction of US$4824 in the total expenditures for advanced illness management hospice enrollees and of US$6127 in inpatient payments. This was in addition to fewer hospitalizations and hospital days.

The study also reveals that savings to Medicare are present for both cancer patients and non-cancer patients. Moreover, these savings appear to grow as the period of hospice enrollment lengthens, with the observed study period of 1–105 days. The authors suggest that investment in hospice care translates into overall savings for the Medicare system, noting that "if 1000 additional beneficiaries enrolled in hospice 15–30 days prior to death, Medicare could save more than $6.4 million"* (Kelley et al., 2013: 2).

A retrospective study of Medicare decedents between 1993 and 2003, by Taylor et al. (2007), showed significant savings of approximately US$2309 per beneficiary to Medicare associated with hospice utilization prior to death, with costs averaging around US$7318, as opposed to the US$9600 average for the non-hospice control. These findings were consistent with those of Weckmann et al. (2012) and Kelley et al. (2013), who found reduced total costs and lower rates of hospital utilization and in-hospital death among hospice patients compared with the control. Interestingly, this effect was observed across a variety of lengths of stay. Once the hospice facility is established by the HMC, economic analyses should be undertaken to find out if the cost differences are of a similar nature in Qatar.

Other Risks

The correct number of staff with specialist palliative care experience should be employed. Staff without the requisite experience or less than the recommended number of staff could put patients at risk.

Conclusion

Qatar's health system must respond to shifts in social and demographic norms brought about by rapid urbanization, as well as increasing numbers of older citizens. The provision of premium hospice care will ensure that financially viable, safe, high-quality care is available to vulnerable patients and their families.

---

* US$1.75 million.

44

The United Arab Emirates

Improving Healthcare through a
National Unified Medical Record

Subashnie Devkaran

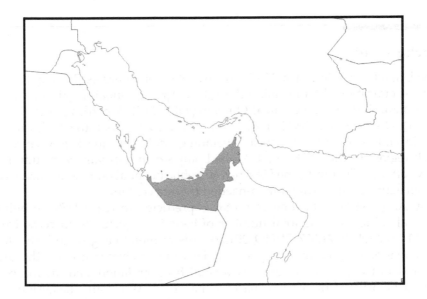

CONTENTS

Emirati Data

- Population: 9,269,612
- GDP per capita, PPP: $72,418.6
- Life expectancy at birth (both sexes): 77.1 years
- Expenditure on health as proportion of GDP: 3.64%
- Estimated inequity, Gini coefficient: Unavailable

Source: All data are from the World Health Organization and World Bank. Latest available data
            used as at October 2017.
GDP = Gross Domestic Product
PPP = Purchasing Power Parity

Background

The United Arab Emirates (UAE) consists of seven emirates. The capital and largest emirate is Abu Dhabi, followed by Ajman, Dubai, Fujairah, Ras al-Khaimah, Sharjah, and Umm al-Quwain (Table 44.1). According to the World Health Organization (WHO), the UAE health system is ranked 27th in the world and is one of the fastest-developing healthcare markets within the Gulf region. In the 1970s, the UAE had only seven hospitals with rudimentary services. Currently, the UAE has 40 public hospitals spread out across its seven emirates, including much-needed trauma centers.

By 2018, the UAE healthcare sector is predicted to reach US$18.6 billion in total value, and the total number of beds is projected to increase from 19,811 (in 2014) to 21,357 (WHO, 2015g). This tremendous growth is due to a combination of government spending derived from hydrocarbon—the UAE has one of the largest oil reserves, a youthful population, and the influx of contract workers from various parts of the world. As a result, the UAE population is expected to reach 10.98 million by 2030. Shifts in lifestyle habits, an increased prevalence of several chronic conditions, and a growing aging population are driving the demand for general, geriatric, and specialist care. The UAE also has a growing medical tourism industry.

Therefore, as part of the UAE's Vision 2021 (Table 44.2), ambitious targets were set to increase the quality of healthcare to international best practice standards (Ministry of Health and Prevention [UAE], 2017). Among the key strategies outlined in the national healthcare agenda, population health management, payment for quality, private–public partnerships, mandatory health insurance, international accreditation, and electronic healthcare record integration are progressing rapidly, with a national unified medical record (NUMR) to be initiated by the UAE's Ministry of Health and Prevention (MOHAP).

TABLE 44.1

UAE Demographics (July 2016)

Indicator	Value
UAE population total	9,267,701
Ratio of Emirati to expatriate	1:9
Deaths per day	47
Births per day	261

Source: United Nations, UN data: United Arab Emirates, 2017c. Retrieved from http://data.un.org/Country Profile.aspx?crName=United%20Arab%20Emirates.

TABLE 44.2

UAE 2021 Vision Indicators

Indicators	2012 Result	2021 Target
Deaths from cardiovascular diseases per 100,000 population	211	158.1
Prevalence of diabetes (%)	19.02	16.28
Deaths from cancer per 100,000 population	78	64.2
Average life expectancy (years)	67	73
Physicians per 1000 population	1.5	2.9
Nurses per 1000 population	3.5	6
Prevalence of smoking	21.6% (men)	15.7% (men)
	1.9% (women)	1.66% (women)
Accredited health facilities (%)	46.8	100

Source: Vision 2021, UAE Vision 2021, n.d, Retrieved from https://www.vision2021.ae/en/our-vision.

The National Unified Medical Record

At the time of writing, the information technology (IT) infrastructure in the UAE consists of independent healthcare information systems with a combination of archaic paper–electronic systems. Implementation of the NUMR project will have multiple benefits. It will transform the UAE healthcare landscape by providing healthcare leaders with accurate data on the burden of disease in the country. It is envisaged that the patient's appointments; queue management; radiology, pharmacy, and laboratory results; dental records; and biomedical information from devices such as dialysis machines will be integrated through the electronic medical record (EMR).

Due to the federal healthcare system being devolved into emirate regulatory authorities in the early 2000s, public health information is fragmented. A unified healthcare record will also facilitate the continuity of care for chronic conditions such as diabetes and cardiovascular disease. In addition, the quality data that are collected can be used in pay-for-performance schemes and to support national benchmarking. This will facilitate the enforcement of regulations and allow compliance to be measured electronically. Patient safety alerts and systems can also reduce the probability of error if implemented appropriately.

There are widespread efforts across the emirates to bring the NUMR to fruition. In 2010, Abu Dhabi implemented an EMR—Cerner—for its public healthcare facilities. Cerner has been integrated across government hospitals, and ambulatory and diagnostic centers. Private facilities, however, have not been part of the EMR integration in Abu Dhabi. Dubai is also making inroads into integrating EMRs in both the public and private sectors through the Dubai Smarter, Dubai Healthier project launched in February 2017 by the Dubai Health Authority (DHA). This deployment of a unique medical record system will integrate all 2770 public and private healthcare facilities in Dubai, providing a single electronic file for each resident (Dubai Health Authority, 2017). This system will eventually be linked to the NUMR. This is significant in that the UAE has a two-tier healthcare system—a universal free healthcare insurance for its nationals and private insurance for non-nationals.

This EMR integration in Dubai will occur in two parts. The EMR system—Salama—will create one electronic record for patients across all public health hospitals, primary health centers, and other facilities, while the National Analysis and Backbone Integration for Dubai Health (NABIDH) system will extend this medical record across all private hospitals. NABIDH, which means "pulse" in Arabic, will create big health data that will be used as an analytical tool in crisis management and the management of disease outbreaks. It will also play a role in implementing preventive healthcare.

Salama (Epic) has been rolled out in three phases or "waves." Wave 1 was launched in April 2017, and Wave 2 was completed in August 2017. In November 2017, the Dubai Health Authority (DHA) launched the third and final phase of the unified electronic system Salama, in Latifa Hospital, Hatta Hospital, Thalassemia Center, Dubai Gynecology and Fertility Center and all DHA Medical Fitness Centers (Figure 44.1).

Following this, Salama will integrate with the NABIDH system. By 2020, every Dubai resident will have one integrated medical record. This will not only ensure accuracy in treatment protocols but also facilitate health insurance coverage.

In the future, any patient presenting to a DHA hospital will have to register using his or her emirate identity card and provide details of his or her health insurance. Using these details, an electronic medical file with a unique number is generated. All medical data related to the patient are attached to the file electronically and accessible wherever the patient receives care.

The system will provide real-time alerts, warnings, and flags to draw attention to changes in the patient medication or patient condition. Patients will be able to access their medical record through a patient portal. A central

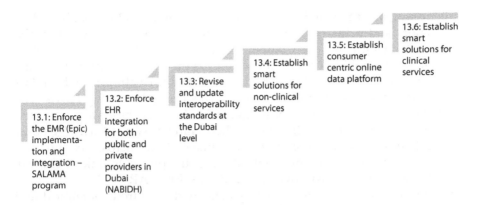

FIGURE 44.1
Dubai Health Strategy 2016–2021: Medical and informatics program 13. (Adapted from Dubai Health Authority, Dubai health strategy 2016–2021, 2016. Retrieved from https://www.dha.gov.ae/Documents/Dubai_Health_Strategy_2016-2021_En.pdf.)

command center to be manned by health professionals will be set up to attend to medical emergencies 24/7. The center will be connected to all hospitals, Dubai Police, and Dubai Ambulance, as well as the Dubai Civil Defense. The NUMR is expected to be fully adopted by all of the UAE's health and medical facilities in 2021.

Promoting Success

The integrated electronic health record should have core capabilities that provide immediate access to key information, such as patients' diagnoses, allergies, lab test results, and medications. This will improve caregivers' ability to make sound clinical decisions in a timely manner. If health information exchange is to be secure, the UAE must develop robust health information privacy regulations. The development of patient portals will support healthcare literacy and self-care, especially for chronic conditions, such as diabetes. Electronic data storage that employs uniform data standards will enable healthcare organizations to respond quickly to regulatory reporting requirements, including those that support patient safety and disease surveillance.

Implementation and Transferability

The success of the Dubai integrated medical record is considered an exemplar and is gradually being replicated within the other emirates. In 2010, Cerner

Millennium, an integrated EMR, was implemented in 12 hospitals and 55 ambulatory and primary health clinics owned and operated by SEHA, the Abu Dhabi Health Services Company. SEHA is responsible for all curative activities of the public hospitals and clinics in the emirate of Abu Dhabi, the largest of the seven emirates in the UAE. The Abu Dhabi private hospitals have followed suit and are in various stages of implementing EMRs. The MOHAP has embraced the information technology revolution and launched a large-scale improvement initiative called Wareed. Wareed will digitalize 15 public hospitals and more than 86 affiliate clinics across six emirates. Cerner Millennium is the core application for the project to link all clinical, operational, and administrative data under the MOHAP.

Digital hospitals are complex ecosystems with hundreds of clinical and business processes made up of thousands of subprocesses. When properly integrated, these processes should seamlessly unite patients, clinicians, staff, assets, and information throughout the hospital, delivering the right information and resources at the right time to the point of care. The path to hospital modernization is not without obstacles, both technical and cultural, and successful reform requires bold leadership, as well as careful planning and execution. In particular, establishing early and clear program management that interlocks the vision, business architecture, technical architecture, and a roadmap for future growth is critical. Hospitals without a substantial foundation in clinical and business process management must invest in design and planning from the outset.

The Next Decades

A number of interventions and policies could sustain progress and assist the UAE in achieving its 2021 targets. A transparent public reporting system is currently being developed, with the Abu Dhabi Department of Health (DOH) working on an Abu Dhabi Healthcare Quality Index, which could guide public decision-making about service providers (Koornneef et al., 2012). At present, the data used for the Quality Index is self-reported. In the future, a unified EMR would support data collection, and identify overutilization and misuse while reinforcing payment for outcomes rather than the current fee-for-service structure. Technology can enhance the patient experience by extending and developing the care process and relationship. EMRs combined with patient portals can facilitate care outside of the clinical setting and enhance self-care.

While regulatory devolution allows each emirate to change based on population needs, it also creates federal fragmentation. A unified EMR will mitigate regulatory fragmentation and standardize compliance reporting. This will reduce the logistical burden on providers who must currently navigate different regulatory and payment systems in a relatively small market. Alignment of policies on healthcare information will support the interoperability of systems. Use of the big data generated from the unified patient databases could provide valuable insights on public health priorities, healthcare expenditure, and capacity planning.

Conclusion

Across the care continuum, healthcare organizations are looking for solutions to help them meet the challenges of providing better patient care at a lower cost. The rapid transformation of the healthcare environment in the UAE, with declining reimbursement values, increasing competition, patient selectivity, stricter regulations, and scarce resources, jeopardizes the survival of hospitals. Thus, healthcare leaders are driven toward technological solutions that offer actionable information that drives effective decision-making, operational effectiveness, improved patient safety, and better clinical outcomes. These technology enablers are just one piece of the UAE's patient-centered business strategy that is focused on efficiency, quality of care, and patient safety. The likely success of the NUMR project brings the UAE closer to its vision of providing world-class healthcare by 2021.

45

Yemen

Integrating Public Health and Primary Care: A Strategy for the Health System of the Future

Khaled Al-Surimi

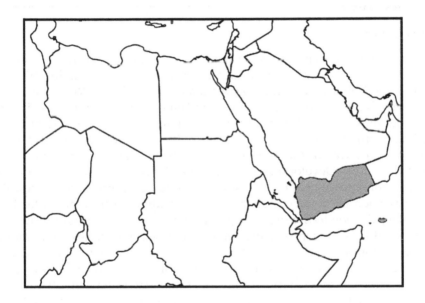

CONTENTS

Yemeni Data

- Population: 27,584,213
- GDP per capita, PPP: $2,508.1
- Life expectancy at birth (both sexes): 65.7 years
- Expenditure on health as proportion of GDP: 5.64%
- Estimated inequity, Gini coefficient: 37.7%

Source: All data are from the World Health Organization and World Bank. Latest available data used as at October 2017.
GDP = Gross Domestic Product
PPP = Purchasing Power Parity

Background

Yemen's health system has experienced many reforms aimed at improving health system performance and its health outcomes (Al-Surimi, 2017; Ministry of Public Health and Population [Yemen], 2000). While there have been significant improvements in some Yemeni health indicators, including a decrease in infant mortality rates and an increased number of health facilities, Yemen is still classified as one of the least developed countries in the world in terms of life expectancy, income per capita, and literacy rates (United Nations Development Programme, 2014). A prominent feature of Yemen's epidemiological profile is the continued presence of communicable diseases, such as malaria, tuberculosis, and upper respiratory infections. This is in addition to the increasing non-communicable disease burden and public health–related problems, such as injuries, disabilities, and trauma. These are compounded by the chronic public health problems of malnutrition and anemia among children and pregnant women (WHO, 2006a), especially in rural areas with limited access to safe water and sanitation, where the majority of the population live (Akala and El-Saharty, 2006; WHO, 2006a).

The many health challenges, in combination with Yemen's political instability, ongoing civil war, and low socioeconomic status, have had an unsurprisingly negative impact on health system performance and population health. As it is likely that the Yemeni healthcare system will face further challenges, health system leaders need to make strategic choices about how best to improve health system performance, particularly when it comes to providing wider access to high-quality basic care. One of the most promising reform possibilities lies in the integration of public health and primary care—moving from hospital-based care to community- and population-based healthcare. In this context, the National Health Strategy for Yemen 2011–2025 recommends activating intersectoral collaboration and coordination to promote healthy lifestyles (EMRO-WHO, 2017).

Integrated Healthcare Services: Rationale and Justification

While public health and primary care share a common goal of building a healthy population, with a focus on health protection, health promotion, and disease prevention, both health professionals and the public still perceive public health and primary care as separate entities with different functions (Institute for the Future, 2003; Lubetkin et al., 2003; Stevenson Rowan et al., 2007). Adopting an integrated care approach can assist in the necessary restructuring, coordination, and collaboration between these two essential aspects of healthcare services. Kodner and Kyriacou (2000) have defined the concept of integrated care as a discrete set of techniques and organizational models designed to create connectivity, alignment, and collaboration within and between the cure and care sectors at the funding, administrative, and/or provider levels. More specifically, Delnoij et al. (2002) explain the concept of integration at the macro-, meso-, and micro-levels of the healthcare system as follows:

- At the macro-level, there will be "functional integration," which involves the financing and regulation of cure, care, and prevention activities of both sectors.
- At the meso- or community-level, there will be "organizational integration and professional integration."
- At the micro-level, there will be "clinical integration," which involves continuity, cooperation, and coherence of healthcare delivery to individual patients.

The World Health Organization (WHO) emphasizes the role of community participation and intersectoral collaboration where health issues cannot be solved effectively by the health sector alone, suggesting a new approach to integrating health systems through strengthening public health functions in primary healthcare settings, in order to improve and enforce both local public health and primary care (WHO, 2003).

Public Health and Primary Care: Concepts, Interlinks, and Joint Functions

The American Public Health Association (APHA) defines the objective of public health as promoting and protecting the health of people and the communities where they live, learn, work, and play (American Public Health Association, 2017). Likewise, Last (1995) has defined public health as the combination of sciences, skills, and beliefs that are directed to the maintenance and improvement of a community's health through collective or social actions, in combination with the prevention of disease and the health needs of the population as a whole. This public health concept has been articulated well in the Yemen 2011–2025 National Health Strategy (EMRO-WHO, 2017).

The Alma Alta Declaration of 1978 defines primary care as the first level of contact for the individual, the family, and the community within a national health system, and outlines the main elements of the care continuum (WHO, 1978). Thus, primary care services should be based on "practical, scientifically sound and socially acceptable methods and technology made universally accessible to individuals and families in the community through their full participation and at a cost that the community and country can afford to maintain at every stage of their development" (WHO, 1978: 1).

According to Hall and Taylor (2003), the WHO's definition of primary healthcare incorporates basic health services, such as education on methods of preventing and controlling health problems; promotion of proper nutrition; sanitation; maternal and child health; vaccination, prevention, and control of endemic disease; appropriate treatment of common disease and injuries; and the provision of essential drugs. Thus, public health and primary care share the same principles and objectives: both strive to promote and protect population health, preventing diseases and injuries and providing basic health services for individuals and the population as a whole. Similarly, the Australian National Public Health Partnership (as cited in Stevenson Rowan et al., 2007) summarized the common features and responsibilities of primary care and public health, highlighting the joint functions, and providing a conceptual framework for integrating the six major functions of public health systems with primary care functions. These include

1. Population health assessment
2. Health protection
3. Health surveillance
4. Health promotion
5. Prevention of disease and injury
6. Disease management

Such a conceptual framework, combining public health and primary care functions, is well suited to countries with limited resources such as Yemen, and can be used to guide both policy-makers and practitioners in their strategic thinking and interventions, decision-making process, and practical decisions and actions.

Research shows conclusively that the integration of public and primary care delivers better health outcomes with lower costs, and generates significant public satisfaction (Starfield, 1994; Van Lerberghe, 2008). As Yemen urgently needs to improve the population's access to basic healthcare services, integrating public health and primary care should be a top priority. The WHO (2003) suggests the following strategies to facilitate the integration of primary care and public health:

- Giving more prominence to the public health functions within primary care
- Using leaders to promote intersectoral collaboration
- Using evidence to demonstrate that important health and social outcomes can be achieved only through intersectoral collaboration
- Encouraging intersectoral stakeholders to agree on health goals and priorities
- Building the mechanisms for collaboration at every level, from national to local
- Integrating health into definitions and processes of wider community development
- Developing appropriate attitudes toward collaboration and power sharing
- Developing influencing skills among primary care professionals and managers at the local level

Implementation Considerations

If public health and primary care are to be successfully integrated in Yemen, the scope of primary care must be broadened. Such an expansion would shift the emphasis to population health at large, supporting the outreach services of primary care centers and incorporating elements such as self-care, first contact care, chronic disease management, health promotion, and primary care management (Gillam, 2013).

A radical shift is needed in the way the relationship between public health and primary care is regarded—by policy-makers, health professionals, and practitioners, as well as the public—if what is known as community-oriented primary care (COPC) is to be established. The COPC approach aims to integrate public health practice by delivering primary care based on community needs assessments (Mullan, 1984; Mullan and Epstein, 2002), and by redirecting the necessary financial incentives toward COPC even within hospital-oriented health systems (Gillam, 2013).

TABLE 45.1

Public Health and Primary Care Core Competencies

Public Health	Primary Care
• Care for populations	• Care for individuals on practice lists
• Use of environmental, social, organizational, legislative interventions	• Use of predominantly medical, technical interventions
• Prevention through the organized efforts of the society	• Care of the sick as their prime function with the consultation as central
• Application of public health sciences (e.g., epidemiology/medical statistics)	• Application of broad clinical training and knowledge about local patterns of disease
• Skills in health services research, report and policy writing	• Skills in clinical management and communicating with individuals
• Analysis of information on population and their health in large areas	• Analysis of detailed practice, disease registers and information on individuals
• Use of networks that are administrative: health and social care authorities and voluntary organizations	• Use of networks that are less bureaucratic: frontline health and social care providers, other primary care teams

Source: Guest, Ricciardi, Kawachi & Lang (Eds). (2013) *Oxford Handbook of Public Health Practice* (3rd ed): Table 3.9.1 (p. 264). Oxford: Oxford University Press. By permission of Oxford University Press.

Prospects

To ensure the successful integration of public health and primary care in Yemen and other developing nations, the capacity and competencies of policy-makers and practitioners must be developed at both the policy and practice levels. Table 45.1 shows the core competencies that could be used as a conceptual framework for building the future capacity of public health and primary care professionals.

Conclusion

While research is still underway, there is sufficient evidence that the adoption of an integrated care approach can significantly strengthen health systems. The integration of public health and primary care, widening the scope of primary care, and moving from hospital-based to community-based healthcare will benefit countries with limited resources, such as Yemen.

46

Middle East and North Africa (MENA)

Health Systems in Transition

Jeffrey Braithwaite, Wendy James, Kristiana
Ludlow, and Subashnie Devkaran

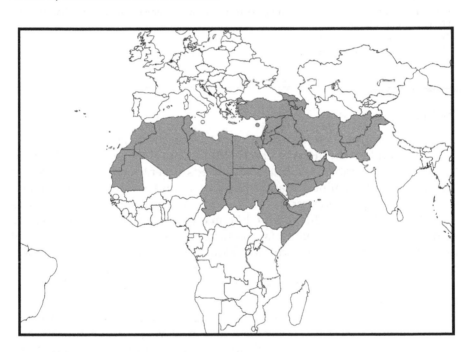

CONTENTS

Middle East and North Africa (MENA) Data[a]

- Population: 921,756,386
- GDP per capita, PPP: $23,729.0[b]
- Life expectancy at birth (both sexes): 71.4 years
- Expenditure on health as proportion of GDP: 5.6%[c]
- Estimated inequity, Gini coefficient: 59.2%[d]

Source: All data are from the World Health Organization and World Bank. Latest available data
used as at October 2017.
[a] Data unavailable for Palestine and Western Sahara.
[b] Data unavailable for Somalia and Syria.
[c] Data unavailable for Somalia.
[d] Data unavailable for Bahrain, Eritrea, Kuwait, Lebanon, Libya, Oman, Somalia, and the
United Arab Emirates.
GDP = Gross Domestic Product
PPP = Purchasing Power Parity

Prologue

Abdul, a 12-year-old boy from a Bedouin tribe in the Negev region of Israel, is traveling on foot through the desert to the city of Be'er Sheva with his family. It is a considerable trek, but Abdul needs urgent medical help. He has developed a strange illness that may be a virus, and is suffering from a high fever and fast-spreading rash on his chest. There are no medical services in Abdul's village, and his family, like many others in his isolated region, cannot afford a car, so Abdul's family have been forced to manage his viral infection according to old traditions. However, there has been no improvement in his condition and now they fear for his life. They worry because once they get to a medical facility in Be'er Sheva, they do not know if their meager family resources will be enough to cover Abdul's treatment.

Meanwhile, Fatima, a 14-year-old girl living in Bahrain, is stressed because of her midschool exams. She is also a little overweight and compares herself unfavorably to her slimmer friends, who sometimes tease her, with the occasional stinging comment, about her size. She has, of late, been thinking negative thoughts about her situation, imagining that she could easily harm herself. Her mom has chatted to her about this and has arranged via their family physician for Fatima to see a child and adolescent psychiatrist who is an expert in early teenage disorders of this kind. In preparation for Fatima's visit, the psychiatrist has discussed her case with a colleague after reviewing the family doctor's notes in the electronic database. She has seen many cases like Fatima's, and is confident that she can work with her and her family to sort out this situation. Fatima is in safe hands, with a team already mobilized around her even in these early stages of her involvement with the health system.

Background

This chapter examines the region known by the acronym MENA—the Middle East and North Africa. MENA centers on the Mediterranean Sea and commonly includes a range of countries and territories surrounding it. Table 46.1 lists the countries usually defined in its core, as well as a second set of countries that are often included in more extended definitions of MENA.

Roughly synonymous with the concept of the "Greater Middle East," the population of the region is approaching 440 million people, or about 5.8% of the world's population. Using the acronym MENA to describe this region is geographically accurate but masks the incredible diversity of the area: MENA includes very rich countries, such as Qatar, Saudi Arabia, and Kuwait, as well as much poorer ones, including Yemen, Sudan, and Eritrea. The primary distinguishing feature of the economies of these countries is their reliance on oil and gas resources. It is estimated that the MENA region has two-thirds of the world's oil reserves and almost half of its natural gas.

While some of the countries are stable, health systems and the health status of populations across MENA are all affected by the region's past and continuing conflicts. These include the wars between Israel and the Arab countries, as well as with Palestine; the disputes, battles, and full-blown external wars that have at various times involved Iran, Iraq, Saudi Arabia, and Syria; the civil wars in Yemen and Libya; and the recent Arab Spring uprisings. While these hostilities often manifest as conflicts over ownership of oil and gas resources, they stem, in part, from many religious, racial, ethnic, and historical differences that characterize the region, which straddles three continents: Asia, Africa, and Europe.

Ongoing political instability and conflict naturally have an effect on socioeconomic conditions, which in turn play a significant part in determining the health status of the MENA region's populations. MENA countries vary widely in this regard: while some nations, such as Djibouti and Sudan,

TABLE 46.1

MENA Countries and Territories

Core Definition of MENA		Extended Definition of MENA	
• Algeria	• Morocco	• Afghanistan	• Pakistan
• Bahrain	• Oman	• Armenia	• Somalia
• Egypt	• Palestine	• Azerbaijan	• Sudan
• Iran	• Qatar	• Cyprus	• Turkey
• Iraq	• Saudi Arabia	• Djibouti	• Western Sahara
• Israel	• Syria	• Eritrea	
• Jordan	• Tunisia	• Ethiopia	
• Kuwait	• The United	• Georgia	
• Lebanon	Arab Emirates	• Mauritania	
• Libya	• Yemen		

devote 10.6% and 8.4%, respectively, of their GDP to healthcare, others, like Pakistan and Kuwait, contribute a much more modest 2.6% and 3.0%. As a consequence of this and other factors, the life expectancy of the MENA population ranges from 56 to 84 years.

Organizational Reforms

Against this backdrop, we want to look across the health systems of the countries in MENA, the organizational reforms that are in train to support their improvement, and the prospects for their success over time. Structures and arrangements differ greatly across the region, but also share some commonalities, as they do with the health systems of other regions. Rich or poor, the MENA group's health systems try to make provision for their populations' acute, primary, aged, community-based, and public health needs. The differences are mainly of degree and emphasis, and based chiefly on factors such as each country's wealth and income distribution, efficiency of the systems of care, access to technology, levels of infrastructure, extent of corruption, commitment to equity, ability to retain healthcare professionals, support for social justice, rule of extant law, and the public–private nature of the system.

Bousmah et al. (2016) suggest that any increase in health expenditure has to be made carefully, because it may not necessarily lead to better health outcomes. Spending money wisely and in an organized way is crucial for the countries of the MENA region. There will doubtless be increases in resources devoted to healthcare over time in most MENA countries, but the organizational structures and the service arrangements predicated on where and how people work and the ways money flows are determining factors in whether health outcomes are enhanced in response to expanding resources. Thus, Bousmah et al. (2016) argue that increasing health expenditure in the region is a necessary but not sufficient condition for advancing population well-being. A huge surge in demand for healthcare in the region is projected in the next decades; as wealth increases, living standards will likely improve and the middle class expand (Bousmah et al., 2016).

All MENA countries, rich or poor, in effect have health systems that are in transition (Grant Thornton, 2009). Shahraz et al. (2014) make the point, however, that while countries of the region share some political, geographical, and cultural aspects, they are not homogenous. As we discussed at the outset of this chapter, MENA is made up of oil-rich, wealthy countries (e.g., Saudi Arabia, the United Arab Emirates [UAE], and Iran), as well as middle- and low-income countries, such as Yemen, Eritrea, Ethiopia, and Jordan. Yemen, for example, has very poor health infrastructure and patchy coverage of services to its population. Implementing a one-size-fits-all set of organizational reforms is clearly not an effective match to the MENA group's needs (Younis, 2013).

Simply classifying countries as rich or poor does not tell the whole story either. Even within MENA's "rich countries" cohort, there has been greater budgetary pressure following recent declines in oil and gas prices. Further, all reputable forecasts predict that more renewable energy will increasingly be made available worldwide, thus reducing reliance on fossil fuels. This will mean that funding wealthy MENA countries' health systems will become problematic. Reorganizing, reforming, and transitioning the health systems of the MENA countries so that they are fit for purpose becomes a crucial issue in rich and poorer countries alike.

With such regional variations, it is unlikely that reform agendas or the pace of change within MENA nations will be in any way homogenous (Braithwaite et al., 2017a,b). So, while in theory all MENA members should be in a position to take advantage of improved technology, the genomics revolution, new public health measures, and the like, the organizational structures and arrangements of all countries and territories in the region need attention.

There is also the problem in the region of current instabilities, or potential and projected threats to stability. Countries that are currently suffering or have recently suffered greatly from hostilities and associated instabilities include Yemen, Syria, Iraq, and Libya. Conflicts pose huge problems for delivering services because when they manifest, the organizational structures and workforce, and the capacity to provide care, become hugely compromised (Younis, 2017). Life expectancies have fallen during times of war in the region (Mokdad et al., 2016), and health systems have been destroyed (as in Yemen and Syria) in whole or part. These systems must eventually be rebuilt, but it is not clear how resources will be found to pay for this.

Some specific challenges to population well-being that healthcare and associated systems in the MENA region have to deal with include a high number of road accidents, climate change, pollution, droughts, and lack of water. Outdated information systems in many MENA nations mean that there is a lack of data on which to base good decisions (Al-Abbadi, 2009). For those populations within the region that have grown wealthier, as in Qatar, the UAE, and Kuwait, life expectancies have risen along with social expectations and aspirations. Some segments of these populations now suffer from creeping "lifestyle diseases"—conditions of the wealthy. In lower-income countries in the region, such as Ethiopia and Somalia, however, people are more likely to suffer from communicable diseases. As in other parts of the world, as communicable-style diseases recede, this is accompanied by a rise in non-communicable diseases, such as cancer, cardiovascular disease, diabetes, and obesity.

Organizational and Reform Success

As these trends continue, and hopefully as conflicts reduce in both number and severity, the imperative to organize care more efficiently, and transition the health systems into more streamlined structures, becomes more important—and feasible (Braithwaite et al., 2015, 2017a; Chan et al., 2009; Truitt et al., 2013).

For the low- and middle-income countries in the region, the creation of integrated, robust, community-based public health systems, with care distributed on as equitable a basis as possible, is the priority organizational matter. In these countries, with less absolute and per capita wealth and a comparatively lower proportion of GDP going to healthcare, there is a need to ensure that resources are spent wisely. One key lesson is not to waste money on risky new technologies, or those that benefit only a few. Instead, it is usually a good investment for governments of less wealthy nations to support basic maternal and infant care, healthcare for infants and young people, and healthcare for those without the capacity to pay themselves. Effective community-based health systems and primary care are the key to strengthening society and improving the health status of the population. This is happening in countries like Oman and Qatar, but in war-torn or otherwise compromised parts of the region, such as Syria and Sudan, there is far less progress than anyone would care to see.

For higher-income countries, it is also a matter of spending money wisely, but at least there is more to spend. Such funding supports the development of technology-enabled, integrated primary–secondary–tertiary referral networks, employing a bigger, well-trained, information-savvy workforce, and advanced research across the four pillars—basic, clinical, public health, and health services research—in order to underpin next-generation progress. The goals are to take advantage of clinical advances and new techniques, harness modern information and communications technology, and capture the benefits inherent in new developments, such as the genomics revolution. This is happening in places such as Saudi Arabia, the UAE, Qatar, and Bahrain. However, if oil and gas are the main sources of revenue or these economies are fragile in other ways, these health systems may not be sustainable in the long run. There are clear threats over time as the demand for fossil fuels irrevocably curves downward. To ensure that their health systems remain sustainable in the face of future challenges, countries such as Saudi Arabia, Bahrain, and the UAE are moving toward private insurance payment models.

Exemplars

Qatar, one of the higher-income countries in the region, is currently undergoing significant demographic change due to rapid urbanization. While palliative care is provided in hospitals, and support for home care is provided, the growing aging population, and the decrease in the availability of traditional home care, means that dedicated hospice care is now needed. The first Qatari hospice service, funded by charitable donations, is due to be opened at Hamad Medical City in 2018. Staff and caregiver training has been implemented, and a public education campaign has begun, and it is envisaged that the service will be beneficial to patients and their caregivers. While the initial infrastructure costs are expensive, dedicated palliative care services are usually thought to be more efficient than hospital palliative care.

Yemen, one of the poorest countries in the region, suffers from political instability and ongoing civil war. The numerous public health challenges, such as communicable diseases, malnutrition, poor sanitation, and unsafe drinking water, are compounded by famine and the impact of warfare. Non-communicable diseases, such as asthma, are also on the rise. However, there have been serious improvement efforts. Infant mortality figures have decreased, and there has been an increase in the number of health facilities. One of the most promising reform possibilities (and one that is also being pursued in wealthier countries) lies in the integration of public health and primary care—moving from hospital-based care to community- and population-based healthcare, including the promotion of a healthy lifestyle. This will involve integration across and within the various care levels, along with community participation.

Epilogue

When he finally arrives at the local public hospital in Be'er Sheva, Abdul is diagnosed with a treatable condition. He is given the necessary medication and spends the first night under close observation. His prognosis is good. His mother will stay with him, and his father and older brother will stay with a distant cousin on the outskirts of the town. On the second day, Abdul is much improved. His father has brought fruit to supplement his diet, and they are all pleased to see that Abdul has recovered his usual huge appetite. Abdul's mother will stay with him in the hospital until he is discharged, after which they will stay with their family until he is well enough to make the journey home.

Meanwhile, in Bahrain, Fatima's visits to her counselor have helped her enormously. Her problems have been discussed and she has been given strategies to manage her anxiety. Fatima's sleeping has returned to normal, her moods have stabilized, and she is generally more relaxed. Her teachers report that her academic results have improved, and she is far happier at school—she is calmer and a little more confident, and her friendship group has expanded. Although there is still work ahead for Fatima, her parents are relieved that she is doing so much better.

Part V

South-East Asia and the Western Pacific

Jeffrey Braithwaite and Yukihiro Matsuyama

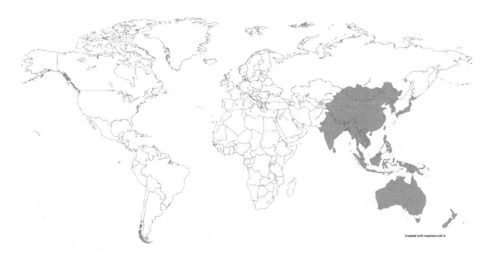

Moving across the planet, we now arrive at the final part of the book, and present chapters on two World Health Organization (WHO) regions: South-East Asia and the Western Pacific. This corner of the globe covers a large part of the earth. It is the most populous; China and India, for example, have a combined 2.7 billion people. The region's range is extensive, and it contains a multitude of countries, including some very small ones, such as Brunei Darussalam and Cocos (Keeling) Islands, with only 423,000 and 1,400 people, respectively. It is diverse, too, in terms of different cultures, historical traditions, and stages of economic development. Singapore, Australia, and Hong Kong are among the wealthiest of nations, whereas Papua New Guinea, Cocos (Keeling) Islands, and Cambodia are low income and thus have limited resources to allocate to healthcare services.

The topics explored by the region's authors are similarly diverse. Ken Hillman, Fakhri Athari, Steven Frost, and Jeffrey Braithwaite from Australia open with a discussion of aging and its effects. They make the very important observation that as people age, they require more healthcare, but often that care is provided, particularly in developed health systems, in the last 12 months of life. Some of this care, and often much of it, is futile—or at the very least non-beneficial. Figuring out ways to deliver the right sort of care in the right circumstances, and learning to not deliver care when it is not called for or not needed, is crucial.

Moving north from Australia to China, Hao Zheng looks at how the integration of different components of the health system is an important consideration for the future. Zheng makes the telling point that creating a more integrated health system is not just a technical challenge, but requires social change of some magnitude, including transforming the cultures of clinicians and institutions—and patients—and how they view the health system.

Eliza Lai-Yi Wong, Hong Fung, Patsy Yuen-Kwan Chau, and Eng-Kiong Yeoh, writing from Hong Kong, also consider integrated care—this time from a patient-centric perspective. They discuss the need to empower people, strengthen systems, alter models of care, and improve ways to coordinate care across sectors, arguing that we must shift our thinking from hospital-based, acute care models to much more patient-oriented approaches.

Shifting to the second most populous country in this region and the world, namely, India, Girdhar Gyani opens with an acute question: How do you build a first-world health system on a third-world budget? He argues that among the key solutions are universal health coverage and strong, capable institutions, such as those that provide accreditation and stimulate quality of care and patient safety.

Japan has a rapidly aging population and an economy that has plateaued over several decades. Yukihiro Matsuyama, in his chapter, looks at dementia care in Japan, observing how the growing burden of care underscores the importance of maintaining quality of life for those suffering from dementia. Measures to support this demand include social initiatives and technologically novel supports, such as using robots and information and communications technology (ICT).

Changing pace and shifting countries yet again, the chapter on Malaysia, written by our long-standing colleagues Ravindran Jegasothy and Ravichandran Jeganathan, takes a specific problem, that of antenatal care, and extrapolates it into the future: that of antenatal care. Malaysia has developed an integrated antenatal care system that focuses on access, affordability, availability of skilled care, and accreditation of providers. Providing ongoing political and institutional support, including setting an agenda for improvement in care over time, is a key consideration.

Moving to Mongolia, a large country in terms of geography, but small by population, the issue of financing of healthcare into the future is tackled by Tumurbat Byamba and Tsolmongerel Tsilaajav. Important measures, according to the authors, include stabilizing health expenditure across time, getting an appropriate mix of resources (state provided, health insurance, and copayments) into the system, and spending the money available in the right places and in an effective way.

New Zealand, another wealthy health system, covered in a chapter authored by Jacqueline Cumming, has recently been focusing on ways to strengthen its primary healthcare capability. Primary care is a key platform for all health systems wishing to improve. Providing adequate funding is a necessary consideration, and placing sustained attention on continuous improvement, although hard to do, is a key feature of New Zealand's primary reform program.

Penultimately, and in contrast to the wealthier countries in the region, Paulinus Lingani Ncube Sikosana writes about how Papua New Guinea (PNG) is currently facing a reduction in resources available to healthcare due to its deteriorating economic situation. Despite this, PNG has been focusing on eHealth solutions as a gateway to improving care delivered to a complex, distributed citizenry located predominantly in rural settings. The hurdles are obvious, but the benefits, if e-enabled care can be achieved, will be considerable.

Shifting to the theme of big data, Yu-Chuan (Jack) Li, Wui-Chiang Lee, Min-Huei (Marc) Hsu, and Usman Iqbal discuss advanced thinking in Taiwan about health improvement based on making big data available, unusually among countries, on an open platform. This is a lesson from which many health systems could learn.

Finally, Jeffrey Braithwaite, Wendy James, Kristiana Ludlow, and Yukihiro Matsuyama have provided a regional perspective in order to discuss not only the specific country exemplars above, but those countries in South-East Asia not covered in their own chapter. These additional countries are Brunei Darussalam, Cambodia, Timor-Leste, Indonesia, Laos, Myanmar, the Philippines, Singapore, Thailand, Vietnam, Christmas Island, Cocos (Keeling) Islands, and Andaman and Nicobar Islands. This chapter discusses initiatives to tackle infectious diseases, both endemic and emerging. In reviewing the literature in this area, Braithwaite and colleagues argue that strengthening institutions, providing resources in the right place, and building on existing successes are crucial factors in combating infectious diseases over the next 5–15 years. Although this work targets South-East Asia, other countries susceptible to infectious diseases, outbreaks, and pandemics might, we hope, find it of value.

47

Australia

The Silver Tsunami: The Impact of the Aging Population on Healthcare

Ken Hillman, Fakhri Athari, Steven Frost, and Jeffrey Braithwaite

CONTENTS

Australian Data

- Population: 24,127,159
- GDP per capita, PPP: $46,789.9
- Life expectancy at birth (both sexes): 82.8 years
- Expenditure on health as proportion of GDP: 9.4%
- Estimated inequity, Gini coefficient: 34.9%

Source: All data are from the World Health Organization and World Bank. Latest available data used as at October 2017.
GDP = Gross Domestic Product
PPP = Purchasing Power Parity

Background

The development of acute hospitals has had a profound impact on the way we deliver healthcare. Up until the 1960s, care was largely delivered by family practitioners and community nurses, but over the last half century, hospitals have become the focus of our concentrated healthcare technology and expertise (Braithwaite et al., 1995; Hillman, 1999). More than 40% of Australia's recurrent health budget is now allocated to acute hospital services (Australian Institute of Health and Welfare, 2017).

Over the same period, however, there has been an insidious incremental change in the demographic and physiological makeup of those who require treatment. Hospitals and medical specializations were initially designed to care for younger patients with a single diagnosis; however, one-third of hospitalized patients in Australia are now elderly, aged 65 years or over (Hillman, 1999; Tinetti and Fried, 2004). This group accounted for 19.6% of all emergency department presentations, and almost half (48%) of total hospital bed days in 2013–2014 (Australian Institute of Health and Welfare, 2017). The proportion of elderly in hospital admissions increased from 35% to 40% between 2004–2005 and 2013–2014, respectively (Australian Institute of Health and Welfare, 2017). Treatment of older patients means that a significant portion of acute hospital resources are now being used to manage elderly frail people, who are often near the end of their lives (Goldsbury et al., 2015).

The information we have on the increasing use of acute hospitals by the elderly frail is compelling. The global population is aging, and many more elderly patients are therefore being admitted to acute hospitals worldwide (Amalberti et al., 2016; Beard et al., 2016). Many require life support during the last few days or weeks of their lives (Hoffman et al., 2016). One-third of urgent calls to patients in acute hospitals are for patients who are near the end of life, and for whom treatment limitations have not been set (Jones et al., 2012). The outcome for these patients is commonly poor: almost a third of hospitalized elderly patients die in the 12 months following discharge (Clark et al., 2014). Almost half of the elderly patients admitted to intensive care units (ICUs) die within 12 months, and those who survive experience a significant decline in their quality of life (Sacanella et al., 2001; Yende et al., 2016).

It would appear that acute hospitals find it difficult to recognize that patients are at the end of life and have little idea of the poor outcomes experienced by the elderly frail after they are discharged from the hospital. Rather than regarding repeated and unplanned hospital readmissions as a failure to recognize and effectively treat the problem, it may be that repeated admissions of the elderly frail are simply a marker of people being near the end of life.

The future of our health system will reflect the future of the aging population. The teaching and practice of medicine, which has historically focused on younger patients with a single diagnosis, has become somewhat anachronistic. Most patients currently being managed in acute hospitals have chronic conditions or comorbidities. Most of these comorbidities are age related and rarely amenable to a "cure." At best, comorbidities such as ischemic heart disease, stroke, and diabetes can be modified to some extent by lifestyle and medical interventions.

Chronic conditions will not only worsen with age, but the number of comorbidities will also increase. Aging is also accompanied by functional deterioration resulting in frailty (Rodríguez-Manas and Fried, 2015). The functional state of a patient is a greater predictor of outcome than the "medical" conditions associated with aging (Beard et al., 2016). Aging and frailty are irreversible and ultimately terminal conditions that do not lend themselves readily to the miracles of modern medicine (Braithwaite, 2014).

The fact that to up to one-third of medical interventions carried out during the last 12 months of life are non-beneficial presents an enormous challenge for the health system (Cardona-Morrell et al., 2016). Continuous admission of elderly patients to the hospital has disadvantages for patients and staff alike: many aged patients would prefer to have their conditions managed at home, and hospital staff gain no satisfaction from delivering treatment that has no effect. Such treatment also contributes to the increasingly unsustainable costs of healthcare. While advances in what a healthcare system can treat continue apace, clear parameters for when such treatment is acceptable and appropriate have not yet been established.

Success Story

The Australian-based Simpson Centre for Health Services Research of the University of New South Wales is currently working on a different system of care for the elderly frail. The research project is focused on circumventing the medicalization of aging and dying, and is working to engage communities and individuals in supporting the frail aged in a way that it is consistent with the reality of these inevitable conditions.

In order to engage individuals, the center has developed a prognostic indicator—the CriSTAL tool—to identify those elderly frail people who are nearing the end of their life (Cardona-Morrell and Hillman, 2015). The tool includes variables such as age, degree of frailty, and chronic conditions, as well as the severity of the acute condition. Of all these variables, frailty and age are the most important. The tool also gives an estimate for longevity, an estimate with inherent uncertainty. Nevertheless, the condition of the elderly frail person is terminal, and the chronic conditions associated with extreme aging are progressive and not reversible. The prognostic tool becomes a "flag" for identifying elderly patients who are near the end of life, and triggers action to share information and engage the patients and their supporters in the caring process. This radical change to current treatment protocols involves a two-step process:

1. An honest and empathetic discussion is conducted with the elderly person and their carers.
2. Patients and carers are empowered to make informed choices about how they wish to live the remainder of their life. Given an informed choice, many elderly people do not want to be repeatedly admitted to acute hospitals, especially when the resources of the hospital will have little effect on the chronic conditions.

The second step represents an important shift in current medical practice, which assumes that these chronic conditions in the elderly are treatable.

Impact

The realities of aging and the quantification of prognosis in the frail elderly will significantly change the way we deliver healthcare. The impact of the approaching "silver tsunami" on the health system can be modified by recognizing chronic conditions in the elderly as predictable and irreversible. This will shift the emphasis from complex acute care in hospitals to community care. The new approach will be based on individual choices and be supported by government and non-government organizations. It is envisioned that this will lead to a better quality of life for the patient, as well as enormous savings in healthcare delivery.

Implementation and Transferability

The system is being trialed in several Australian and international hospitals, and evaluation of the system will be conducted by researchers with a track record in health services research (Chen et al., 2016; Hillman et al., 2005).

The strength of the system is based on being embedded in healthcare delivery by clinicians in acute hospitals. Clinicians will use the prognostic tool to conduct an appropriate response to patients identified by the tool. The system will operate when patients attend the emergency department, when urgent calls are made regarding deteriorating patients, and when patients are admitted to the ICU. It will be operated by staff in the general wards and in any outpatient clinics associated with the hospital.

Prospects

The drivers for uptake of the system include

- Being embedded in routine clinical practice
- Having intuitive appeal to clinicians, patients, and funders of the health system
- Being consistent with the wishes and built around the needs of patients and their carers
- Addressing one of the major contributors to the unsustainable costs of healthcare (Braithwaite et al., 2017c)

In order to change the attitudes and beliefs associated with the current medicalization of aging and dying, successful implementation of the new system will also require widespread dialogue and community interaction. This has already begun: research literature has attracted not only widespread international attention by academics, but also extensive coverage in the lay media (Cardona-Morrell and Hillman, 2015; Cardona-Morrell et al., 2016). A book and film, both aimed at the general public, have been produced, using narrative to outline the often futile medicalization of the aging process (Hillman, 2017; Tyson, 2017).

It is expected that the system's strongly embedded clinical acceptance will drive change. This coincides with patients' increasing desire to be involved in their own care goals as they approach the end of life (Hillman, 2009). The system will be implemented on a wide scale in Australia. Many developed countries have comparable challenges, and collaboration is already occurring internationally.

Conclusion

Reducing the incidence of inappropriate and often deleterious interventions that elderly people are subjected to in the last months of life is arguably the most important challenge for contemporary healthcare delivery. A new approach has been developed that simultaneously identifies the elderly frail near the end of life and empowers them to establish their own goals of care. The system is currently being evaluated, and strategies are in place to implement the new approach on a wide scale.

The approach will shed new light on the way we currently view chronic conditions in the elderly frail as treatable and amenable to acute hospital management. Such a radical change in approach will need to occur in conjunction with open and vigorous dialogue between multiple stakeholder groups.

48

China

Integrated Stratified Healthcare System

Hao Zheng

CONTENTS

Chinese Data

- Population: 1,378,665,000
- GDP per capita, PPP: $15,534.7
- Life expectancy at birth (both sexes): 76.1 years
- Expenditure on health as proportion of GDP: 5.6%
- Estimated inequity, Gini coefficient: 42.2%

Source: All data are from the World Health Organization and World Bank. Latest available data used as at October 2017.

GDP = Gross Domestic Product
PPP = Purchasing Power Parity

Background

China's healthcare system has experienced considerable change since the 2009 healthcare reforms. Since then, a public health system has been established: basic health coverage has been provided to 95% of the population, including health insurance for serious illness; essential medicine lists have been put in place; and a series of reforms in public hospitals have been launched. According to a recent study by the Chinese Academy of Social Sciences, however, these healthcare reforms have had little impact on two significant system challenges: the difficulty of accessing care due to a lack of primary health services, and the unaffordability of healthcare due to deficiencies in the newly established essential drug system (Peilin, 2015). The study recommended that an integrated stratified healthcare system be established to make healthcare more accessible and affordable.

China has what can be described as an "inverted triangle" pattern of healthcare delivery—care is hospital-centric, and primary healthcare (PHC) is both insufficient and of poor quality. Healthcare resources tend to be allocated to hospitals, especially Class III tertiary hospitals in major cities. The central government has committed more than 6.5 trillion RMB (around US$1 trillion) to healthcare since 2009; 80% of this money went to hospitals, while only 10.6% was invested in PHC facilities. In 2012, revenue generated by PHC facilities was only 20.5% of that generated by hospitals, of which 50.5% came from Class III tertiary hospitals, which make up 12% of all hospitals (National Health and Family Planning Commission [China], 2013). By the end of 2016, out of a total 992,000 health institutions, China had 29,000 hospitals, 930,000 PHC facilities, and 30,000 public health institutions. The number of patient visits reached 7.09 billion from January to November 2016; 2.9 billion patients (40.9%) were treated in hospitals, an increase of 5.6%, while 3.93 billion were treated in PHC facilities, a decrease of 0.6% (National Health and Family Planning Commission [China], 2016a,b). A serious shortage of general practitioners (GPs) has further weakened service delivery at the PHC level; most health workers at the PHC level have only limited post–high school training (World Bank, 2016a), and patients would rather wait for hours in enormous lines to see a specialist at a tertiary hospital than step into a neighborhood PHC facility to see a GP.

In the 13th 5-year plan (2016–2020), China's state council built on the success of the 2009 health reforms, reiterating its goal of providing greater access to efficient and quality care to all citizens, both rural and urban. The state council endorsed a national strategy known as Healthy China 2030, which was designed to "deepen the health reforms of the medical and health systems, promote the interaction of medical services, health insurance and pharmaceutical supply, implement the tiered delivery system and establish primary care and modern healthcare systems that cover both urban and rural areas" (World Bank, 2016a: xxix; National Health and Family Planning Commission [China], 2016c).

Implementation

While an integrated stratified healthcare system is an integral component of China's healthcare reforms from a policy perspective, there has been little in the way of implementation. From 2013 to 2016, the number of hospitals increased by 25%, with Class III tertiary hospitals increasing by 37.4% between 2013 and 2015. The number of PHC facilities, by contrast, increased by only 1.9% (National Health and Family Planning Commission [China], 2013, 2015, 2016a).

The successful implementation of these much-needed reforms faces many challenges. These include

- The fragmented nature of China's healthcare system (World Bank, 2016a; Zhao and Huang, 2010; Swedish Agency for Growth Policy

Analysis, 2013): Under the leadership of the National Health and Family Planning Commission (NHFPC), healthcare governance has been decentralized to local health bureaus in 31 provinces or autonomous regions. The latest reforms under the leadership of the state council involve as many as 16 government agencies in the management, supervision, development, and implementation of various policies, guidelines, and action plans, making it difficult to reach consensus.

- The demographic imbalance in health resource allocation: 80% of health resources have been allocated to hospitals in cities, of which 80% were allocated to Class III tertiary hospitals (National Health and Family Planning Commission [China], 2013).

- China's inefficient health insurance system: Health insurance reforms have not kept up with other reforms, or with the need for healthcare services. For example, patients need to pay twice for the same illness when referred to higher-level health facilities—so why not go to tertiary hospitals directly? Moreover, health insurance bodies lack control over Class III tertiary hospitals due to their dominant position in the system.

- The absence of service integration (World Bank, 2016a): No strategies or plans have been put in place to improve integration; consequently, there is little communication between providers at different levels, services are not effectively coordinated, and there is no structured referral system.

- Reform-resistant tertiary hospitals: Even though they are publicly owned, all tertiary hospitals function as independent profit-driven entities that compete for patients; thus, many hospitals refuse to transfer patients back to lower-level facilities.

- The public perception of GP care: Chinese medical education focuses on producing specialists rather than GPs, and patients prefer tertiary hospitals due to the perceived poor-quality care in PHC and the serious shortage of qualified and trusted GPs. Further, rural areas have difficulty in recruiting and retaining qualified health professionals, as the opportunities available to city-based doctors are superior (World Bank, 2016a).

In light of these challenges, the NHFPC introduced further reforms in 2016, in which 270 cities were selected to run pilots establishing a localized hierarchy for diagnosis and treatment (National Health and Family Planning Commission [China], 2016d). Accordingly, local governments introduced relevant policies and endorsed models and strategies based on local contexts to implement action plans.

Vertically Integrated Regional Medical Alliances

This refers to the integration of regional health resources by creating an alliance among tertiary, secondary, and community health facilities, led by the regional tertiary hospitals (Jing, 2014; Huang and Yi, 2015; Liu, 2016). For example, in Zhenjiang, Jiangsu Province, two tertiary hospitals, five secondary hospitals, and 10 community hospitals have formed an alliance through asset integration since 2009 to coordinate service delivery across the continuum of care. In 2011, Shanghai launched the Ruijin-Luwan Medical Alliance (led by Ruijin Hospital) and the Xinhua-Chongming Medical Alliance (led by Xinhua Hospital). Both medical alliances brought together secondary hospitals and community health facilities in an effort to provide more efficient and better-quality care to residents in these districts. Under the leadership of a newly established executive board, these alliances centralized financial management, integrated health resources, and allowed health professionals mobility within participating institutions. The possibilities of establishing a shared purchasing platform and payment system are still being explored.

The Shanghai-Minhang Model

The Shanghai Municipal Commission of Health and Family Planning initiated an information-sharing Internet platform for tertiary, secondary, and PHC facilities at the municipal, district, and community levels (Yang, 2016). The platform was piloted in the Minhang District in 2015. PHC facilities were newly built or expanded, and residents were each assigned a GP, who was responsible for initial diagnosis and treatment, referral, health management, and continued treatment when the patient was referred back to the community. In contrast to the medical alliance model, the Minhang model has no binding requirements among different tiers of health facilities. Shanghai expects the model to spread at the municipal level.

Health Insurance Leveraging

Beijing initially promoted the stratified healthcare system by introducing new health insurance policies whereby 90% of medical expenses at PHC facilities would be reimbursed, but only 70% at tertiary hospitals (Zhu, 2015; Zhang and Yu, 2016).

Impact

By establishing an integrated stratified healthcare system, with well-structured referrals among primary, secondary, and tertiary care, these reforms to China's healthcare system are expected to

- Improve access to affordable healthcare, especially for people in rural areas
- Use healthcare resources more efficiently in order to reduce costs and improve health outcomes
- Reduce the quality gap at different levels of healthcare services

There are some reservations about the integrated stratified healthcare system being implemented as a government-led "political system" or an administrative order simply to enforce the roles of different healthcare facilities. This would result in a drift away from the original intention of providing patients with easily accessible, affordable, well-coordinated quality care across the healthcare continuum.

Prospects

In September 2015, the state council released a guide to advance the integrated stratified healthcare system, specifying targets and implementation plans (State Council [China], 2015). The guide outlined two major objectives in relation to PHCs:

1. By 2017, policies at the national, provincial, municipal, and county levels pertaining to the stratified healthcare system should be well established; efforts should be made to clarify roles of health facilities at different levels, and to optimize the distribution of health resources. These efforts should strengthen GP capacities, improve efficiency and outcomes of healthcare, and increase the use of PHC services.

2. By 2020, an integrated stratified healthcare system should be in place, including initial diagnosis and treatment at a PHC facility, a well-structured referral system, continuing care for acute and chronic conditions, vertical integration to promote sharing of health resources within different levels of healthcare facilities, and enough family doctors to cover the entire population.

China currently has about 180,000 GPs in practice, with only 80,000 registered with the NHFPC. In order to reach a ratio of one GP per 2000 residents, China needs 700,000 GPs by 2020; however, there are currently no plans underway to ensure that the required additional 500,000 GPs receive training by then (Guo, 2004).

A degree of social change will also be necessary if the stratified healthcare system is to be successfully implemented. Since the 1980s, Chinese citizens have been free to choose their own care and are likely to be resistant to the imposition of new controls. Additionally, the Chinese public, including many health professionals, traditionally consider GPs to be less qualified than specialists; they do not trust GPs or PHC facilities, and believe that only tertiary hospitals can provide quality care. Consequently, newly opened PHC centers find it difficult to attract patients, and it will take a long time to build the public's trust in reformed primary care and change their habits when it comes to medical care.

Conclusion

The World Bank has estimated that it will be 10 years before these proposed health reforms are fully implemented (World Bank, 2016a). Various strategies and approaches need to be designed to restructure healthcare delivery and to provide the sticks and carrots needed to achieve policy goals. If the reforms are to be widely effective, social and economic disparities between different regions must also be considered and addressed.

49

Hong Kong

Integrated Health Services: A Person-Centered Approach

Eliza Lai-Yi Wong, Hong Fung, Patsy Yuen-Kwan Chau, and Eng-Kiong Yeoh

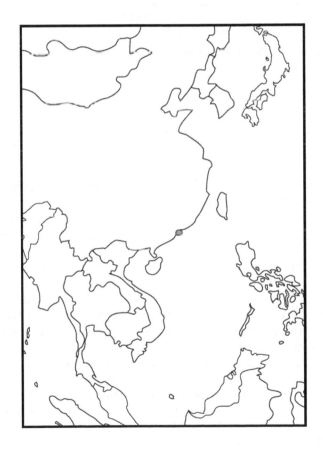

CONTENTS

Hong Kong Data

- Population: 7,346,700
- GDP per capita, PPP: $58,552.7
- Life expectancy at birth (both sexes): 84 years
- Expenditure on health as proportion of GDP: 5.7%
- Estimated inequity, Gini coefficient: 53.7%

Source: All data are from the World Health Organization and World Bank. Latest available data used as at October 2017.
GDP = Gross Domestic Product
PPP = Purchasing Power Parity

Background

With the dual burden of emerging and re-emerging communicable disease and the increasing prevalence of chronic diseases, nations are facing challenges in providing timely and equitable access to health services, shortages of trained health workers, and rising public expectations. The World Health Organization (WHO) advocates an integrated, people-centered approach as crucial in enabling the development of a responsive and sustainable health system (WHO, 2016d). Well-coordinated health service throughout the life course is the core principle of integrated health services. WHO defines integrated health services as "health services that are managed and delivered so that people receive a continuum of health promotion, disease prevention, diagnosis, treatment, disease-management, rehabilitation and palliative care services, coordinated across the different levels and sites of care within and beyond the health sector, and according to their needs throughout the life course" (WHO, 2016d: 2). Incorporating the concept of people-centered care (PCC) has the greater potential to contribute to the health of all people because the health system has to take into account the preferences and values of people it serves in the design and delivery of health services.

PCC has been increasingly emphasized since the turn of the century. The term evolved from, and is still used interchangeably with, person-centered care and patient-centered care. While definitions of PCC vary depending on different theoretical perspectives, it is most commonly described as a relationship between doctors and patients whereby the doctor engages with the patient as a holistic person rather than focusing primarily on the patient's disease (Pelzang, 2010). Considering the patients' perspective, the International Alliance of Patients' Organizations (IAPO) (2016) emphasizes the importance of patient engagement in different dimensions of healthcare, including health policy, mutual respect, patient choice, patient empowerment, accessibility to healthcare, and information about self-care. The WHO describes PCC as encompassing clinical and non-medical needs, in combination with the crucial role people play in shaping health policy and health services, which is a broader concept than patient-centered care and person-centered care. Subsequently, the WHO provided a more detailed definition of PCC as an approach to care that "consciously adopts individuals' cares, families' and communities' perspectives as participants in, and beneficiaries of, trusted health systems that are organized around the comprehensive needs of people rather than individual diseases, and respects social preferences" (WHO, 2016d: 2).

People-Centered Care in Hong Kong

In Hong Kong, the term patient-centered care was first used during the 1990s by the Hospital Authority, the statutory body responsible for Hong Kong's public hospital services. Patient-centered care is currently a core value of Hong Kong's health system, which acknowledges the "importance of having a caring heart" and provides "good two-way communication" in its efforts to approach patients' needs more holistically (Hong Kong Hospital Authority, n.d.). In 2008, the Hospital Authority launched a patient empowerment program in collaboration with non-government organizations to improve understanding and enhance self-care management in patients with chronic disease (Hong Kong Hospital Authority, n.d.).

In 2009, the Hospital Authority launched the first patient experience exercise to integrate the patient's view into a review mechanism to improve the quality of healthcare. E. L.-Y. Wong and her team (at the JC School of Public Health and Primary Care of The Chinese University of Hong Kong) were commissioned to develop a tool for measuring patient experience, and to conduct the first public hospital system survey (Wong et al., 2012). Locally developed and validated, the tools were based on analysis of data from patient and staff focus groups using the Care Quality Commission's General Inpatient Questionnaire framework and validation survey (Wong et al., 2015). Subsequent to the first benchmarked patient experience exercise, the Hospital Authority implemented patient experience exercises on a regular basis, surveying hospital inpatients every 3 years, and conducting specialty-based surveys in between (Hong Kong Hospital Authority, 2015).

Specialty-based patient experience surveys were conducted in outpatient settings in 2014 and in accident and emergency departments in 2016.

The surveys elicit patients' overall impressions and experiences through the care journey. The survey covers sequential aspects of the care journey, from receiving a medical service to discharge from care. Patients rate each experience from 0 to 10, where the higher score reflects a better experience. Survey averages included 7.8 for inpatient services, 7.7 for specialist outpatient services, and 7.7 for accident and emergency services (Hong Kong Hospital Authority, 2015, 2016, 2017a). Improved patient engagement to support self-care and decision-making was required in outpatient, inpatient, and accident and emergency settings. The first benchmark inpatient experience survey in 2010 was encouraging, with a mean score of 7.4 for the overarching item "Overall, how would you rate the care you received?" (Wong et al., 2012). Most elements of the care journey rated between 7 and 8 and were regarded as moderate. Thus, we identify the comparatively better experiences as those receiving a mean score of >9, and areas with a comparatively worse experience as those with a mean score of <7 in Table 49.1.

The most negative experiences related to patient engagement on choice, information, and involvement in decision-making. To further investigate the PCC experience, particularly in reference to patient engagement, the Hospital Authority commissioned E. L.-Y. Wong and her team (Wong et al., 2017) to explore the operational understanding and experience of patient engagement from the perspectives of patients as well as healthcare staff (doctors and nurses). The operational framework of patient engagement was described by both patients and healthcare staff as 12 elements in the areas of "communication and information sharing," "involvement in decision-making," and "self-care and safety." The study found that there was a significant gap between the experiences of patients and healthcare staff, particularly in the areas of communication and information sharing and involvement in decision-making. Organizational factors, such as heavy workloads, time constraints of staff, and the hospital setting itself, were identified as challenges to achieving patient-centered care.

Impact

There have been a number of systematic reviews of PCC measurements, intervention, and outcomes. For the PCC measurements, the IAPO reviewed the literature and identified existing patient-centered healthcare domains and measurement indicators at the health system level, and in hospital or primary care settings (International Alliance of Patients' Organizations, 2012). The term patient-centered first appeared in family medicine; however, the tools for measuring patients' perception of PCC in family medicine

TABLE 49.1

2010 Inpatient Experience Survey Results

Care Journey	Comparatively Better Experience (Mean Score > 9/10)	Comparatively Worse Experience (Mean Score < 7/10)
Admission to hospital	NA	• Providing a choice of admission dates for planned admission
Environment, food, and facilities	• Feeling bothered by noise at night from hospital staff	• Providing a place to keep patients' personal belongings while in the ward • Taste of hospital food • Offering a choice of food types or amount
Hospital staff	• Having confidence and trust in doctors and nurses	NA
Patient care and treatment	• Hospital staff providing consistent information to the patient • Having enough privacy when patients were being examined or treated • Getting enough help from staff when needed	• Involvement in decisions about patients' care and treatment • Having enough opportunity for family members or someone else close to patients to talk to doctors • Healthcare staff comforted patients or discussed their condition if they had worries or fears
Leaving hospital	• No delayed discharge for any reasons • Providing a clear and understandable explanation of the purpose of at-home medications • Providing clear explanations of how to take medications • Providing clear written or printed information about medications	• Involvement in decisions about patients' discharge • Describing danger signs to watch for after discharge • Providing information needed for patient care and recovery to a family member or someone else close to the patient in care and recovery • Providing contact information when patients were worried about their conditions or treatments
Overall impression	• Being treated with respect and dignity in the hospital	• Asking patients' views on the quality of care • Providing a drop box for the collection of patients' opinions and complaints

NA = No item with score > 9/10

were limited to four key dimensions: disease and illness experience, whole person, common ground, and patient–doctor relationships (Hudon et al., 2011). In terms of intervention and outcomes, the literature published between 1998 and 2013 summarized the positive impacts of person-centered approaches on patients' peer support, information understanding, health-care experience, and shared decision-making. Most studies showed that patient-centered care had a positive impact on patient satisfaction, as well as increasing perceptions of quality of care (McMillan et al., 2013). Among a target population, such as those suffering from dementia, intensive activity-based person-centered intervention resulted in a significant reduction in agitation over the short-term (Kim and Park, 2017). Meaningful analysis of cost-effectiveness, including research over the long-term, was very limited.

Future Directions of Integrated People-Centered Care

While the terms patient-, person-, and people-centered care are often used interchangeably, they can be differentiated in that they refer to distinct levels of care: patient- or person-centered care targets individual patients at a micro-level, whereas PCC targets specific population groups within the community at a macro-level and forms the overarching framework driving patient- or person-centered care. These important conceptual frames help to define the dynamic processes needed to build an integrated people-centered health system.

The WHO identified five independent strategies for strengthening integrated people-centered health systems:

1. Empowering and engaging people and communities
2. Strengthening governance and accountability
3. Re-orienting the model of care
4. Coordinating services within and across sectors
5. Creating and enabling the environment in four domains: population and individuals, service delivery processes and health practitioners, healthcare organizations and system enablers, and management and stewardship (WHO, 2016d,e)

Echoing the WHO's advocacy of PCC, the Hong Kong Hospital Authority made patient-centered care a priority in its 2017–2022 strategic plan (Hong Kong Hospital Authority, 2017b). The report emphasized the importance of giving patients timely access to high-quality and responsive services through multifaceted strategies relating to efficiency, effectiveness, and patient engagement. A coordinated team of healthcare providers working across different settings and levels of care and committed to PCC is a fundamental transformation in the way we deliver care. Engagement of different stakeholders in the process of implementation and management is also crucial for people-centered integrated care. In order to monitor the process for continuous improvement, policy measures and indicators should first be developed by experts and then validated by health professionals, patients, and, in certain contexts, the general public.

Conclusions

Hospital-based, disease-based, curative care models are unlikely to meet the health system challenges presented by rapidly aging populations and limited resources. Compartmentalized, uncoordinated care provisions are inadequate in such a context. In 2016, the WHO published a framework for integrated people-centered health services to improve health system performance in terms of accessibility, equity, affordability, quality, and responsiveness. Interconnected strategies were outlined for building up more effective health services that embrace all stakeholder perspectives, from individuals to communities. There are many challenges ahead; however, the implementation of an integrated patient-centered health system, where patient engagement and public participation are encouraged, is the key to creating a successful and sustainable health system.

50

India

How to Build a First-World Health System on a Third-World Budget

Girdhar Gyani

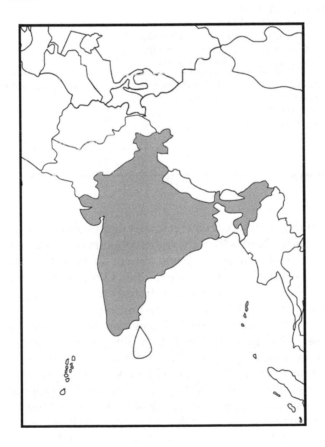

CONTENTS

Indian Data

- Population: 1,324,171,354
- GDP per capita, PPP: $6,572.3
- Life expectancy at birth (both sexes): 68.3 years
- Expenditure on health as proportion of GDP: 4.7%
- Estimated inequity, Gini coefficient: 35.1%

Source: All data are from the World Health Organization and World Bank. Latest available data
used as at October 2017.
GDP = Gross Domestic Product
PPP = Purchasing Power Parity

Background

Universal health coverage is being pursued aggressively by the government of India. Currently, about 50% of the Indian population is covered under some or other government insurance scheme. Patient safety and the affordability of health services have emerged as key priorities. Other issues currently in focus include making healthcare services uniformly available across the country, particularly in rural areas, and reducing regional disparity in health indicators like the infant mortality rate (IMR) and maternal mortality rate (MMR).

At the time of independence in 1947, healthcare was predominantly provided by the government, with only 8% delivered by private hospitals. The public health network was well established, with each district being served by a 200-bed secondary care hospital, supported by an appropriate number of community hospitals (30 beds) and a number of primary health centers and subcenters to cater for primary healthcare services. While this system worked adequately during the initial years, it failed to respond to the needs of the growing population. The private sector seized the opportunity, and from 1980 onward, India witnessed the corporatization of the healthcare sector. Presently, more than 70% of India's outpatient department (OPD) and 60% of its inpatient department (IPD) healthcare services are provided by the private sector.

Until recently, healthcare was not a prominent feature of the Indian political agenda; this is rapidly changing, however. Central and state governments are aggressively promoting their plans for health insurance coverage. The biggest push, in the form of the National Health Protection Scheme (NHPS), is currently being launched by the central government. The NHPS, which will cover about 500 million people, will be a cashless scheme under which specified categories of impoverished populations will be able to access quality healthcare from empaneled private hospitals for secondary and tertiary healthcare services. It will be by far the most extensive social health coverage, in terms of sheer numbers, anywhere in the world.

Success Story

India successfully established the National Accreditation Board for Hospitals and Healthcare Providers (NABH) in 2006, primarily to meet the demand for medical tourism. In order to achieve global recognition, the NABH's standards were accredited by the International Society for Quality in Health Care (ISQua) in 2008. Five hundred Indian hospitals are now accredited by NABH (National Accreditation Board for Hospitals and Healthcare Providers, 2017).

Since accreditation, India has emerged as a favored destination for patients from the Middle East, Central Asia, Africa, and South Asian Association for Regional Cooperation (SAARC) countries who wish to access first-world health services at third-world prices. The growth of medical and wellness tourism in the country has been between 23% and 25% over the past couple of years. In 2015, the total number of overseas patients visiting India on medical visas was 134,344; this has risen to about 200,000 in the year 2016–2017 (Chowdary, 2017). A large number of Indians working in overseas countries (non-resident Indians) also receive medical treatment during their visits to India. The availability of accredited quality care at a significantly low cost when compared with other countries has boosted medical tourism (Table 50.1).

In a 2016 initiative, the Indian Insurance Regulatory and Development Authority (IRDA), which represents 33,000 healthcare organizations (HCOs), most of which are smaller hospitals with less than a 100-bed capacity, has mandated that private insurance companies empanel only those hospitals that have NABH accreditation within the next 2 years. The Central Government Health Scheme (CGHS), which provides cover to central government employees (including pensioners), provides cash incentives to the NABH-accredited tertiary care hospitals, by which the accredited hospitals are reimbursed at a higher rate (15% more) than unaccredited hospitals. Eight Indian states have also made NABH accreditation a criterion for their discrete government-run health insurance schemes. These states have jurisdiction over more than 2000

TABLE 50.1

Treatment Cost Comparison

Procedures	United States ($)	Costa Rica ($)	India ($)	Korea ($)	Mexico ($)	Thailand ($)	Malaysia ($)
Heart bypass	$144,000	$25,000	$5,200	$28,900	$27,000	$15,121	$11,430
Angioplasty	$57,000	$13,000	$3,300	$15,200	$12,500	$3,788	$5,430
Heart valve replacement	$170,000	$30,000	$5,500	$43,500	$18,000	$21,212	$10,580
Hip replacement	$50,000	$12,500	$7,000	$14,120	$13,000	$7,879	$7,500
Hip resurfacing	$50,000	$12,500	$7,000	$15,600	$15,000	$15,152	$12,350
Knee replacement	$50,000	$11,500	$6,200	$19,800	$12,000	$12,297	$7,000
Spinal fusion	$100,000	$11,500	$6,500	$15,400	$12,000	$9,091	$6,000
Dental implant	$2800	$900	$1,000	$4,200	$1,800	$3,636	$345
Lap band	$30,000	$8,500	$3,000	N/A	$6,500	$11,515	N/A
Breast implants	$10,000	$3,800	$3,500	$12,500	$3,500	$2,727	N/A
Rhinoplasty	$8,000	$4,500	$4,000	$5,000	$,500	$3,901	$1,293
Face lift	$15,000	$6,000	$4,000	$15,300	$4,900	$3,697	$3,440
Hysterectomy	$15,000	$5,700	$2,500	$11,000	$5,800	$2,727	$5,250
Gastric sleeve	$28,700	$10,500	$5,000	N/A	$9,995	$13,636	N/A
Gastric bypass	$32,972	$12,500	$5,000	N/A	$10,950	$16,667	$9,450
Liposuction	$9,000	$3,900	$2,800	N/A	$2,800	$2,303	$2,299
Tummy tuck	$9,750	$5,300	$3,000	N/A	$4,025	$5,000	N/A
Lasik (both eyes)	$4,400	$1,800	$500	$6,000	$1,995	$1,818	$477
Cornea (both eyes)	N/A	$4,200	N/A	$7,000	N/A	$1,800	N/A
Retina	N/A	$4,500	$850	$10,200	$3,500	$4,242	$3,000
In vitro fertilization treatment	N/A	$2,800	$3,250	$2,180	$3,950	$9,091	$3,819

hospitals. These reforms have led to great improvements in patient safety and quality of care.

While the major push for quality and accreditation has so far come from government health insurance schemes and IRDA-empaneled private health insurance companies, media campaigns are currently underway in an effort to increase citizen awareness of the benefits of accreditation. This will motivate and empower patients to choose accredited hospitals as their preferred healthcare destinations. This will further accelerate the process and encourage the vast majority of small hospitals, even outside the jurisdiction of IRDA, to seek accreditation.

In order to strengthen India's regulatory framework, the Indian government introduced the Clinical Establishments Act (CEA) in the year 2010, under which HCOs are required to comply with minimum infrastructural requirements. Under the act, HCOs are also required to adhere to standard treatment guidelines (STGs). While compliance is mandatory in hospitals located within the seven union territories (as these are directly under administrative control of central government), individual states have the option to adopt the CEA 2010 or draft their own regulatory framework. By March 2017, nine out of 29 states had adopted the act. The Indian government is encouraging the remainder of the states to follow suit so that all regulatory standards are aligned. This will progressively improve quality even among the non-accredited hospitals.

Impact

Demand for quality accreditation has largely been driven by advocacy from health insurers and government. Various incentives, along with the regulatory push, are aimed at creating a large pool of accredited HCOs over the next couple of years. India's success story thus has two drivers—mandatory regulation through the CEA, and voluntary accreditation based on global standards and incentivized by payers, including government and private insurance companies. Combined governmental and private health insurance schemes are likely to cover about 70% of the Indian population by the end of 2018. The beneficiaries will receive quality health services through a cashless scheme. While hospitals will experience an increased volume of patients, the reimbursement will be fixed by the government, which in all probability will be lower than expected. Hospitals will therefore be required to improve their operational efficiency so as to remain sustainable.

In order to help hospitals comply with both the regulatory framework under the CEA and the new accreditation standards, large numbers of training and counseling professionals will be required. A number of institutes, including the Association of Healthcare Providers (India) (AHPI) Institute of Healthcare Quality, the Quality Council of India, and the Consortium of Accredited Hospitals, have taken the lead in preparing competent training and consulting agencies to take up this role. As a result, more than 300 HCOs have been accredited, and it is expected that this will continue to improve as more and more staff are trained.

Implementation and Transferability

The Indian model is easily transferable to other countries in general and developing nations in particular. The Indian experience has highlighted the importance of encouraging a good regulatory framework. Once regulation becomes effective, it becomes easy to introduce accreditation. As accreditation is voluntary, a multipronged approach must be taken to promote the process. First, citizens must be made aware that accreditation is directly linked to patient safety, thus empowering them to demand accreditation and to only seek medical treatment from accredited hospitals. NABH has initiated a campaign in the print media in which the benefits of accreditation are regularly publicized in each region. Each promotion includes a list of accredited hospitals in the local area. Print media is accessible to citizens at the grassroots level, and the campaign is beginning to show positive results. Financial backers, including government and private health insurance agencies, will continue to incentivize quality by linking accreditation with reimbursement at a higher rate. The Indian example also illustrates the importance of establishing training and consultancy resources prior to accreditation and the imposition of regulations to facilitate implementation.

The Indian initiative to improve patient safety through national accreditation could be adapted to suit most developing countries. The Indian model of private–public partnership (PPP), whereby the government provides incentives to the private sector to invest in the setting up of hospitals and nursing homes and then procures healthcare services under their social welfare programs, could be helpful to those countries that lack public resources and infrastructure.

Prospects

India's accreditation and patient safety initiatives have been launched and are running successfully. India is a big country with diverse terrain and cultures, and additional efforts will be required to achieve uniform results. There are a number of states that are short of revenue resources, and unless central government makes good on the shortfall, these regions may not be able to match the achievements of leading states.

Although patient safety is a priority, the implementation of the accreditation program is still widely perceived as a cost that does not provide a return on investment, rather than a benefit. Case studies focusing on the "cost of poor quality" model are currently underway and should provide evidence that quality may cost money in the initial phase, but ultimately helps to reduce medical errors and save lives.

Cost-saving tools like Lean, 5S, Kaizen, and Six Sigma, which originated in the Indian manufacturing and service sectors (automotive industry and information technology sector), are also being utilized to increase economic efficiency.

Conclusion

Healthcare accreditation is more widespread in developed nations, where it is a requirement for insurance coverage. It is also used as a marketing tool for attracting medical tourism. India has launched an accreditation program, adopting a multipronged approach, to ensure patient safety and improve health outcomes. These include linking accreditation with regulation, incentivizing accreditation, and empowering Indian citizens to make informed choices. Simultaneously, the use of management tools is being encouraged to improve efficiency and bring down cost. The Indian healthcare sector has made impressive advances, and India is increasingly regarded as a destination that provides first-world services at a third-world price.

51

Japan

Toward a Community-Friendly Dementia Strategy

Yukihiro Matsuyama

CONTENTS

Japanese Data

- Population: 126,994,511
- GDP per capita, PPP: $41,469.9
- Life expectancy at birth (both sexes): 83.7 years
- Expenditure on health as proportion of GDP: 10.2%
- Estimated inequity, Gini coefficient: 32.1%

Source: All data are from the World Health Organization and World Bank. Latest available data used as at October 2017.
GDP = Gross Domestic Product
PPP = Purchasing Power Parity

Background

According to a report published by the World Health Organization (WHO) in 2012a, "Dementia—A Public Health Priority," the number of people currently living with dementia globally is estimated to be 35.6 million, a number that is predicted to double by 2030 and more than triple by 2050. The report warns that "dementia is a costly condition in its social, economic, and health dimensions. Nearly 60% of the burden of dementia is concentrated in low- and middle-income countries and this is likely to increase in coming years" (WHO, 2012a: v). In other words, regardless of a nation's stage of economic development, dementia is a problem that needs to be addressed.

Japan has the highest proportion of elderly citizens in the world. As the prevalence of dementia increases with age, an increase in dementia is set to become both a major health and a social problem. The social consequences of such an increase can already be seen in the rise in the number of people missing due to wandering, and traffic accidents in which dementia patients are victims or perpetrators. The Japanese government launched a comprehensive dementia policy called the New Orange Plan in January 2015 (Ministry of Health, Labor and Welfare [Japan], 2015).

This chapter discusses the basic data on dementia in Japan and the outline of the New Orange Plan, which replaces the Japanese Health Ministry's

5-year Orange Plan of 2013. In addition to promoting the use of advanced technologies, such as information and communication technology (ICT) and robots, as a measure against dementia, aspects of the Japanese social system also need to be reformed. It seems that some of these efforts will be successful and others will fail. However, even failed initiatives will provide a useful example to those planning dementia measures in other countries.

The New Orange Plan

The New Orange Plan is based on 2012 estimates of dementia prevalence by age group (Table 51.1). The prevalence of dementia increases sharply in both males and females over the age of 75, and by age 85, 47.09% of males and 58.88% of females will be suffering from dementia. While the proportion of elderly will continue to rise, from 17.4% of the population in 2000 to 30.3% in 2025 and 33.4% in 2035, the number of elderly over 85 will be more than 10 million—around 9% of the total population by 2035 (Table 51.2).

Even if the dementia prevalence percentage remains the same as in 2012, it is estimated that the number of people with dementia will be 8 million in 2035, accounting for 21.4% of the elderly population (Table 51.3).

The growing number of patients with dementia has not only increased the economic cost for medical and nursing care but also caused various social problems. According to the National Police Agency, the number of missing persons with dementia increased from 9,607 in 2012 to 12,208 in 2015. Most of these people, who have usually wandered, have been discovered successfully within a week, but there are also a number who have never been found or who have died. The incidence of dementia sufferers who continue to drive has led to a serious increase in traffic accidents.

Because much of the Japanese labor force moved from rural to metropolitan areas during periods of strong economic growth (from 1955 to 1973), both the concentration of elderly population and the problems associated with

TABLE 51.1

Dementia Prevalence Percentage in 2012

Age Range	Male (%)	Female (%)
65–69	1.94	2.42
70–74	4.30	5.38
75–79	9.55	11.95
80–84	21.21	26.52
Over 85	47.09	58.88

Source: Ninomiya, T., The study on projections of the elderly population with dementia in Japan 2015. Tokyo: Ministry of Health, Labor and Welfare, 2015.

TABLE 51.2

Elderly Population Projection (Thousands)

		2000	2015	2025	2035
Total population (1)		126,926	127,095	120,659	112,124
Elderly population (2)		22,0 41	33,866	36,573	37,407
(2) ÷ (1)		17.4%	26.6%	30.3%	33.4%
Male	65–69	3,365	4,723	3,407	3,863
	70–74	2,676	3,625	3,622	3,308
	75–79	1,629	2,817	3,797	2,806
	80–84	917	2,016	2,505	2,599
	Over 85	655	1,478	2,378	3,447
Female	65–69	3,753	5,036	3,665	4,095
	70–74	3,234	4,161	4,094	3,687
	75–79	2,528	3,537	4,600	3,375
	80–84	1,701	3,010	3,522	3,525
	Over 85	1,582	3,467	4,984	6,702

Source: National Institute of Population and Social Security Research [Japan], Population projections for Japan (January 2012): 2011 to 2060, 2012. Retrieved from http://www.ipss.go.jp/site-ad/index_english/esuikei/ppfj2012.pdf.

TABLE 51.3

Number of Dementia Patients (Thousands)

	2012	2025	2035
Number of dementia patients	4620	6760	7990
Ratio to the elderly population	15.0%	18.5%	21.4%

Source: Ninomiya, T., The study on projections of the elderly population with dementia in Japan 2015. Tokyo: Ministry of Health, Labor and Welfare, 2015.

the increase in dementia are greater in cities. Between 2010 and 2035, the population over 75 is set to increase by 82.5% in metropolitan areas, compared with 46.2% in rural areas (Table 51.4). The proportion of single-person households among those over 75 was 25.5% in 1980; it is estimated that this will be around 40% by 2035 (Table 51.5). In terms of dementia prevention and support, it is a risk factor for elderly people to live alone because they tend to lose opportunities for communication and may become quite isolated.

The Japanese government has implemented the New Orange Plan in an effort to alleviate some of the negative consequences of the projected increase in dementia sufferers. The following seven policy goals have been set to create a friendly community for elderly people with dementia:

1. Promotion of education to deepen understanding of dementia
2. Provision of timely and appropriate medical and nursing care depending on the condition of dementia

TABLE 51.4

Comparison of Elderly Population in Metropolitan Area Prefectures and Others (Thousands)

		2010 (1)	2025	2035 (2)	(2) ÷ (1)
Total	Population	128,057	120,659	112,124	0.876
	65–74	15,290	14,787	14,953	0.978
	Over 75	14,194	21,786	22,454	1.582
Metropolitan area prefectures	Population	52,254	50,925	48,264	0.924
	65–74	6,126	5,535	6,537	1.067
	Over 75	4,683	8,415	8,548	1.825
Other prefectures	Population	75,803	69,734	63,860	0.842
	65–74	9,164	9,252	8,416	0.918
	Over 75	9,511	13,371	13,906	1.462

Source: National Institute of Population and Social Security Research [Japan], Population projections for Japan (January 2012): 2011 to 2060, 2012. Retrieved from http://www.ipss.go.jp/site-ad/index_english/esuikei/ppfj2012.pdf.

Note: Metropolitan area prefectures include Tokyo, Kanagawa, Saitama, Chiba, Osaka, and Aichi.

TABLE 51.5

Increase of Households Whose Head is Over 75 and Living Alone (Thousands)

	Number of Households Whose Head is Over 75	Number of Households Whose Head is Over 75 and Living Alone	
	A	B	B ÷ A
1980	1,070	273	0.255
1990	2,176	640	0.294
2000	3,943	1,393	0.353
2010	7,308	2,693	0.368
2025	11,867	4,473	0.377
2035	11,736	4,660	0.397

Source: National Institute of Population and Social Security Research [Japan], Household projections for Japan 2010–2035, 2013. Retrieved from http://www.ipss.go.jp/pp-ajsetai/e/hhprj2013/t-page_e.asp.

3. Enhancement of measures against early-onset dementia

4. Support for those who care for dementia patients

5. Promotion of communities that are friendly to the elderly, including people with dementia

6. Promotion of research and development of the prevention, diagnosis, and treatment method, and the rehabilitation and nursing care model, and dissemination of the result

7. Focus on the point of view of people with dementia and their families

Developing Infrastructure

The New Orange Plan has set several numerical targets in order to achieve these policy goals. For example, 8 million people are to be trained as dementia supporters by March 2018. Dementia supporters are volunteers who are correctly informed about dementia, warmly watch over people and families with dementia, and provide support while family members undergo a training course. In 2018, an initial intensive support team for dementia will be established in all municipalities. Under this program, medical and nursing care professionals will visit dementia patients and people suspected of having dementia and consult with their family members, introducing and adjusting medical and nursing care as required. Thus far, it appears that the implementation of this program is proceeding smoothly. The challenge now is whether each program can improve its function further.

To be effective, dementia control measures require the involvement of local residents as well as trained care workers. In view of this, the government has established a number of regional comprehensive healthcare delivery systems within populations of 10,000 people. These systems will be expanded across the country by 2025. Since the emphasis in such a system is not medical care but long-term care and living support, it is not hospitals but medical practitioners and social welfare corporations that play a central role. Social welfare corporations are non-profit organizations that provide relief activities for the disadvantaged. There are about 20,000 social welfare corporations; approximately 18,000 operate elderly facilities, disabled facilities, child care facilities, nurseries, and so on. In order to help coordinate care provided by social welfare corporations, which are generally only small organizations, and to enable them to play a more significant role in elderly care, the government revised the social welfare law in March 2016.

Transferable Exemplars

As a successful example of support for dementia patients by social welfare corporations, Share Kanazawa, which was created by the social welfare corporation Bussien in Ishikawa Prefecture, has attracted attention. Share Kanazawa is a model community where people with disabilities and healthy people, including children, students, working generations, and the elderly, live together. There are facilities for children, senior citizen facilities, a laundry operated by disabled workers, an outdoor school, a sports club, a restaurant, a jazz cafe, hot spring facilities, and so on.

There is a much-told anecdote about dementia patients in Bussien: a grandmother with dementia was trying to feed spoonfuls of jelly to a young man who was unable to move his neck, but her hand was trembling so much that the jelly spilled. Two weeks later, her trembling stopped, but in the meantime, in an effort to anticipate her movements, the young man had learned to move his neck. While the grandmother had formerly tended to remain at home, she soon began to actively attend the facility in Bussien. In supporting the young man, she had been given a reason to live. This story reinforces the importance of non-drug care in the treatment of dementia. It also acts as a caution to healthy people who tend to underestimate the capabilities of dementia patients as well as their determination. And like most people, whether well or ill, people suffering from dementia place great value on interacting with others. If communities such as Share Kanazawa are replicated across the country, with local residents involved in their management, dementia patients, and their families, will be greatly supported.

In order to ensure that all sufferers are able to remain in their communities, the Japanese government aims to establish initial intensive dementia support teams in all municipalities by 2018. According to the New Orange Plan commentary material, this will involve medical and nursing professionals visiting people with dementia and their families to provide necessary medical and nursing care and family support.

Prospects

In dementia care, there are two stages. The first stage aims to maintain a good quality of life for dementia patients in the initial stages of the disease, and the second stage to provide an appropriate environment for those who are in the final stages of life. Because patients with dementia generally develop various other diseases, all doctors and nurses, regardless of their own specialization, will need to be able to treat dementia patients. However, dementia specialists are also necessary. Currently, there are only about 1000 doctors accredited by the Japan Society for Dementia Research. Given the expectation that the number of dementia patients will be 8 million by 2035, around 10,000 specialists will be required.

The use of robots and ICT is effective for improving the care and security of dementia patients and supporting care providers, and various tools, such as communication robotics and the wanderer discovery system, have already been developed in industrialized countries. Japan, which has the largest clinical field when it comes to dementia, can make more contributions in this area, including the development of dementia drugs.

Conclusion

Improvement in public health and advances in medical technology have led to an increase in life expectancy in many countries; this increased life expectancy has, in turn, led to a rapid increase in the number of dementia patients in all developed nations. In Japan, the development of a regional comprehensive healthcare delivery system that enlists community support rather than just offering medical services will not only provide an effective solution, but also help ensure quality of life for all dementia patients.

52

Malaysia

The Future Malaysian Antenatal Care System: Building upon the Old

Ravindran Jegasothy and Ravichandran Jeganathan

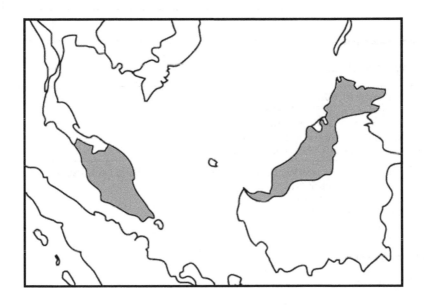

CONTENTS

Malaysian Data

- Population: 31,187,265
- GDP per capita, PPP: $27,680.8
- Life expectancy at birth (both sexes): 75.0 years
- Expenditure on health as proportion of GDP: 4.17%
- Estimated inequity, Gini coefficient: 46.3%

Source: All data are from the World Health Organization and World Bank. Latest available data used as at October 2017.
GDP = Gross Domestic Product
PPP = Purchasing Power Parity

Background

Since independence in 1957, Malaysia has developed a far-reaching network of primary care through the creation of 2869 health and community clinics that provide midwifery services. In addition, there are mobile and Flying Doctor teams that supplement the service. This achievement was possible due to the government's long-term commitment to healthcare, with a particular emphasis on maternal and reproductive health.

Primary care is considered the fulcrum of the health system. Patients can access the healthcare system at the primary care level or by referral to a specialist at the hospitals. However, primary care exercises only a partial gatekeeper function since public-sector patients can obtain a referral from a general practitioner and, for a small additional fee, can go directly to specialists and hospitals. Private-sector specialists can be accessed through a fee-for-service mechanism. The Malaysian system is currently dichotomous:

public-sector services increasingly serve the poor, while private-sector services serve the more affluent people who live in urban areas (Jaafar et al., 2013). Three patterns can be recognized in healthcare delivery at hospitals:

1. Strengthening specialty care at large public hospitals and spreading the reach of specialty services to identified district centers
2. Introducing subspecialty care in major disciplines, expanding top-end private hospitals with private-sector investment to cater to the medical tourism market
3. Increasing the number of ambulatory care centers

Malaysia's current healthcare policies emphasize both a wellness and a disease perspective (Pathmanathan and Liljestrand, 2002). Malaysia is undergoing an epidemiological transition, with causes of mortality shifting from communicable to non-communicable diseases. Diabetes, hypertension, metabolic syndrome, and obesity are now a permanent feature of the Malaysian healthcare scene. Although there are episodes of communicable disease outbreaks, such as dengue fever, the health budget increasingly focuses on lifestyle and nutritional diseases. Thus, operational efficiency is essential to ensure that priorities in healthcare are met and that there is value for money spent.

Since independence, Malaysia has achieved a significant decrease in the maternal mortality rate: from 280 deaths per 100,000 live births in 1957 to 27 per 100,000 in 2015. The initial decline was rapid, driven by improved access to healthcare and the professionalization of midwifery services. Current rates have plateaued, due to the complexity of care required when dealing with increasingly indirect causes of maternal morbidity and mortality, such as cardiac and respiratory diseases in pregnancy.

Current System

Antenatal care essentially monitors the well-being of the mother and her unborn child. The quality of care reflects the capabilities of the health and obstetric sectors of the health service. The quality and spread of care are essential to the reduction of maternal and infant mortality given that many health problems in pregnant women can be prevented, detected, and treated during antenatal care visits. In 1990, 78.1% of pregnant women attended an initial antenatal care visit at a health or midwives' clinic. By 2007, this number had increased to 91.9%, and in 2013 the number reached 98%. Between 2007 and 2013, 93% of pregnant Malaysian women attended at least one antenatal care visit; this compares well with the global average of 83% from 2007 to 2014 (WHO, 2015a).

Based on a review of the effectiveness of different antenatal care models, the World Health Organization (WHO) currently recommends a minimum of four antenatal care visits. These visits should include interventions such as tetanus toxoid vaccinations, screening and treatment for infections, and identification of warning signs during pregnancy. The average number of antenatal visits to public and private health facilities increased from 6.6 visits per mother in 1990 to 8.5 visits in 2000, rising to 10 visits in 2012, with an average of 9 visits per mother since the year 2004.

In Malaysia, the quality of the four antenatal care visits has not been evaluated in a national study. In the audit of maternal deaths, it has, however, been noted that the number of antenatal visits does not necessarily equate to the quality of care. Many maternal deaths have occurred despite multiple antenatal visits. Ineffectual and unnecessary visits must be reduced if quality of care is to be improved, and finite resources conserved.

A recent study attempted to analyze adherence to recommended antenatal care schedules, specifically with regard to examining the extent of adherence to recommended antenatal care content, and to determine the factors associated with the care content from the obstetrics, provider, and utilization perspectives. The findings indicated non-adherence to the recommended content in many cases (Yeoh et al., 2015). This indicates that monitoring and audit of care content is vital to the success of any effort to improve the system. Adhering to proposed timing of visits and content of care is essential due to resource and maternal health implications.

In 2013, the Ministry of Health developed a Perinatal Care Manual, which provides guidelines on adequate initiation of care, number of visits, and content of routine (Ministry of Health [Malaysia], 2013). The adequacy of antenatal care is measured by the application of indices that assess initiation of care and number of visits. In addition to measuring the initiation and number of visits, the care content should be assessed.

New System

The proposed new antenatal care system (Figure 52.1), which has been phased in since 2016, will utilize the following building blocks, which have been developed over the years:

- Standard operating procedures will be consolidated through use of the Perinatal Care Manual, which is now in its third edition. This manual is meant for both the health and hospital sectors and takes into account the frequent referrals between both sectors.

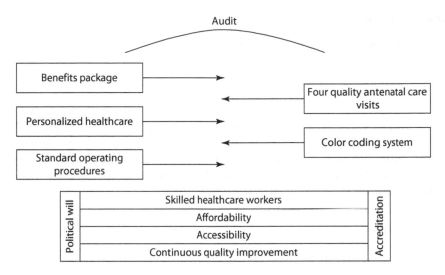

FIGURE 52.1
Proposed Malaysian antenatal care system.

- The obstetric color coding system, which empowers midwives to move mothers to specialist units when required, will be strengthened (Ravindran et al., 2003).

- Personalized healthcare will be upscaled throughout the country, prioritizing the rural and semiurban areas. Every antenatal patient will be assigned a particular midwife who will monitor the patient throughout her pregnancy. Pilot studies in selected health districts have shown that such care has a good impact on patient satisfaction and care. The challenge will be to implement such a system in urban areas due to higher patient–staff ratios.

- The four-antenatal-visit model will be implemented so as to minimize resource wastage and improve quality of care. There will be no compromise in the care provided by specialists, as referrals will still be possible on a needs basis.

- The new system will be tied to a national health insurance scheme. The proposed benefits package ensures that antenatal care is affordable and emergency obstetric care will always be available, irrespective of a mother's ability to pay.

- An audit of the quality of care will be done through the quality assurance framework, which includes a number of predetermined standards for obstetric care, as well as an incident reporting system. The previously established and highly successful Confidential Enquiries into Maternal Deaths and Maternal Death Review Frameworks will also contribute to the audit (Ravichandran and Ravindran, 2014).

Projected Strengths of the New System

Accessibility

Accessibility is the cornerstone of any successful healthcare system. Accessibility will not be an issue, as infrastructure expansion and strengthening was achieved during the early phase of development. The proximity of facilities to patients' homes will be enhanced by the merging of the public- and private-sector facilities under a national health financing scheme. This will also serve to address the issue of overcrowding in public facilities.

Affordability

Antenatal care has always been provided inexpensively in Malaysia. For the cost of RM1 (US$0.23), the full spectrum of antenatal care is provided by a trained midwife or medical officer. If a referral is made to a specialist clinic for an opinion, RM30 (US$6.8) will be charged for a first visit, and subsequent follow-ups cost only RM5 (US$1.1). Additional investigations, such as ultrasound scans, cost only RM50 (US$11.4), and if the patient is unable to pay, a welfare officer may waive the charges. Under the proposed national health financing scheme, charges will be standardized across the sectors. Public servants and the underprivileged will be covered by the state, and those who can afford to will contribute a fixed sum each month (dependent on income) to access medical care.

Availability of Skilled Care

The Malaysian antenatal color coding system is a managerial tool that empowers midwives who are considered professionals under the Midwifery Act to utilize a risk assessment system that enables them to rapidly escalate care for high-risk patients to a specialist if required. Despite criticism that there is no valid risk prediction model in obstetrics, a decision has been made to retain the color coding system. Most patients are still managed by trained midwives who ensure that care remains at the appropriate level. An essential cog in this system is the established referral system, which ensures safe and appropriate transfer of a pregnant mother when required.

Accreditation

A national system of accreditation has been in place since 1991, and the fifth edition of Malaysian standards is underway. The Malaysian Society for Quality in Healthcare (MSQH), which runs the accreditation system, has itself been accredited by the International Society for Quality in Health Care

(ISQua) for organization, standards, and surveyor training. A Malaysian accreditation standard that covers hospitals and clinics has thus been internationally benchmarked and has listed quality indicators in obstetric care that need to be adhered to and complied with.

Political Will

Successive Malaysian governments have placed great emphasis on female and infant health. Since independence, there have been specific ministries responsible for women's affairs. The Ministry of Health initially ran a Maternal and Child Health division, which has since evolved into a Family Development Division with a wider focus, adopting a "womb-to-tomb" approach to women's care.

Setting an Agenda That Suits Our Needs

Developing a system that fits Malaysia's unique political and sociocultural environment has been crucial to recent reforms. Malaysia has never been dependent on external funding sources for health; however, external expertise and overseas studies have been used in policy development.

What Is New?

The system features an increased emphasis on audit, accountability, and development of key performance indicators. Efforts are focused on determining what is required, instead of unnecessary antenatal visits, to provide an empowering system that escalates care to specialists only when required. Work guidelines that emphasize teamwork have been inculcated. We hope that these modifications serve to overcome the plateauing of maternal and infant mortality in Malaysia. Like a marathon, the last mile will be the most difficult. Only an increasing emphasis on quality will help.

Conclusion

The proposed Malaysian antenatal healthcare system is a composite system that builds on elements of the system that has been established since independence. After adapting the system to local needs, it should be possible for any country to successfully replicate the Malaysian antenatal care system.

53

Mongolia

Health System Financing

Tumurbat Byamba and Tsolmongerel Tsilaajav

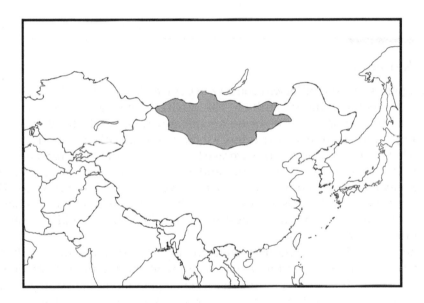

CONTENTS

Mongolian Data

- Population: 3,027,398
- GDP per capita, PPP: $12,220.4
- Life expectancy at birth (both sexes): 68.8 years
- Expenditure on health as proportion of GDP: 4.73%
- Estimated inequity, Gini coefficient: 32.0%

Source: All data are from the World Health Organization and World Bank. Latest available data
used as at October 2017.
GDP = Gross Domestic Product
PPP = Purchasing Power Parity

Background

Mongolia is a vast landlocked country in Central Asia with a relatively small population of just under 3 million. The harsh climate and conditions have a negative impact on both the socioeconomic situation and the health status of the population. While life expectancy in Mongolia is still comparatively low, rates have been increasing. The average life expectancy at birth reached 69.57 years in 2016 (75.10 years for women and 65.58 years for men), indicating that longevity has increased by approximately 6 years over the last 20 years. The birth rate has also increased over the last few years; in 2016, the number of children aged 0–4 was the highest in the population pyramid (Center for Health Development, 2016).

The country has undergone an epidemiological transition since the 1990s. As a result of increased immunization coverage and the implementation of national programs, infectious disease is no longer among the leading causes of death. Instead, lifestyle- and behavior-dependent diseases, such as circulatory system diseases, cancer, and injuries, have become the leading causes of morbidity and mortality (Bolormaa et al., 2007; Tsilaajav et al., 2013).

Health services in Mongolia are provided at primary, secondary, and tertiary levels through a nationwide delivery system. Primary healthcare in rural areas is delivered by state-owned *soum** health centers (SHCs). The secondary level of care, or general hospital care, is through general hospitals as well as through private specialty hospitals and clinics (Figure 53.1). In urban areas, primary healthcare facilities are privately run family health centers (FHCs).

Since the late 1990s, the primary healthcare system has been significantly reformed. Centers that were previously designated *Soum* hospitals were

---

* *Soum* are the smallest administrative units within a rural area; the population ranges between 5,000 and 10,000.

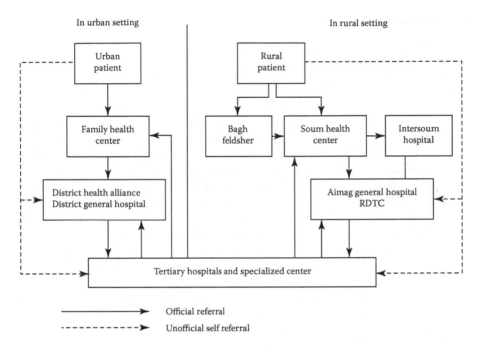

FIGURE 53.1
Patient pathway. (From Tsilaajav et al., *Mongolia Health System Review: Health Systems in Transition*. World Health Organization Regional Office for the Western Pacific, Geneva, 2013.)

transformed into health centers in 2011, and many previously solo practitioners have now formed private cooperative family health centers or practices with groups of family doctors and other medical personnel. Further, respective national standards and guidelines have been revised, and the operational focus of health facilities has been redirected from curative to preventive care.

While the number of primary healthcare facilities continues to increase, the hospital sector still dominates the Mongolian health system: there are 328 hospitals, both public and private, providing inpatient and outpatient services. The public- or government-owned hospitals are outdated and poorly maintained, provide limited services, and do not meet modern requirements for a high-quality, risk-free care environment (Center for Health Development, 2016; Tsilaajav et al., 2015).

Recent policy changes involve the possible application of public and private partnerships and the structural reorganization of the district hospitals into multispecialty facilities, which will provide inpatient services for internal medicine, pediatrics, surgery, obstetrics and gynecology, neurology, infectious disease, and dental care. Other reforms include establishing governing boards in the government-owned secondary- and tertiary-level hospitals, in order to provide organizational autonomy in management and decision-making (Bolormaa et al., 2007; Tsilaajav et al., 2013).

Health System Financing

Mongolia's system of health financing has also experienced a number of challenges over the last three decades. During this time, a number of reforms have been implemented in order to transition from the Soviet-era Semeshko model,* first established in the 1920s, to address challenges and also changes in the health system. After the collapse of the socialist system, the Mongolian health system faced significant problems due to inadequate funding, leading to reduced accessibility, availability, and quality of services; infrastructural problems; an inadequate supply of equipment and drugs; and a severe shortage of trained staff. While the current health financing system is much improved, there is still a long way to go. Major achievements include the introduction of social health insurance as one of the main sources of health financing, resource allocation to service providers through mixed payment methods, private-sector performance, and government-funded primary care services (Chimeddagva, 2011; Tsilaajav et al., 2013, 2015).

Tsilaajav et al. (2013) conclude that while total health expenditure (THE) in Mongolia has fluctuated for the last three decades, it has been relatively stable since 2005, ranging from 4% to 5.4% of the GDP. Total government health expenditures as a proportion to GDP remained between 2.6% and 2.8% over the last 10 years. As in other post-Semashko countries, more than 50% (53.5% in 2005 and 60.8% in 2016) of health expenditure goes to inpatient services (Center for Health Development, 2016).

Funding for the Mongolian health system currently comes from central government revenues, social health insurance contributions, and out-of-pocket (OOP) payments, as well as some very limited private (voluntary) health insurance contributions, and donor support. Table 53.1 shows funding sources and their percentage of total expenditure on health.

These funding sources have been managed in a fragmented manner by different government institutions. The government (Ministry of Finance and Ministry of Health) regulates the funding from the state budget according to the budget law. Social health insurance (SHI) is regulated by the Ministry of Social Welfare and Labour as stipulated in health insurance law.

The majority of SHI revenue is generated from payroll tax contributions from formal sector employees. The contribution rate is set at a very low level—4% of monthly payroll and 1% of (reported) monthly income. These contributions provided 66.3% of total health insurance fund revenue in 2000 and 88% in 2013. Those who are categorized as vulnerable have their contributions subsidized; however, this subsidy decreased from 27.4% of total revenue in 2000 to 6.6% in 2013.

---

* The Semeshko model is a Soviet-designed healthcare model: government funded and run, integrated and centralized. The model focuses on communicable diseases and hospital-based care.

TABLE 53.1

Sources of Revenue as a Percentage of Total Expenditure on Health, 2000–2012

Sources of Revenue	2000	2001	2002	2003	2004	2005	2006	2007	2008	2009	2010	2011	2012
Government expenditure	62	58	50	43	44	44	46	50	49	46	47	50	49
Social health insurance	20	27	29	12	12	12	12	10	13	14	15	14	13
OOP expenditure	12	10	17	42	41	41	39	38	35	37	35	34	35
Other private expenditure	6	5	4	3	3	3	3	3	3	3	3	3	3

Source: WHO, National health accounts, 2015c, Retrieved from http://www.who.int/nha/en/.

While central government expenditure and SHI combined account for most health revenues, these have decreased substantially from 82% to 62% of THE for the period of 2000–2012 (Table 53.1). The main factor influencing this trend is the increasing OOP payments in the THE due to the various user fees charged by health facilities (Dorjdagva et al., 2016; Tsilaajav et al., 2015).

Most healthcare expenditure is for secondary and tertiary hospital care. Over the last 6 years, 25.1–29.3% of the healthcare budget was paid to providers of primary healthcare services, 35.0–36.7% to providers of secondary healthcare services, and 35.3–39.2% to providers of tertiary healthcare services (Center for Health Development, 2016).

The SHI primarily covers inpatient care; however, the benefits package was expanded to include diagnostics, outpatient visits and daycare, and some primary healthcare provider services since 2016. The state budget funds services (as stipulated in the Health act) including primary care, maternal health, child health, infectious diseases, cancer care, and mental health. The fragmented nature of health funding has created a highly disjointed payment system for healthcare providers.

There are currently four main provider payment mechanisms operating in the Mongolian health system:

1. Budget line items from the state budget
2. Case-based payments from the social health insurance fund (HIF), using diagnostic-related groups (DRGs)*
3. Capitation payments from the state budget
4. User charges and OOP payments

Table 53.2 lists the types of healthcare organizations and their funding mechanisms, including revenue sources. All public hospitals receive funding from the state budget depending on their service provision scale for maternal and child health, infectious diseases, cancer, and mental health services. Some specialist hospitals, such as maternity hospitals, are fully funded by the state. These services are paid by a line-item payment system. Since 2007, hospitals have been paid under a DRG system using the 115 locally designed DRGs from social health insurance fund. The SHI funded hospital services include inpatient, outpatient, diagnostic, and daycare services, high cost surgical devices and some cancer treatments.

While public hospitals are funded from the state budget, the HIF, and patients' OOP contributions, private hospitals are funded by the HIF and patients' OOP payments. Due to the unequal funding, there are vast differences in service provision, quality, and workloads. There is a need to rationalize services and develop a better definition of diseases; this is reflected

---

* DRGs are used to classify hospital cases into clinically similar patients likely to use similar resources.

TABLE 53.2

Payments to Service Providers

Level of Care	Types of Provider	Purchaser/Payment Methods (% of Revenue)		
		State Budget	Health Insurance Fund	OOP Payment
Tertiary level	General hospitals and special centers at the national level	Line-item budget (12–83%)	Case-based DRG (7–83%)	Fee-for-service (4–10%)
Secondary level	District health centers, hospitals, and maternity hospitals	Line-item budget (17–100%)	Case-based DRG (0–80%)	Fee-for-service (0–30%)
	Regional diagnostic and treatment centers	Line-item budget (60%)	Case-based DRG (34%)	Fee-for-service (6%)
	Province/Aimag general hospitals	Line-item budget (58–60%)	Case-based DRG (30–40%)	Fee-for-service (1–10%)
	Sanatoria	—	Case-based DRGs (19–90%)	Fee-for-service
	Private hospitals	—	Case-based DRGs[a] (10–30%)	Fee-for-service (70–90%)
	Private clinics	—	—	Fee-for-service (100%)
Primary care	*Soum*/*intersoum* hospitals	Line-item budget (75–96%)	Case-based DRGs (4–20%)	—
	Soum health centers	Line-item budget or capitation (100%)	—	—
	Family group practices	Capitation (100%)	—	—
All levels of care	Outpatient pharmacy	—	Reimbursement by reference price[b] (12–83%)	Fee-for-service (100%)

Source: World Bank, Assessment of systems for paying health care providers in Mongolia: Implications for equity, efficiency and universal health coverage (Report No. 98790-MN). 2015a, Retrieved from http://documents.worldbank.org/curated/en/docsearch?query=98790-MN.

[a] Private hospital reimbursements are less than those of public hospitals, even when DRGs are the same.

[b] The number of pharmacies, drugs, and reimbursement percent; those entitled to HIF reimbursement are few.

in recent amendments to the health law (2017) and health insurance law (2016), and in other policy documents.

FHCs and *soum* health centers are paid on the basis of capitation for their catchment area population. This method is evolving to suit the Mongolian context in terms of determining catchment population and their health risks, as well as conditions where facilities operate. The original FHC payment was determined by age group and socioeconomic status. It was subsequently decided that this was ineffective, and from 2007 the socioeconomic adjustment was made according to places of residence: *ger* (traditional housing) district or apartment district. From 2012, the FGP risk-adjusted capitation payment model was set at US$7 (per capita) for patients living in *ger* districts and US$6 for patients living in an apartment district (Tsilaajav et al., 2015).

The Ministry of Health has attempted to regulate the level of OOP payments and has approved and revised lists of service fees at public health facilities. This is seen as a way of monitoring charges and discouraging informal payments. However, patients are still expected to make OOP payments for many services other than primary care. The share of OOP payments in THE shows a sharp increase from 12.1% in 2000 to 35% in 2012 and 41% in 2014. Some studies suggest that OOP payments for healthcare have increased, and estimate that health-related expenses have pushed approximately 0.7% of the total population into poverty (Dorjdagva, 2017; Tsilaajav et al., 2015).

Conclusion

The financing of Mongolia's health system has experienced a number of successful reforms since the 1990s, with increased social protection, more equitable access, and the provision of higher-quality care. Further work is needed, particularly when it comes to eliminating the current funding fragmentation and the inequities of public and private hospitals, and ensuring that vulnerable sections of the population are not impoverished due to OOP health payments.

54

New Zealand

Strengthening Primary Healthcare

Jacqueline Cumming

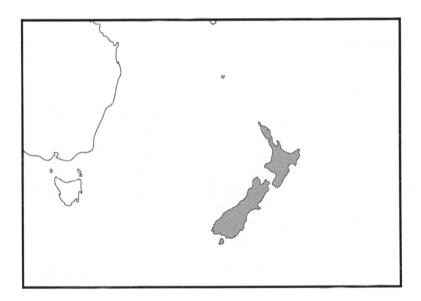

CONTENTS

New Zealand Data

- Population: 4,692,700
- GDP per capita, PPP: $39,058.7
- Life expectancy at birth (both sexes): 81.6 years
- Expenditure on health as proportion of GDP: 11.03%
- Estimated inequity, Gini coefficient: 36.2%

Source: All data are from the World Health Organization and World Bank. Latest available data used as at October 2017.
GDP = Gross Domestic Product
PPP = Purchasing Power Parity

Background

Although New Zealand's (NZ) healthcare system is regarded as successful, providing high-quality care at a reasonable cost, like all systems, it faces challenges. The NZ population is growing as well as aging. The number of people with long-term conditions is rising, and key concerns include heart disease, diabetes, depression, dementia, and musculoskeletal conditions. There are numerous new technologies that might be funded to improve health and well-being. Those using services appear to be increasingly vocal when high-quality care is not delivered in a timely way, including in relation to new technologies. Finally, there are significant inequities in health status in NZ. All these factors are driving an increase in demand for health services (Cumming et al., 2014).

With more than 80% of funding from government sources, the NZ government maintains a high degree of control over the NZ healthcare system, and is able to direct resources to services it believes will best meet the needs of New Zealanders and achieve key health goals (OECD, 2015a). The direction of the system is laid out in the New Zealand Health Strategy (NZHS), first introduced in 2001 (King, 2000) and "refreshed" in 2016 (Minister of Health [New Zealand], 2016a,b).

A system-level measures performance framework is in development to track performance toward high-level "aspirational" health goals (Ministry of Health [New Zealand], 2016). These goals include ambulatory-sensitive hospitalizations for 0- to 4-year-olds, acute hospital bed days, patient experience of care, amenable mortality rates, smoke-free households for babies, and improved youth health. A large number of contributory indicators support these system-level measures. For example, the amenable mortality rate measure has contributory indicators relating to immunization and cervical smear testing, and offers of assistance to quit smoking.

Strengthening Primary Healthcare

Key to achieving the goals of the 2016 NZHS will be the further strengthening and extension of primary healthcare (PHC) services in NZ.

The 2016 NZHS builds on the 2001 New Zealand Primary Health Care Strategy (NZPHCS) (King, 2001). The NZPHCS was developed in response to long-held concerns over poor access to PHC services in NZ (Cumming et al., 2014), and pointed to international evidence that strong PHC can deliver better overall health outcomes, and greater equity in outcomes, at a reasonable cost (Starfield, 2004; Starfield and Shi, 2002; Starfield et al., 2005). The key characteristics of PHC that support these healthcare system outcomes include promoting ease of access to services; a focus on generic health as opposed to a service user's specific condition; prevention (e.g., through screening); early management; continuity of care, comprehensiveness, and coordination; and less unnecessary or inappropriate specialty care (Starfield et al., 2005).

Implementation of the NZPHCS led to significant increases in funding for PHC services, the introduction of primary health organizations (PHOs) to oversee the development of PHC services and support PHC providers, a reduction in the fees that New Zealanders pay when they use PHC services, and a considerable increase in the numbers of consultations in PHC, including those delivered by nurses (Cumming and Mays, 2011; Cumming et al., 2008).

By the late 2000s, however, concerns were being raised that service delivery had not changed as much as had been desired (Smith, 2009). There was more to be achieved in delivering proactive, population health services emphasizing prevention through a larger PHC team, which might include, for example, nurse practitioners, pharmacists, social workers, and case managers. There was also a perceived need to increase the role of PHC in delivering more integrated care for service users across PHC and community providers, between PHC and hospital providers, and between health and social service providers. This might be achieved by PHC taking on a greater coordinating role and arranging the best mix of services to meet service user needs, including social service providers delivering housing, work and income, or employment services.

The refreshed NZHS therefore reemphasizes the NZPHCS aims of delivering more preventive services, with care increasingly delivered closer to home; of delivering more patient and family-centered services; and of achieving more integrated care, through greater collaboration and teamwork.

The 2016 NZHS is supported by a roadmap of actions that sets out 27 actions to be undertaken over the next 5 years. These goals will be achieved by, among other things, an emphasis on information and support to help people make healthy choices and better manage their own health; strengthened community engagement in decision-making; people-led service design; clear referral pathways; an emphasis on using the full skills of health

professionals; an increased focus on prevention, early intervention, and rehabilitation; greater collaboration across government agencies; an emphasis on
a "one-team" approach to health; the development of leadership capability;
and the improvement of data quality and analytical capability (Minister of
Health [New Zealand], 2016b).

Features Supporting the Likelihood of Success

The likelihood of successfully achieving a strengthened PHC in NZ is supported by the stability of the healthcare system, a well-established PHC sector, and an accountability framework that supports enhancement of the PHC
sector.

Healthcare System Stability

NZ's 20 district health boards (DHBs)—responsible for the health and
health services of those living in their geographical region—have been in
place since 2001, and the PHOs they contract with have existed in their
current form since around 2010. This stability enables those working in
the system to focus on key service delivery challenges, and has allowed
key relationships to evolve within well-recognized funding and planning
arrangements.

Established PHC Sector

The NZ system has been emphasizing PHC services for a number of years,
enabling sector leaders to develop plans, implement change, learn how to
successfully bring about change, and adapt their plans based on assessments
of progress.

Accountability

The 2016 NZHS and system-level measures appear to be driving accountability in a far more coordinated way than has occurred in recent years.
Improvement plans are required of DHBs, through the local decision-making alliances they have formed with PHOs. Funding is provided to support
PHO capacity and capability, and to reward performance (New Zealand
Government, 2016). Alliances can include local contributory indicators in
their plans; for example, the Canterbury DHB includes referrals to health-
promoting lifestyle services and the completion of motivational conversations training as contributory factors to be tracked (Canterbury Health
System, 2017).

Challenges to Success

While there are some positive factors supporting the likelihood of achieving strengthened PHC services in NZ, there is more that needs to be done.
A significant issue relates to the funding available to PHOs. PHOs (and the
practices they contract with) are funded on a weighted capitated basis that
takes into account age, gender, high user status, and rurality for a large proportion of services. Practice contract status also in part determines funding,
with additional capitation funding for those practices agreeing to low service user fees through very low-cost access (VLCA) and zero fees for those
aged under 13 years. The weighting is unlikely to be sufficient to support
those with higher health needs. Some evidence suggests that PHOs working
with higher-needs populations struggle financially, yet these are the very
populations that need the greatest support (Hefford et al., 2010; Love and
Blick, 2014).

Little new funding has been made available since 2007, and service user
fees are rising once again. For adults aged 45–64 years and those aged
65 years and over, fees remain low on average at NZ$15.48 (US$10.14)
and NZ$14.26 (US$10) in VLCA practices (with a maximum of NZ$17.50
[US$12.26]). However, in non-VLCA practices they now average NZ$41.40
(US$29) and NZ$39.35 (US$27.57), respectively, with the maximum fees set at
NZ$69 (US$48.34) and NZ$60.50 (US$42.39), respectively. A significant proportion of NZ adults are reporting unmet need due to cost—that is, they are
unable to see a general practitioner when they feel they need to, due to cost.
In 2015–2016, 14.3% of adults—or 533,000 people—reported such unmet need
(Ministry of Health [New Zealand], 2017). There are much higher rates of
unmet needs for women, Māori and Pacific peoples, and those living in disadvantaged areas. To many, an NZHS that seeks to enhance the role of PHC
services and bring services closer to home is difficult to reconcile within an
overall policy setting where service user fees for adults are now well above
where they were in 2001, prior to the release of the NZPHCS (Raymont et al.,
2013), and where significant proportions of the highest-needs communities
report high levels of unmet need.

It remains unclear whether funding will be allocated in the future to support the achievement of the 2016 NZHS goals in delivering services closer to
home. Much news reporting on the health sector continues to focus on hospital care—with elective services, maternity care, mental health services, the
quality of hospital care, hospital food, and hospital parking being the focus
of recent attention. A key challenge for the government will be to better
support PHC services in delivering a greater range of services; this should
involve workforce development (which isn't as well supported in the PHC
as in the hospital sector), along with increased capital investment. There is,
however, a need for the PHC sector to demonstrate that expanded PHC services are actually making a difference; the lack of recent research on this

makes it difficult to argue successfully for new funding to support the further development of PHC services.

Achieving transformation in service delivery is extremely difficult (Best et al., 2012), and some NZ evidence also points to the need for sustained attention to reform processes and funding to support people's time as they develop new ways of delivering care (Cumming, 2011; Lovelock et al., 2014; Middleton and Cumming, 2016).

The 2016 NZHS and system-level measures do not appear to sufficiently emphasize reducing inequalities in health, particularly for Māori and Pacific populations. Although the actual measures include data on inequalities, there is no requirement for the gaps to be closed between the different ethnic groups; nor is there an emphasis on improving the health of those with disabilities (mentioned in the NZHS) or of those on lower incomes (where the actual measures are not reported at all).

Conclusion

NZ is on the cusp of prioritizing PHC policy and services to a far greater extent than has occurred in the past. If we are to achieve the expectations of the NZHS, and build on nearly 16 years' worth of work, NZ governments must demonstrate sustained investment and support for the sector.

55

Papua New Guinea

Strengthening the Collection, Analysis, and Use of Health Data through eHealth Solutions

Paulinus Lingani Ncube Sikosana

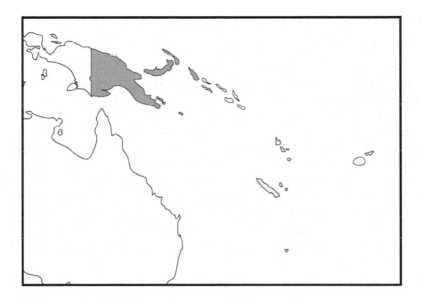

CONTENTS

Papua New Guinean Data

- Population: 8,084,991
- GDP per capita, PPP: $2,760.8
- Life expectancy at birth (both sexes): 62.9 years
- Expenditure on health as proportion of GDP: 4.26%
- Estimated inequity, Gini coefficient: 43.9%

Source: All data are from the World Health Organization and World Bank. Latest available data used as at October 2017.
GDP = Gross Domestic Product
PPP = Purchasing Power Parity

Background

Following the midterm review of Papua New Guinea's (PNG) national health strategy, the National Health Plan (NHP) 2011–2020, in 2015, a more focused version of the NHP—Health Sector Strategic Priorities 2016–2020—was developed to guide the sector during the last 5 years of the plan. The Health Workforce Enhancement Plan, which describes short-term measures to address PNG's health workforce crisis, was extended from 2016 to 2019. Medical supply reforms continued to strengthen the quantification, procurement, quality assurance, and distribution of essential medicines to front-line health facilities. The Provincial Health Authorities reform has been accelerated to fast-track integration of the management of rural health services and curative services nationwide. In 2014, the government introduced District Development Authorities (DDAs) to consolidate decentralization and local decision-making.

In line with objective 3.4 of the NHP (National Department of Health [Papua New Guinea], 2010), information and communication technology (ICT) has taken center stage in the country's health reforms to improve data use in performance monitoring. In 2014, an electronic health information system (eNHIS) was introduced to improve the quality, time-lines, flow, and use of health data in decision-making; however, there are currently several stand-alone mobile health (m-health) initiatives that do not interface with the National Health Information System (NHIS) (Loring et al., 2013). The Bloomberg Data for Health (D4H) Initiative is a public–private partnership that builds capacities of countries to better collect and use health-related data. The initiative is assisting the National Department of Health (NDoH) to build capacity to implement information technology (IT) strategies to compile and manage data, and to develop data-sharing policies and the strategic use of data in policy development, advocacy, and public communications.

Due to a slump in commodity prices, PNG's economic growth declined from 9.9% in 2015 to 4.3% in 2016, and to 2.2% in 2017. Per capita expenditures on health are projected to decline from US$92 in 2014 to about US$69 in 2016 and US$53 in 2017 (Flanagan, 2015). For m-health initiatives to unlock other options to improve access to health services, the government must ensure sustainable funding, increase access to electricity, and improve 3G network coverage (Hayward-Jones, 2016). Improved network coverage is particularly important, as increased mobile penetration can overcome some of the challenges of geographic remoteness through e-services and mobile applications. There have already been significant improvements since the introduction of competition in the mobile market in 2007 (BuddeComm, 2016).

Success Story

The goal of eHealth reform is to achieve an efficient and responsive health system that relies on continuous, sustainable, and secure data and information sharing at all levels of the system. The 2014 reform introduced the eNHIS, which was successfully piloted over 2 years in 17 districts. The eNHIS upgrades the paper-based NHIS into an electronic and computerized format. This is a digital health information system that uses mobile-enabled devices to capture, analyze, and share NHIS data. Data are transmitted via mobile networks to centralized servers in near real-time. The system autogenerates customized reports, delivers them to all levels of the health system, and facilitates timely use of the data for planning and management. The eNHIS incorporates a geographic information system (GIS) that is linked to data sets that generate digital maps. According to an independent review, the eNHIS is "capable of timely, reliable health data collection, data analysis and presentation" (National Department of Health [Papua New Guinea], 2016a: 3). The timeliness of data collection and data quality reportedly improved in all pilot districts (National Department of Health [Papua New Guinea], 2016a). The positive pilot project experience has prompted the NDoH to consider rolling out the eNHIS nationally.

The absence of timely and reliable information contributes to inefficient use of health sector resource inputs and ultimately to poor health outcomes. Effective collection and use of health data are vital for policy development, tracking imminent health threats, and monitoring progress toward set health goals. If access to accurate health information can be improved, decision makers can exert more influence when negotiating with donors and other government departments (Cibulskis and Hiawalyer, 2002). Currently, the NHIS records primary data on paper forms that are transported to provincial offices for computer data entry, using an obsolete software application, Visual FoxPro (National Department of Health [Papua New Guinea] and WHO, 2015). The data are often submitted late, inaccurate, and incomplete; they are awkward to analyze manually, and therefore rarely used in decision-making.

eHealth applications can generate real-time, accurate, reliable, and complete health information on service delivery, disease programs, availability of essential drugs and supplies, budgets, and financial expenditures. While vertical programs currently operate separate information systems that use applications that do not interface with each other or with the NHIS, mobile phone applications can consolidate information from a variety of sources and quickly generate an up-to-date picture on population health.

eHealth reforms herald an IT-based communications culture with the potential to replace paper-based data handling. The eHealth Steering Committee, which consists of individuals and organizations involved in eHealth initiatives, has recently been established to oversee the integration, harmonization, and standardization of PNG's eHealth initiatives. The eHealth strategy, health sector ICT policy, and ICT enabling plan will form the basis for the development of common standards and the establishment of an overarching architecture for an integrated eNHIS. A common health IT infrastructure has the potential to create a common platform for health institutions to share information and achieve interoperability. A dedicated national broadband network can connect hospitals and primary healthcare facilities to enable videoconferencing, telemedicine, and e-learning. Appropriate policies will be required to ensure data privacy, confidentiality, and access.

Potential barriers to the adoption of eHealth include weak IT infrastructure and a shortage of IT experts to operate the system. In PNG, mobile phone charges and Internet connectivity are prohibitively expensive. For example, fixed broadband subscriptions are 150% of the average monthly income and wireless broadband ranges from 20% to 80%, well above the international benchmark of 5%. The government has, however, taken steps to address low rural Internet penetration by expanding infrastructure and establishing additional connections to international undersea fiber-optic cables. 3G and 4G networks, domestic fiber cables, microwave links, and satellite systems are being expanded as complementary technologies (Oxford Business Group, 2015).

Impact

The proposed nationwide rollout of the eNHIS builds on successful pilots in 17 districts that demonstrated feasibility and scalability, and improvements in the timeliness of data collection and analysis, presentation, and use

(National Department of Health [Papua New Guinea], 2016a). Because the eNHIS collects data using electronic versions of the original tally sheets used in the paper-based NHIS, it has been relatively easy to introduce the new system to health workers. The existence of a draft ICT policy, ICT enabling plan, and multistakeholder eHealth Steering Committee demonstrates leadership commitment to addressing the governance issues necessary for the deployment of the eHealth solutions. The development of an eHealth strategy, eHealth action plan, and a monitoring and evaluation framework is the next logical step to a successful eHealth development process. The transition to a computerized information system will depend not only on improving systems and software, but also on building in-house capacity to develop and use ICT infrastructure, the promotion of a culture of information use, and sustainable funding.

Implementation and Transferability

According to the review, the eNHIS was feasible at the facility and provincial levels, technically sustainable, and replicable. As a result, the third meeting of the eHealth Steering Committee resolved to roll out the eNHIS nationally (National Department of Health [Papua New Guinea], 2016b). A national eHealth strategy for PNG, eHealth action plan, monitoring and evaluation plan, and regulatory framework for implementing the proposed ICT solutions will be developed. As part of the rollout phase, staff at primary healthcare facilities will be trained to use the technology and provided with tablets to collect, process, store, print out, and display health information. Health workers must be trained to collect accurate data and to use them for planning, management, and other decision-making processes. The paper-based system will be phased out to reduce the workload resulting from operating a dual system. As per the eNHIS review, "the initiative now faces considerable challenges, many of which arise from the core challenges of the PNG health system itself, such as utilization of data by decision makers and capacity development within NDoH to manage the future eNHIS" (National Department of Health [Papua New Guinea], 2016a). The implication is that there is a need for a whole system approach to the introduction of ICT solutions to strengthen health information systems.

The next 10–15 years will likely witness growth in the national IT infrastructure that will in turn enhance interorganizational communication and flow of information. Improvements to the NHIS will be driven by digitization and connectivity, with strong collaboration between private and public institutions and development partners. m-Health will evolve as the technological platform of choice for many components of eHealth, which include telehealth, health informatics, e-learning, and e-commerce. m-Health applications will increasingly dominate web-based disease surveillance systems, human resources for health information systems, telemedicine, electronic district health information systems, electronic medical records, emergency medical response, integrated financial management systems, and supply chain management. The health sector will embrace telehealth programs to provide clinical and patient referral support to health workers in isolated areas. However, because of prohibitive cell phone charges, geographic barriers to connectivity, and practical issues of charging mobile-based devices, victims of the digital divide, including vulnerable groups in rural and remote areas, will continue to be discriminated against.

Prospects

The success of this reform requires the development of leadership that embraces and promotes the adoption of eHealth, allocates sufficient resources, and focuses on health informatics development, education, and training. Sound technical infrastructures, such as telecommunications, electricity, and access to computers, are necessary prerequisites for successful introduction of eHealth solutions. Inequalities in access to and use of ICT and a lack of interest in and motivation to adapt and adopt the use of technology by health professionals are potential social and cultural barriers that must be addressed in the PNG context (Anwar and Shamim, 2011). This reform requires a resilient health system infrastructure and a cultural transformation that embraces ICT and routine use of data in decision-making at all levels of the health system.

Conclusion

Globally, eHealth is growing rapidly, with several projects demonstrating its benefits as well as challenges in measuring e-readiness (Qureshi et al., 2014). PNG's NDoH has taken the bold step of establishing a single health information system with databases that harmonize information requirements across vertical programs and integrate processes of data collection, aggregation, and reporting. The ability of the eNHIS to stimulate increased utilization of health information in policy development, planning, and management at lower levels of the health system will be critical to its sustainability. Successful deployment of eHealth applications requires a mix of skills and commitment to data management responsibilities that are currently not available in PNG. PNG must invest in short- and long-term training strategies to develop the necessary capacities, relevant policies, and an appropriate eHealth regulatory environment.

56

Taiwan

"My Data, My Decision": Taiwan's Health Improvement Journey from Big Data to Open Data

Yu-Chuan (Jack) Li, Wui-Chiang Lee, Min-Huei (Marc) Hsu, and Usman Iqbal

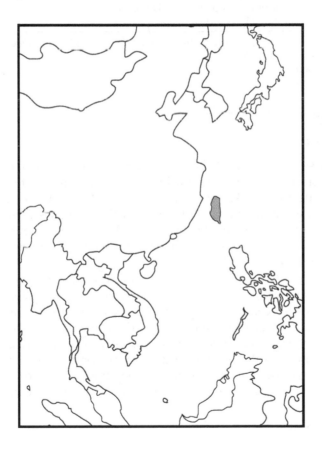

CONTENTS

Taiwanese Data

- Population: 23,508,428
- GDP per capita, PPP: $47,800
- Life expectancy at birth (both sexes): 80.1 years
- Expenditure on health as proportion of GDP: 6.2%
- Estimated inequity, Gini coefficient: 33.8%

Source: All data are from the World Health Organization and World Bank. Latest available data
used as at October 2017.
GDP = Gross Domestic Product
PPP = Purchasing Power Parity

Background

In terms of service, Taiwan's healthcare delivery system is tailor-made for its population of 23 million, embracing a pluralistic distribution of services that echoes its free-market economy. Taiwan's National Health Insurance (NHI), which is estimated to cover 99.9% of the country's total population, provides comprehensive benefit coverage inclusive of inpatient services and ambulatory care, and ample freedom in terms of medical care options (陳育群 [Chen] and 李偉強 [Lee], 2016).

With the advent of groundbreaking communication technology, from high-speed Internet to web-connected personal devices, like computers and mobile phones, data have been easily digitized—collected extensively and at impressive speed. Health data collected from individual patients by healthcare providers have the potential to greatly enhance research (Mitchell et al., 2014). The NHI is now able to return the public's health information to them via "My Health Bank," thus ensuring that individuals can keep track of their own health. This will help to revolutionize the global healthcare medical service model.

Big Data

Since the launch of the NHI in Taiwan 20 years ago, it has become one of the largest administrative healthcare systems in the world, managing to cover the entire Taiwanese population. The National Health Insurance Research Database (NHIRD), which was established by the NHI at the same time, collects data from all patient visits. By 2015, these data, which are provided to Taiwanese scientists for research purposes, had been used in more than 3000 studies published in 656 scientific journals. These studies are all indexed in PubMed, in the National Library of Medicine in the United States. Since 2015, the research has continued growing at a tremendous rate (Yen et al., 2016; 陳育群 [Chen] and 李偉強 [Lee], 2016).

Big data can be defined as an extremely large data set that commonly represents a wide variety of information that can be analyzed computationally. Such data sets are particularly useful in the health context, revealing patterns and trends that may assist in research and/or influence policy directions, as well as helping individual patients. Advances in health information technology (HIT) in Taiwan have expedited the compiling of observational health data. The compilation of such data has been made possible due to the fact that there is a digital record made of all visits to Taiwan's hospitals and clinics, which are now paid exclusively via e-claims. On average, every Taiwanese citizen visits a doctor 15 times each year, between one and five diagnoses are made on each of these visits, and an average of 15 drugs are prescribed to each individual. This all adds up to a mass of potential data that can be used by healthcare organizations.

Taiwan NHI Information Now Accessible: Open Data

The worldwide trend of governments to give their citizens, organizations, and corporations access to governmental data is not without its ethical concerns. After two decades, Taiwan's NHIRD has managed to amass an extensive amount of medical data. In its initial application of these data, the National Health Insurance Administration (NHIA) established 120 data sets on the government's open data platform so that the public could conduct any value-added applications or innovations free of charge. This has further grown to 355 data sets plus two data visualizations that include information about individual hospitals' features and service quality, along with supplementary information on insurance premium coverage. NHIA's data set categories include the following:

- Medical care quality
- Medical information disclosure
- Medical institution
- Important statistics
- Drugs and medical devices

The most popular topics downloaded by citizens include "NHI under-writing-related statistical data," "NHI drug coverage archives," and "disease-related data." The "disease-related data" category is used to facilitate the search for the 20 most common ailments that incur the greatest medical expense, the cost burden and growth rate of these ailments, and changes in their ranking. In 2014, the diseases that incurred the greatest medical expenses were, in descending order: chronic renal failure, dental-related diseases, diabetes, hypertension, and acute upper respiratory infections. Information such as this is significant to the health of the general public and thus the nation (National Health Insurance Administration, Ministry of Health and Welfare [Taiwan], 2015a).

In order to improve the clarity and quality of government-held knowledge, elevate the value of NHI data, fulfill the requirements of medical industries, and increase the economic output of the medical industry, the NHIA has progressively established data sets with multiple dimensions. The objective here is to create constructive and mutually beneficial outcomes for the government and the medical industry, as well as private organizations. For detailed information, see the government's open data platform: http://data.gov.tw.

It is worth noting that within 1 week of the platform's launch, the top three most downloaded data sets included diseases covered by the National Health Insurance System, statistics on accepted NHI claims, and medicines covered by the National Health Insurance System.

My Data, My Health Bank

Preserving a healthy workforce despite low birth rates and an aging population is a challenge that all nations will need to deal with in the near future. Encouraging individuals to take an active role in their own healthcare is regarded as a cost-effective way to ensure optimum health outcomes.

With this goal in view, the NHIA input comprehensive current and historical data into the NHI database and introduced the My Health Bank system in September 2014. This system lets individuals download their personal health information online (see the My Health Bank system at http://www.nhi.gov.tw). The website gives its users immediate access to all data related to their visits to NHI-affiliated medical institutions (Figures 56.1 through 56.4). By fully utilizing information and communication technology, the NHIA delivers health-related data to cloud storage from which citizens can instantaneously obtain comprehensive personal health information. Access to such information will help patients take personal responsibility for their own health decisions.

In 2014, the Digital Opportunity Survey of the National Development Council reported that in Taiwan, 78% of people aged 12 and above are

National Health Insurance Administration, Ministry of Health and Welfare Outpatient Health Bank

ID: Hackathon****
Date of data production: 105/09/30
Period date of data: 105/08/01~105/08/31 (Health ID card Upload)

National Health Insurance Administration Service Unit	Medical Institutions		Visit Date	Prescription Or Exam Or Rehabilitation Date	Health ID Card Number Of Visit	Disease Code	Disease Category Name	Procedure Code	Procedure Name	Copayment Money	NHI Payment Count
	Medical Code	Medical Name									
Taipei	Hackathonclinic		105/08/01		0018	K529	Non-infectious gastroenteritis and colitis, other and unspecified, diarrhea, ileitis, jejunitis, sig moditis			50	229
	A029828100	Kascoal Tablets 40mg(Dimethylpolysiloxane)									12
	AC19337100	LIMODIUM TABLETS (LOPERAMIDE)									6
Taipei	Hackathonclinic		105/08/15		0019	J309	Allergic rhinitis			90	376
	AC41169100	KOJAZYME(Lysozyme) TABLETS 90mg									16
	AC49362100	MINLIFE-P SUSTAINED RELEASE FILM COATED TABLETS									8
	BC22924424	NASONEXR Aqueous Nasal Spray									1
Taipei	Hackathonclinic		105/08/30		0020	J069	Acute upper respiratory infections of multiple or unspecific			50	229
	AC21758100	Medicon-A									12
	AC33939100	ABROXOL TABLETS 30mg									12

FIGURE 56.1
My Health Bank outpatient data sample. (From Authors' conceptualization.)

National Health Insurance Administration, Ministry of Health and Welfare Dentistry Health Bank

ID: Hackathon****
Date of data production: 105/09/30
Period date of data: 105/06/01~105/06/30 (Declare)

National Health Insurance Administration Service Unit	Medical Institutions	Visit Date	Health ID Card Number Of Visit	Disease Code	Disease Category Name	Procedure Code	Procedure Name	Copayment Money	NHI Payment Count
	Medical Code	Medical Name				Tooth Position Code	Tooth Position Name		Medical Quantity
Taipei	Hackathonclinic	105/06/01	0015	K0530	Chronic periodontitis			50	943
	00130C	Enhanced infection control							1
	34001C	Periapical X-ray photography				36	6, left lower permanent		1
	91004C	SCALING FULL MOUTH				FM	FULL MOUTH		1

FIGURE 56.2

My Health Bank dentistry data sample. (From Authors' conceptualization.)

National Health Insurance Administration, Ministry of Health and Welfare Laboratory or Exam Result

ID: Hackathon****
Date of data production: 105/09/30
Period date of data: 104/12/01~104/12/31

National Health Insurance Administration Service Unit	Medical Institutions	Visit Date/ Admission Date	Laboratory or exam Date	Data Upload Time	Medical Code	Medical Name	Exam Items Name	Result	Reference
Taipei	Hackathon hospital	104/12/01	104/12/01	104/12/15	08011C	Full blood test	HCT [37–52%]	45.3%	39.0–52.0
Taipei	Hackathon hospital	104/12/01	104/12/01	104/12/15	08011C	Full blood test	HGB [12–18 g/dL]	15.5	13.0–17.0
Taipei	Hackathon hospital	104/12/01	104/12/01	104/12/15	08011C	Full blood test	MCH [26–34 pg]	27.9 pg	26.0–34.0
Taipei	Hackathon hospital	104/12/01	104/12/01	104/12/15	08011C	Full blood test	MCHC [31–37 g/dL]	34.1	33.0–37.0
Taipei	Hackathon hospital	104/12/01	104/12/01	104/12/15	08011C	Full blood test	MCV [80–99 fL]	81.7 fL	80.0–99.0
Taipei	Hackathon hospital	104/12/01	104/12/01	104/12/15	08011C	Full blood test	MPV [7.2–11.1 fL]	9.08 fL	7.20–11.10
Taipei	Hackathon hospital	104/12/01	104/12/01	104/12/15	08011C	Full blood test	PCT	0.16%	0.16–0.35
Taipei	Hackathon hospital	104/12/01	104/12/01	104/12/15	08011C	Full blood test	PDW	18.1 fL	N/A
Taipei	Hackathon hospital	104/12/01	104/12/01	104/12/15	08011C	Full blood test	PLT [130–400 x 10.e3/uL]	176	130–400
Taipei	Hackathon hospital	104/12/01	104/12/01	104/12/15	08011C	Full blood test	RBC [4.2–6.1 x 10.e6/uL]	5.55	4.20–6.10
Taipei	Hackathon hospital	104/12/01	104/12/01	104/12/15	08011C	Full blood test	RDW [11.5–14.5%]	14.2%	11.5–14.5
Taipei	Hackathon hospital	104/12/01	104/12/01	104/12/15	08011C	Full blood test	RDW-SD	40.3 fL	N/A
Taipei	Hackathon hospital	104/12/01	104/12/01	104/12/15	08011C	Full blood test	WBC [4.0–11.0 x 10.e3/uL]	11.16	4.00–11.00
Taipei	Hackathon hospital	104/12/01	104/12/01	104/12/15	08013C	White blood cell count	%BASO [0–1.5%]	0.2%	0.0–2.0
Taipei	Hackathon hospital	104/12/01	104/12/01	104/12/15	08013C	White blood cell count	%EOS [0–7%]	0.4%	0.0–7.0
Taipei	Hackathon hospital	104/12/01	104/12/01	104/12/15	08013C	White blood cell count	%LYM [19–48%]	7.6%	19.0–48.0
Taipei	Hackathon hospital	104/12/01	104/12/01	104/12/15	08013C	White blood cell count	%MONO [2.0–10.0%]	5.8%	2.0–12.0
Taipei	Hackathon hospital	104/12/01	104/12/01	104/12/15	08013C	White blood cell count	%NEUT [40–74%]	86.0%	40.0–74.0
Taipei	Hackathon hospital	104/12/01	104/12/01	104/12/15	09005C	Glucose in blood	Glucose (Blood) 1 [70–110 mg/dl]	115	70–99
Taipei	Hackathon hospital	104/12/01	104/12/01	104/12/15	09015C	Creatinine (blood)	Creatinine (Blood) [0.5–1.3 mg/dl]	0.8	0.7–1.2
Taipei	Hackathon hospital	104/12/01	104/12/01	104/12/15	09015C	Creatinine (blood)	eGFR	118	N/A

FIGURE 56.3
My Health Bank laboratory exam data sample. (From Authors' conceptualization.)

experienced in using the Internet; 85% of these use their cell phones to access the Internet, with an increasing number making use of online services through their mobile devices (National Health Insurance Administration, Ministry of Health and Welfare [Taiwan], 2015b). In order to take advantage of this, a NHI mobile app was developed. This app comes with two significant new features: My Health Bank and "Mobile Counter." Users can access their personal My Health Bank or check their NHI payment status in Mobile Counter on a cell phone or tablet. All they need to do is register with the NHI smart card online services on the NHIA website. Other features of this app include information relating to

- Medical care
- Service locations
- NHI audiovisuals

National Health Insurance Administration, Ministry of Health and Welfare Chinese Medicine Health Bank

ID: Hackathon****
Date of data production: 105/09/30
Period date of data: 105/07/01~105/07/31 (Declare)

National Health Insurance Administration Service Unit	Medical Institutions	Visit Date	Health ID Card Number of Visit	Disease Code	Disease Category Name	Procedure Code	Procedure Name	Copayment Money	NHI Payment Count
	Medical Code	Medical Name							Medical Quantity
Taipei	Hackathon chinese medicine clininc	105/07/01	0017	S161XXA	Strain of muscle and tendon of head, initial encounter			50	480
	A01	Outpatient examination fee							1
	B42	Acupuncture treatment and disposal fees (including the cost of materials) - unopened medication							1

FIGURE 56.4
My Health Bank Chinese medicine data sample. (From Authors' conceptualization.)

- NHI laws and regulations
- NHI headlines
- Long-term care

An NHI "Q&A" is also featured. The app is available for Android mobile devices and iOS systems, and can be downloaded from the Google Play Store or Apple Store. Automatic updates ensure full functionality of the latest features.

My Health Bank allows people to keep track of their own healthcare records, which include all outpatient medical files, the individual's history of allergies, and any hospice palliative care. All information is updated regularly by a dedicated system. The data are programmed to maintain sequential

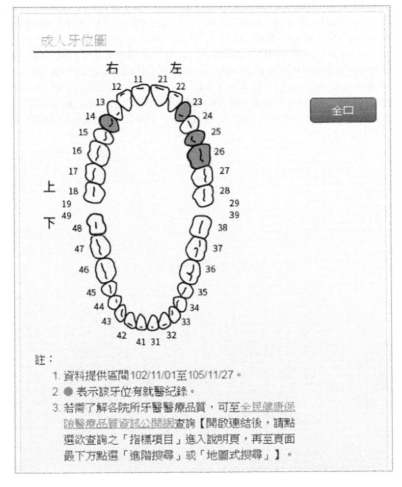

FIGURE 56.5
My Health Bank teeth map sample. (From Authors' conceptualization.)

records by either the date of visit, healthcare facility, or disease, and this not only helps the patient but is also a significant aid for a physician if the patient requires immediate medical attention. The NHIA is committed to continuing its efforts to expand and improve the existing services, making full use of online technology to achieve better health outcomes.

Conclusion

Thus far, the efforts of the Taiwanese Ministry of Health and Welfare to move from big data to open data have been a success, improving the overall efficacy, safety, and quality of medical care. Taiwan's NHIA system is a revolutionary medical service model that provides the public with the power to take their healthcare into their own hands. The My Health Bank system will continue to grow, and, with advances in technology, will become capable of producing increasingly rich and accessible data. Initiatives such as My Health Bank are making "My Data, My Decision" a reality, and have set the country on the path to reach the World Health Organization's 2020 health goal for advanced countries: that of realizing a holistic, people-centered care model.

57

South-East Asia

Taming Communicable Diseases

Jeffrey Braithwaite, Wendy James, Kristiana
Ludlow, and Yukihiro Matsuyama

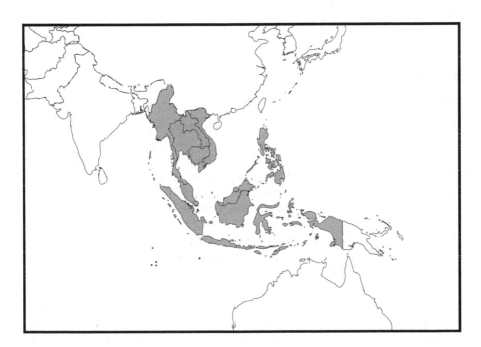

CONTENTS

South-East Asian Data[a]

- Population: 639,895,454
- GDP per capita, PPP: $23,065.5[b]
- Life expectancy at birth (both sexes): 72.1 years[b]
- Expenditure on health as proportion of GDP: 3.8%[b]
- Estimated inequity, Gini coefficient: 43.9%[c]

Source: All data are from the World Health Organization and World Bank. Latest available data used as at October 2017.
[a] Data unavailable for Andaman and Nicobar Islands.
[b] Data unavailable for Christmas Island and Cocos (Keeling) Islands.
[c] Data unavailable for Brunei, Myanmar, Christmas Island, and Cocos (Keeling) Islands.
GDP = Gross Domestic Product
PPP = Purchasing Power Parity

Background

South-East Asia has a rich history and prehistory. Situated in the Southern Hemisphere, a part of the Oceania-Pacific region, it is the third most populous part of the world. The region covers an area of about 4.5 million km^2 (1.7 million square miles) and hosts a population approaching 650 million people. It comprises a range of countries (Table 57.1), which are characterized by considerable diversity in population, culture, racial mix, and social and religious traditions.

South-East Asia's borders are framed by Australia and the Indian Ocean to the south; the countries of East Asia to the north; South Asia, India, and the Bay of Bengal to the west; and the Pacific Ocean and Oceanic countries to the east.

Mainland South-East Asia, which was traditionally known as Indochina, comprises Vietnam, Laos, Cambodia, Thailand, Myanmar, and West Malaysia. Oceanic South-East Asia encompasses Indonesia, East Malaysia, Singapore, the Philippines, Timor-Leste, Brunei, the Christmas Islands, the Andaman and Nicobar Islands, and the Cocos (Keeling) Islands. Historically, this region was known as the East Indies and Malay Archipelago.

As an illustration of the diversity of the countries in the region, a two-hour flight will take you from Cambodia, where the GDP per capita is US$3,737, to neighboring Singapore, where the per capita GDP is US$90,151—24 times that of Cambodia (World Bank, 2016b). This multiplication factor is even more pronounced if you contrast the wealthiest of South-East Asian nations with the most impoverished Christmas Island, the Cocos (Keeling) Islands, and the Andaman and Nicobar Islands.

While the wealthier South-East Asian countries have done very well in creating modern health systems and programs to tackle communicable diseases and

TABLE 57.1

South-East Asia Countries and Territories

Countries	Territories
• Brunei Darussalam	• Christmas Island
• Cambodia	• Cocos (Keeling) Islands
• Timor-Leste	• Andaman and Nicobar Islands
• Indonesia	
• Laos	
• Malaysia	
• Myanmar	
• Philippines	
• Singapore	
• Thailand	
• Vietnam	

non-communicable diseases alike, other countries in the region, whether lacking resources or effective programs, have struggled against various headwinds to make progress (Coker et al., 2011). In this chapter, we discuss aspects of healthcare provision across the region and how these can be strengthened, focusing specifically on current and future efforts to combat communicable diseases.

Challenges

Common issues facing the countries presented in this book include questions of how to strengthen and organize health systems to manage different types of patients across the community–acute–aged–rehabilitation continuum, secure adequate funding for all aspects of care, and ensure that funding achieves maximum effect. While all countries in South-East Asia are aiming for a trajectory of improvement, the region has some notable "big-ticket" problems. The most prominent of these is infectious diseases; indeed, South-East Asia has been described as a "hot spot" for both endemic and emerging contagion (Coker et al., 2011). Regardless of whether an individual country has good programs for the detection, monitoring, and surveillance and treatment of infectious diseases, all countries of the region are susceptible. Singapore and Brunei Darussalam are striving with much success to meet the healthcare needs of their population with good health systems and advanced disease control models, but infectious diseases do not respect borders, and can become a heavy burden on societies very quickly. They can become pandemics very readily: severe acute respiratory syndrome (SARS), by way of an example, exacted a heavy toll on populations across the region, as well as the region's tourist industry, between 2002 and 2003. Within a similar time frame, the influenza A (H5N1) outbreak devastated the poultry industry in countries such as Vietnam, Indonesia, and Thailand.

The countries of South-East Asia are interconnected in myriad complex ways, but particularly through trade and tourism. Despite this obvious transmission factor, it is not always clear why South-East Asia is the epicenter of both common and more exotic infections, and its populations so susceptible. It may be that ecologically the region presents a diversity of niches that microbes can explore. Dense, shifting populations and poor control mechanisms in parts of the region may also play a role.

Exacerbating the problem, governance of infectious diseases at the regional level is patchy, with overlapping institutional responsibilities, complex political relationships, and historical tensions between the countries of the region. While some disease outbreaks have led to political attention and economic investment, such as the Canada-Asia Regional Emerging Infectious Disease (CAREID) Project (CAD\$4.3 million [US\$3.4 million] in funding) in response to the 2003 SARS epidemic, progress to tackle infectious diseases in South-East Asia has generally been slow. There are significant shortcomings in capacity to tackle infectious diseases in some countries—the 2012 outbreak of hand, foot, and mouth disease in Cambodia, for example, killed more than 50 children (WHO, 2012b). A 2009 outbreak of leptospirosis in the Philippines' capital, Manila, killed 178 and hospitalized 2299; this despite the prevalence of the disease during typhoon season and the availability of rapid diagnostic tests (Amilasan et al., 2012).

Such instances are due to numerous system deficits—from inadequate infrastructure to lack of pharmaceuticals to less than optimal monitoring and surveillance. Research capability in infectious disease control and surveillance across the region suffers from underinvestment, and expertise is unevenly distributed.

The diseases that contribute the most burden include those that are drug resistant, such as tuberculosis and malaria; endemic diseases, such as HIV/AIDS and dengue fever; and emerging diseases, such as avian influenza A (H1A1), SARS, and pandemic influenza A (H1N1) (Coker et al., 2011; WHO, 2010).

Success

Where there have been successes, what are they? General health is, of course, a significant factor in combating the most serious effects of infectious disease. Improved nutrition in South-East Asian countries over the last quarter century—the prevalence of undernourishment has decreased from 30.6% in 1990–1992 to 9.6% in 2014–2016—has gone some way to improving the overall health of the regions' population (Food and Agriculture Organization of the United Nations et al., 2015). In terms of infectious disease–specific health improvements, there is an increased awareness across the region of the importance of investment in tackling infectious diseases, making prevention a priority, and putting control measures in place early in any disease cycle. Water and sanitation systems are generally improving, as are surveillance systems. The Mekong Basin Disease Surveillance initiative, for instance, has had some control success across Cambodia, Laos, Myanmar, Thailand, and Vietnam, as well as China (Phommasack et al., 2013). This initiative has been far more rigorously pursued than those in the past and showcases the strength of such collaborative efforts. Many countries of the region collaborate via the Association of South-East Asian Nations (ASEAN). This association, established in 1967 and currently comprising 10 member states, promotes economic, cultural, social, legal, and scientific development throughout the region. The ASEAN Health Cooperation works to address shared public health challenges, particularly those posed by communicable diseases (Association of South-East Asian Nations, n.d.).

In addition, many governments across the region have set up new or strengthened emerging diseases monitoring, protection, and control agencies in response to the emergence of the SARS, H5N1, and H1N1 viruses. Such initiatives have been supported by increased investment and training, and laboratory capacity has been enhanced in order to generate reliable data and research ubiquitous and emerging infections (South-East Asia Infectious Disease Clinical Research Network, 2013).

Impact

While there have been some successes, low levels of funding and poorly structured funding arrangements across the region strain the ability to make steady progress. Environmental changes and economic transitions, including the growing shift to urbanization and industrialization, have further complicated matters. And, while there have been many successful initiatives to improve the education and training of health professionals throughout the region (Castro Lopes et al., 2016; Dhillon et al., 2012), the brain drain of medical professionals to wealthy countries is also an ongoing problem.

Notwithstanding this, the many national, regional, and local initiatives, including the establishment of surveillance systems, are having a positive effect. For example, the Centers for Disease Control and Prevention (CDC) in Cambodia has been working closely with the Cambodian Ministry of Health since 2002 to improve surveillance capacity and strengthen public health systems to combat infectious disease, with a particular focus on HIV/AIDS, tuberculosis, and malaria (Centers for Disease Control and Prevention, 2013a). The CDC has provided strategies and guidance for monitoring immunization programs for measles and hepatitis B and assistance in keeping Cambodia polio-free. The Applied Epidemiology Program, established in 2010, trains Ministry of Health staff in surveillance and effective outbreak response. Since the establishment of the CDC, Cambodia's surveillance capacity has been enhanced, with more than 5600 specimens tested. Seventy percent of tuberculosis patients are now tested for HIV, and the country's 215 HIV counseling and testing centers are subject to external quality control (CDC, 2013b).

There have also been a number of successes in Indonesia, where early warning and response systems for 22 diseases, including the deadly avian influenzas, have been adopted, again with the assistance of the local CDC. Immunization programs have been strengthened, and malaria control has improved, with fatalities in the South Halmahera district decreasing from 226 in 2004 to 11 in 2009 (Centers for Disease Control and Prevention, 2013c). Results from research into disease-carrying mosquitoes in Yogyakarta have been highly influential in the quest to control the disease throughout the region. Field epidemiology training programs have been established throughout the country, with more than 600 graduates now able to assume roles as public health leaders.

Timor-Leste, formerly East Timor, has also had some notable successes, with a significant decrease in malaria and the virtual elimination of leprosy. Rates of immunization have increased, with 77% of 1-year-olds now immunized against diphtheria, tetanus, and polio (WHO, 2016f).

Implementation and Transferability

While these initiatives and others throughout the region are improving the situation in many areas, substantial challenges remain. Coker et al. (2011) made a cluster of recommendations in their analysis of the regional infection challenges facing the South-East Asian countries. These included accelerating high-level regional approaches and the sharing of ideas across the constituent countries. Evening out the imbalances in the capacities of the different countries and providing specific support to less well-off health systems are important considerations. The European Union (EU) has a similar problem, and has established horizontal governance and collaboration structures and institutions for tackling common health system problems, including surveillance and infection control. This EU approach could provide a model for South-East Asia. Further, the spread of disease can be tackled by building capacity and training the next generation of public health specialists, monitoring personnel and outbreak analysts, establishing widespread surveillance capability, and investing in agencies and providing institutional capacity, which can take the lead and stimulate improved systems.

Better data and laboratory capacity are seen as key determinants of future success, with collaborative capacity-building measures of particular value

(Wertheim et al., 2010). The South-East Asia Infectious Disease Clinical Research Network (SEAICRN), a collaborative partnership between hospitals and research institutions in Vietnam, Thailand, and Indonesia, is one such venture that has achieved encouraging results. The SEAICRN was established in 2005 to conduct clinical research to address emerging threats, and to advance the scientific knowledge and clinical management of infectious diseases in the region (South-East Asia Infectious Disease Clinical Research Network, n.d.a). Among its achievements, SEAICRN has conducted interventional studies (a multicenter randomized trial) for oseltamivir to combat severe influenza, and has published research protocols to deal with infectious diseases in member laboratories (South-East Asia Infectious Disease Clinical Research Network, n.d.b).

Prospects

The next steps predicating further progress include accelerating the programs and initiatives already underway and taking advantage of demonstrated successes in other regions. Building momentum with existing measures is also crucial. The Gates Foundation, for example, is investing heavily in research into malaria and typhoid fever that will ultimately benefit the whole region.

As in other parts of the world, South-East Asian nations are making progress toward the Millennium Development Goals (MDGs). The World Health Organization (WHO), World Bank, and International Monetary Fund are funding or supporting initiatives at the global level to enhance international efforts. WHO, for instance, is doing extensive work to support universal healthcare initiatives (WHO, 2016g) and promoting quality of care and patient safety measures internationally (WHO, 2017d). South-East Asian countries and the region are actively participating in these initiatives (WHO, 2015f). Next-generation success will require concerted efforts; political will and support; further investments, even when resources are scarce; learning from past successes in South-East Asia and elsewhere; and doing more of what we know already works well.

Conclusion

Looking from between 5 and 15 years into the future is useful when considering the challenges facing South-East Asia in protecting its communities against infectious diseases. The time to start dealing with a pandemic is before it has taken hold. Preventing or ameliorating the next virulent influenza event or post-typhoon infection is far preferable. Long-term planning is the key: establishing more robust policies, systems, infrastructure, practices, and capacities, and building on what has already been established, is likely to be an investment of great benefit to the region.

Discussion and Conclusion

Jeffrey Braithwaite, Russell Mannion, Yukihiro Matsuyama,
Paul G. Shekelle, Stuart Whittaker, Samir Al-Adawi,
Kristiana Ludlow, Wendy James, and Elise McPherson

CONTENTS

In this compendium, we offer evidence-based predictions (or, more precisely, explorations) about the future of healthcare, resulting from the input of 148 contributing authors drawn from all over the world. These authors have investigated and analyzed multiple settings, topics, professional groups, and ideas drawn from three-quarters of the world's countries. Now that we can see the chapters in context, we can celebrate the range, depth, and quality of the case studies across 52 individual countries and five regional groupings, covering the 152 countries and territories in our sample of stories.

Once the chapters were submitted, representatives of the core editorial group in Sydney, Australia (Braithwaite, Herkes, Ludlow, and McPherson), on behalf of the larger editorial team, gathered around a whiteboard to synthesize the information contained in the case stories. When the team grouped the cases using an affinity clustering technique (based on, e.g., Thomas, 2006), nine themes emerged (Figure D.1).

FIGURE D.1
Whiteboard session outputs.

The chapters mapped to the nine themes are expressed in Table D.1.

TABLE D.1

Broad-Based Themes, Aggregated from the Chapters

Theme	Number of Chapters
Financing, economics, and insurance	5
Patient-based care and empowering the patient	8
Universal healthcare	3
Technology and information technology	10
Aging populations	4
Preventative care	3
Accreditation, standards, and policy	9
Human development, education, and training	7
Integration of healthcare services	8

We then developed a chapter summary providing a concise description of each case study for each chapter within each theme (Table D.2).

These nine themes account for the wide range of initiatives that the authors believe are important in creating a health system for the future. In order to examine the landscape of learnings further, we did a secondary analysis of the chapters.

TABLE D.2

Themes, Country/Region, Chapter Title, and Summary

Theme	Country/Region and Chapter Title	Chapter Summary
Financing, economics, and insurance	**Africa** Equity for All: A Global Health Perspective for the Continent	The high burden of disease that is linked to demography and poverty necessitates reforms, such as healthcare financing, to address the inequality in healthcare access and health system strengthening and resilience through a shift from vertical funding (disease specific or patient group) to horizontal funding (system-level interventions).
	Chile The Struggle for an Integrated Health Insurance System	The health insurance system (private vs. public) is inequitable and requires integration. There is increasing consensus that a social security system covering the whole population will be the most effective solution.
	England Getting Personal? Personal Health Budgets	Personal health budgets–healthcare plans set up between individuals and their health professionals–are proposed to individualize funding, encourage more self-directed support, and encourage the National Health Service to be more responsive to the needs of patients.
	Malaysia The Future Malaysian Antenatal Care System: Building upon the Old	A proposed new antenatal care system, tied to a national health insurance scheme to ensure affordability, aims to improve quality of care by consolidating standard operating procedures, upscaling personalized healthcare nationally, and implementing a four-antenatal-visit model to minimize resource wastage and improve care quality.
	Mongolia Health System Financing	Since transitioning from the Soviet-era Semeshko model, respective national standards and guidelines have redirected the operational focus of health facilities from curative to preventive care. Other reforms include establishing governing boards in the government-owned secondary- and tertiary-level hospitals, to provide organizational autonomy in management and decision-making. Major achievements include the development of several sources of funding, the introduction of social health insurance, resource allocation to service providers through mixed payment methods, and structural reorganization of the district hospitals into multi-specialty facilities.
Patient-based care and empowering the patient	**Canada** The Future of Health Systems: Personalization	The provider-centric system is being shifted to a personalized system, characterized by democratized health information, accountability to citizens, personalized care approaches, value-based resource allocation, and visibility by tracking and traceability
	Denmark Patient-Reported Outcomes: Putting the Patient First	The recent development of a Danish patient-reported outcome (PRO) data strategy, Program PRO, provides a blueprint for the systematic integration of PRO data in quality improvement across the healthcare system to improve clinical care and patient satisfaction.
	Hong Kong Integrated Health Services: A Person-Centered Approach	The need for an integrated people-centered care approach in the design and delivery of health services is articulated in order to meet the health system challenges of creating a successful and sustainable health system in the context of rapidly aging populations and limited resources.
	Namibia Lessons from Patient Involvement in HIV Care: A Paradigm for Patient Activation and Involvement across Health Systems	Formal patient and community involvement in quality improvement activities at various government levels is promoting patient-centered care and health system improvement culture across the public health system.

(Continued)

TABLE D.2 (CONTINUED)

Themes, Country/Region, Chapter Title, and Summary

Theme	Country/Region and Chapter Title	Chapter Summary
	Oman Paradigm Change: Healthy Villages to Meet Tomorrow's Health Needs	The concept of enabling people to choose their own health outcomes is currently being tested by the Healthy Village initiative, which aims to decentralize the universal healthcare structure, and empower local, provisional, and regional sectors by letting them chart their own paths in developing and improving health services.
	Scotland Deliberative Engagement: Giving Citizen Involvement Meaning and Impact	User, carer, and citizen participation is a key principle of supporting and driving improvement in health and social care systems and services in Scotland. The Our Voice initiative is a national program, with commitment across health and social sectors, conducted in partnership with citizens and health system stakeholders to enable dialogue with citizens to help deliver health gains and reduce inequalities.
	Spain How Can Patient Involvement and a Person-Centered Approach Improve Quality in Healthcare? The Patients' University and Other Lessons from Spain	The Patients' University is a patient-centered initiative that aims to meet patient information needs and promote health literacy and self-management to patient and health professionals through educational programs and research projects in response to the complexities of the modern healthcare systems and as a future investment in fostering the patient's voice in decision-making forums.
	The United States of America The U.S. Healthcare System: A Vision for the Future	In the ideal future U.S. health system, • Health status measurement will be essential. • Consumers will know when and how to use the system. • Quality of care will have a central role. • There will be benefits from updated information about appropriateness of care. • Geography will not affect the amount or quality of care a patient receives. • Care for physical and mental health conditions will be integrated. • Health system leaders will understand and address social determinants of health.
Universal healthcare	**Argentina** Achieving Universal Coverage	A strategy is proposed to gradually achieve universal coverage by the gradual targeting of segments of the population and the creation of a technology assessment agency that would evaluate new technology and determine reimbursement rates to ease tensions between insurers and providers.
	Mexico Leveraging Conditional Cash Transfers and Universal Health Coverage to Tackle Non-Communicable Diseases	The rollout of universal health coverage through *Seguro Popular*, accompanied by conditional cash transfers, has reduced the financial burden of healthcare intervention on the poor and has the potential to improve access to healthcare, which is needed to address the rise in non-communicable diseases.
	Central Asia From Russia with Love: Health Reform in the Stans of Central Asia	The future of health reform is covered, including the introduction of health insurance financing, universal primary care, reducing inequities, better organization, and governance.
Technology and information technology	**Estonia** e-Consultation Services: Cooperation between Family Doctors and Hospital Specialists	An e-consultation service system has been developed through collaboration between the Estonian Health Insurance Fund, the Estonian Society of Family Doctors, and hospital specialists in response to extended waiting times for outpatient specialist services and inequitable access to healthcare services, which has increased satisfaction with provider services since the 1990s.

(Continued)

TABLE D.2 (CONTINUED)

Themes, Country/Region, Chapter Title, and Summary

Theme	Country/Region and Chapter Title	Chapter Summary
	Finland A Real-Life Experiment in Precision Medicine	Precision medicine, supported by data, technology, and digitization, is gradually progressing toward integration with healthcare practices, which may achieve better accessibility and equity of services, and improve quality of care and freedom of choice, in the context of an aging population with increasing service needs and limited resources.
	Italy The Introduction of New Medical Devices in an Era of Economic Constraints	The recently established national health technology assessment program aims to make recommendations about the efficacy and adoption of medical devices, and will ensure stricter regulation of the adoption of devices and cost minimization.
	Lebanon m-Health for Healthcare Delivery Reform: Prospects for Lebanese and Refugee Communities	m-Health, as part of a broader eHealth strategy, aims to improve population health and treatment through a Short Message Service–based system providing health promotion, disease awareness, decision support, and treatment follow-up, as part of an emphasis shift toward primary health and service accessibility.
	Papua New Guinea Strengthening the Collection, Analysis, and Use of Health Data through eHealth Solutions	To achieve an efficient and responsive health system, the electronic National Health Information System was piloted to facilitate data accessibility and timely delivery, in replacement of paper records. The success of the pilot has led to the proposal of a nationwide rollout, which will require improvements to systems and software and the promotion of a culture of information use and sustainable funding.
	Sweden The Learning Health System	Digital data and technologies are being integrated into healthcare provision and research via the learning health system to support everyday medical decision-making and longer-term learning and system improvement.
	Taiwan "My Data, My Decision": Taiwan's Health Improvement Journey from Big Data to Open Data	Big data, collected by the National Health Insurance Research Database, has established 120 data sets on the government's open data platform to allow the public to conduct value-added applications or innovations. The National Health Insurance Administration delivers users' health-related data to cloud storage, from which citizens can instantaneously obtain comprehensive health information, allowing them to take responsibility for their health decisions.
	The United Arab Emirates Improving Healthcare through a National Unified Medical Record	Implementation of the national unified medical record (NUMR) will provide healthcare leaders with accurate data. The NUMR will also integrate patient appointments and other health service provider results and records, replacing the independent healthcare information systems, facilitating regulation and compliance, and promoting patient safety.
	Venezuela Learning from Failure and Leveraging Technology: Innovations for Better Care	Learning from the programs *Barrio Adentro* and *medicos integrales comunitarios*, the implementation of well-meaning innovations has the potential to turn failed efforts into assets. Projects that leverage technology, in combination with proposed reforms to pre-existing programs, have the potential to improve access to healthcare in rural communities.
	Wales Realizing a Data-Driven Healthcare Improvement Agenda: A Manifesto for World-Class Patient Safety	A manifesto for a data-driven organization was built on a functioning patient safety incident reporting and learning system able to analyze and visualize data. With the support of trained staff proficient in quality improvement methods, such a system will improve quality in care and build a more complete national picture of preventable harm.

(Continued)

TABLE D.2 (CONTINUED)

Themes, Country/Region, Chapter Title, and Summary

Theme	Country/Region and Chapter Title	Chapter Summary
Aging populations	**Australia** The Silver Tsunami: The Impact of the Aging Population on Healthcare	The CriSTAL tool was developed to identify frail elderly in order to facilitate a shift from complex acute care in hospitals to community care within the context of population aging, thereby improving support and decision-making, and circumventing the medicalization of the aging and dying.
	Japan Toward a Community-Friendly Dementia Strategy	In response to population aging, a dementia policy, the New Orange Plan (2015), has been developed to maintain quality of life and provide an appropriate environment for those in the final stages of life. The emphasis is on long-term care and living support, with medical practitioners and social welfare corporations playing a central role, rather than hospitals.
	The Netherlands Reform of Long-Term Care	To support the challenges of an aging population and the need for long-term care, the responsibility for long-term care has been transferred to local municipalities, which has in turn created administrative, capacity, complexity, and financial challenges. Improvement initiatives in response include strengthening administrative capacity, developing support structures, improving quality and efficiency in nursing homes, and developing elderly-friendly hospital environments.
	Qatar Hospice Palliative Care	Rapid aging necessitates a reform of end-of-life care, which is currently provided by limited palliative care services in the main hospital network. Proposed hospice services are aimed at increasing patient-centered care and improving quality and safety.
Preventative care	**Portugal** Prevention of Antimicrobial Resistance through Antimicrobial Stewardship: A Nationwide Approach	A nationwide approach to the prevention of antimicrobial resistance was facilitated by the mandatory implementation of a national antimicrobial stewardship program in hospitals by the end of 2014. Program interventions included restrictive prescribing rules and educational advice and feedback for physicians and have succeeded in reducing consumption of some types of antibiotics, with other types still requiring priority in terms of hospital management goals.
	Rwanda Embracing One Health as a Strategy to Emerging Infectious Diseases Prevention and Control	A One Health approach has been adopted to reduce the threat of infectious disease epidemics. It is expected to be a successful initiative because it has strong political backing and emphasizes collaborative action across sectors and governing systems.
	South-East Asia Taming Communicable Diseases	Governments across the region have set up new or strengthened emerging diseases monitoring, protection, and control agencies in an attempt to tame communicable diseases. Successes have come in the form of strengthened immunization programs, and enhanced strategies to combat infectious disease outbreaks.
Accreditation, standards, and policy	**India** How to Build a First-World Health System on a Third-World Budget	The 2006 National Accreditation Board for Hospitals and Healthcare Providers (NABH) was implemented to improve patient safety through national accreditation, and a 2016 initiative has mandated that private insurance companies empanel only hospitals with NABH accreditation. The successful accreditation initiative is creating a first-world service at a third-world price.

(Continued)

TABLE D.2 (CONTINUED)

Themes, Country/Region, Chapter Title, and Summary

Theme	Country/Region and Chapter Title	Chapter Summary
	Iran Hospital Accreditation: Future Directions	The 2012 Accreditation Standards for Hospital program has had a positive impact on hospital services, but has not translated into enhanced quality and safety of services and patient satisfaction. It needs national legislative support to maximize benefits and achieve further reforms. The processes and methods of hospital accreditation also need reforming to enhance credibility.
	Malta The National Cancer Plan: Strengthening the System	The 2011 national cancer plan, One Health, a concerted approach to improve cancer detection and care, has improved cancer outcomes through political commitment and a multisectorial approach in part by emphasizing public health actions related to the environment and lifestyle habits.
	Middle East and North Africa (MENA) Health Systems in Transition	Organizational structures across Middle East and North Africa (MENA) a nations need to take advantage of improved technology, the genomics revolution, and new public health measures. Effective community-based health systems and primary care are the key to strengthening society and improving the health status of the population.
	New Zealand Strengthening Primary Healthcare	A road map of actions emphasizing patient-centered care, community engagement, and disease prevention and early intervention, as laid out in the New Zealand Health Strategy, has been developed to cope with the growing and aging population. Sustained investment and government support are necessary for success.
	Nigeria Doing More with Less: Lean Thinking in the Health System	Reductions in allocated health management budgets will require administrators to become more resourceful in the redesign of systems. Lean thinking principles suggest that the components that have made systems weak and volatile will provide the ingredients for a highly integrated and resilient system. The principles of Lean thinking focus on the elimination of waste in all its forms from the system. Improvements based on Lean thinking have seen health organizations reduce costs without compromising quality.
	Norway Bridging the Gap: Opportunities for Hospital Clinical Ethics Committees in National Priority Setting	Clinical ethical committees provide a forum for hospitals, individuals, and government to discuss how best to manage decision-making and priority setting processes; they are proposed as a potential solution for managing scarce healthcare resources, improving the coordination of messages, and identifying and resolving problems.
	South Africa Regulated Standards: Implementation and Compliance	The Office of Health Standards Compliance (OHSC) acts as an independent regulator of health services, guiding, monitoring, and enforcing safety and quality standards. The OHSC develops regulated health standards and provides a framework against which service delivery at health establishments can be evaluated. The next steps for the OHSC include an advanced information technology system for the capture, collation, and analysis of data procured during health service inspection. The system will also calculate certification status by means of a predetermined algorithm and generate inspection reports.
	Turkey Moving Quality in Healthcare Beyond Hospitals: The Turkish Accreditation Model	Quality improvement through accreditation is one of the main principles of the Health Transformation Program, which seeks to achieve an overall transformation of healthcare service quality and patient safety. Standardization and accreditation of health services outside the hospital system is necessary to ensure the delivery of high-quality healthcare, e.g., improved and accredited school health services.

(Continued)

TABLE D.2 (CONTINUED)

Themes, Country/Region, Chapter Title, and Summary

Theme	Country/Region and Chapter Title	Chapter Summary
Human development, education, and training	**Brazil** Patient Safety: Distance-Learning Contribution	A patient safety specialization course has been developed and delivered for health professionals to improve the quality of healthcare and patient safety.
	Jordan Improving Quality of Care by Developing a National Human Resources for Health Strategy	A formal national human resources strategy, focusing on strengthening planning and information systems, workforce recruitment, and training and professional development, and aligned with global initiatives, is required for quality care to be maintained and the system strengthened in the context of recent rapid population growth and refugee influx.
	Northern Ireland Developing a Framework to Support Building Improvement Capacity Across a System	In response to a model of care that is neither affordable nor sustainable, a quality improvement approach, the Attributes Framework for Health and Social Care, has been developed to support leadership in quality improvement and building improvement capacity across a complex system and workforce.
	Pakistan The Way Forward	To reach its health goals, Pakistan needs to improve the quality of its medical education and training, inculcate a culture of medical research in support of evidence-based care, and increasingly utilize technologies to give access to rural populations.
	Switzerland Teamwork and Simulation	Team training and non-technical skills training are promoted through simulation for quality improvement and patient safety in healthcare.
	Trinidad and Tobago Nurse Training: A Competency-Based Approach	Implementing competency-based models to address deficiencies in the supervision of nurses and improve the quality of care requires validating supervisory and managerial competencies, enhancing the curriculum of the institutes that train nurses, and training existing staff to deliver the new supervisory and management models.
	Russia The Future of Physicians' Specialization	The side effects of increasing specialization in post-Soviet countries are reflected on, as well as the need to retain generalists. An increase in the subspecialization of physicians is predicted in successful healthcare systems.
Integration of healthcare services	**Austria** Primary Healthcare Centers: A Silver Bullet?	A new primary care model—multidisciplinary ambulatory outpatient care—was introduced to help shift costly inpatient care to outpatient care to balance utilization and improve coordination of care.
	Central and Eastern Europe Strengthening Community-Based Family Care and Improving Health Equities	Despite the fall of the Soviet Union, health systems are frequently influenced by Soviet-inspired politics. Evaluation of reform is inconsistent and infrequently conducted. Initiatives to tackle inequities and improve family and community-based care have improved the integration of services.
	China Integrated Stratified Healthcare System	Regional pilot trials of an integrated stratified healthcare system (tertiary, secondary, and community health facilities) were supported by information-sharing Internet platforms and health insurance leveraging, to improve access to and quality of primary healthcare in a hospital-centric system.
	France Horizon 2030: Adopting a Global-Local Approach to Patient Safety	A "global-local" health system reform approach was outlined that would meet the needs of an aging population and regional disparity by prioritizing the entire patient journey and implementing a flexible regionalized system, in place of the existing centralized system.

(Continued)

TABLE D.2 (CONTINUED)

Themes, Country/Region, Chapter Title, and Summary

Theme	Country/Region and Chapter Title	Chapter Summary
	Germany Health Services Research and Future Planning in Pediatric Care	To ensure quality pediatric care and address economic inefficiency in rural regions, a series of research projects were run to examine the integration of hospitals, the efficacy of telemedical triage, cooperation between healthcare specialties, and delegation. Concepts should be re-evaluated in larger-scale implementations in all pediatric settings.
	Greenland Everyday Life with Chronic Illness: Developing a Democratic and Culture-Sensitive Healthcare Practice	A new professional focus on well-being, not just health, is needed to address the growing number of people with chronic illnesses. Home care, or "home rehabilitation," is a potential solution currently being trialed, which involves integrated, cross-professional care to create a democratic and culturally sensitive healthcare practice.
	Guyana Paradigm Shift: From Institutional Care to Community-Based Mental Health Services	Mental health is an area of emerging priority in the decentralized healthcare system. A community-based mental health model to close the treatment gap would emphasize primary care and promote comprehensive and community-based psychosocial rehabilitation. Coordination of all government sectors and community healthcare services is required.
	Yemen Integrating Public Health and Primary Care: A Strategy for the Health System of the Future	Integrating public health and primary care into "community-oriented primary care"—moving from hospital-based care to community- and population-based healthcare—is a promising reform possibility for the resource-limited country, which will require a radical shift in the way all key stakeholders regard the relationship between public health and primary care.

In what follows, we provide word clouds (http://wordart.com) of the key concepts explored throughout the book. These are images that distinguish the most frequently occurring words across large volumes of text. To generate the frequently occurring concepts, we made some adjustments to the word clouds by taking out overlapping themes, essential to each section and the book in its entirety. We deleted functor words, such as *and, but, if,* and *the,* and frequently occurring concepts that occurred in all the word clouds, for example, *system, patient, services, health,* and *hospitals,* and words stemming from these concepts. Creation of word clouds with these deletions enables readers to focus on the essential nature of large volumes of text.

Background of Initiatives and Enhancement Projects In-Country

In this first part of their chapters, the contributing authors presented background information and the context of the situation in which their case study was provided, framing the story and the challenges faced by the country, highlighting what is missing and the obstacles faced. Information about their past and current improvement projects and the key health issues, where relevant, was provided in each chapter. This background section attempts to answer the question, "What were the projects being delivered and who were they for?" Some authors highlighted specific sectors in the health services their respective country or region was working to improve, such as aged care delivery or mental health services.

Word Cloud 1 provides a diagram to illuminate the contextual constructs. As we have seen, the context allowed the authors to introduce their chapter and present the key topic areas and where projects were focused. Many authors discussed the context of their interventions and introduced providers and settings in relation to the projects. This was evident through the most popular words, which include *public, population,* and *private.* Additionally, words that related to the specific areas being targeted led to the frequent occurrence of words such as *medical, social,* and *aging.*

WORD CLOUD 1
Key words relating to the context and setting of chapters.

Building on Previous Successes and Their Challenges

The authors were asked to discuss the core elements of their chosen case study's successes and, where relevant, how this was measured. The purpose of this section was to discuss where projects had been successful or where success could be extrapolated and accomplishments extended. The variations and differing perspectives of the authors were most evident throughout this section of the book, with the authors delivering a mixture of reflective and prospective stories depending on where projects were situated in terms of progress and perceived success, and whether this had already occurred to some degree. Some authors focused mostly on challenges and how these needed to be addressed to achieve future success.

Discussions of where their health system was in terms of progress, and the capacity of the system or project to achieve advancements, were based on elements of the respective systems and examples drawn from previous interventions. Some authors discussed initiatives that were in the early stages of delivery and the progress that had been made so far, while others discussed initiatives that were well established, the nature of their success, and their potential impact on the health system or the health of their populations.

While there is considerable socioeconomic inequity between countries with varying health needs, and differing health systems showcased, the authors who described the most significant challenges were not necessarily from the developing world. All countries, including the wealthiest, articulated perceived failures, gaps, and resistance that were occurring or would occur during the implementation of projects.

Word Cloud 2 illuminates the fact that despite the widely varying stories, there were many commonalities: it is clear that all countries, regardless of their progress, have similar goals. Reoccurring words, such as *quality, safety, development*, and *standards*, were a reflection of common objectives and what constitutes success. The recurrence of words such as *data, accreditation*, and *training* suggests that these were key improvement mechanisms globally, whether or not they had been achieved.

WORD CLOUD 2
Key words discussing the criterion for success and what a successful intervention delivers.

Impact of Enhancement Projects

In this section, the authors were asked to describe the effects their proposed successes have had or were projected to have on their corresponding health system, and the overall health of target populations resulting from system projects. The response of the authors was dependent on whether the success story had been written in a reflective or prospective context. Notwithstanding that, a similar set of core themes was identified, evident through the reoccurrence of words such as *quality, care, training,* and *safety.* This emphasized that these are the key elements of successful health system improvements for many authors. Some authors used case studies of specific clinical applications of systems improvement, which was evident through the repetition of words such as *HIV, rural, elderly, inpatient, outpatient,* and *dementia.* Additionally, words such as *support, time, development,* and *established* were a reflection of what elements are needed to ensure meaningful and sustainable impact in the future.

As Word Cloud 3 shows, the impact of the success stories comes in many forms. The most important aspects were whether groups' or populations' health was improved, or whether care, or how it is organized, was enhanced. The impact of success, for case studies that had been successful in one way or another, was measured in terms of new programs, better provision of services, the development of infrastructure to improve things, and enhanced policy, services, or implementation of new initiatives.

WORD CLOUD 3
Impact of the success story on health systems and health outcomes.

Transferability or Implementation of Enhancement Projects

We turn to the authors' contributions to diffusion and implementation. The authors broadly conceptualized this in line with the current trend of "implementation science" (Rapport et al., 2017). This involves the investigation of methods designed to promote the adoption of research results into routine healthcare delivery. It encompasses clinical, organizational, or policy considerations. Our authors homed in on two themes relating to transferability and implementation. First was whether they had borrowed elements of their proposed projects from elsewhere, whether it was a health system different from their own or one that had similarities. Second, some authors described whether their project could be transferred or used in other health systems and whether this was likely to be successful, and why. Many authors discussed whether their projects could be implemented effectively in other systems or if their program was entirely unique and based on the exclusive needs of their population.

Word Cloud 4 provides a synthesis of these reflections. As the word cloud highlights, there are a wide range of opinions on transferability with some reoccurrence of leverage words, such as *use, practice, provide,* and *need*. Frequently occurring words, such as *more, new, primary, level, training, public,* and *management,* suggest that respective countries foresee elements of their interventions being transferable under the right circumstances or, if adjusted, to suit the differing needs and requirements of other systems.

WORD CLOUD 4
Key words relating to the perceived transferability of interventions.

Next Steps for Prospective Strategies

The final word cloud, Word Cloud 5, is a reflection of our request that the authors discuss whether the building blocks of their prospective strategies are in place, and what still needs to be done to achieve this ideal. Some authors listed what had already been planned but not yet implemented, and others discussed what needed to be planned, changed, or delivered to make progress. The recurring presence of words such as *quality, data, national,* and *level* mirrors the concepts in Word Clouds 1–4, identifying the need for the focus to shift, and for changes to be made at a systems level, to ensure consistency and achieve equitable systems, with quality care being the cornerstone of enhancement projects. The repetition of words such as *social* emphasizes the growing consensus that health systems need to move from a curative to a preventative outlook, highlighting the significance of people. Other key words summarizing the elements required for future success include *support, training, provide,* and *needs.* These demonstrate the similarities between Word Clouds 2 and 5, based on commonalities between "what will make a successful system?", "the challenges that may arise and what are the next steps in addressing those challenges," and "where do you see your health system in an ideal future?"

WORD CLOUD 5
Key words relating to the next phase of projects.

Conclusion

Leland R. Kaiser once said, "The hospital is a human invention and as such can be reinvented at any time." We agree, although we would substitute *health system* for *hospital*.

One key message to be drawn from the book is that there is a universal need to realign health systems to reflect the changing needs of populations and to promote improvement wherever possible. Countries are recognizing the importance of doing more with limited resources and maneuvering around the many other constraints they face. Health systems everywhere are optimizing funding through strategies to ensure that wasteful spending is identified and reduced, and that identifiable areas are improved, often through inexpensive and relatively simple initiatives that look at health systems and population health in a social and environmental context.

We are mindful that there is much in this book to reflect on, absorb, and apply. For an inspiring way to conclude, we like the words of Jonathan Bush and Michael Chernew: "If our society can seize this moment to unleash the power of innovators to create better models, which allow them to profit from removing waste and finding new efficiencies, we can put ourselves on the path to a satisfying and sustainable health care system" (Chernew and Bush, 2016).

References

Aagaard, T. (2015). Hverdagsliv med sygdom—patienters kulturelle perspektiver på sundhedspraksis i Grønland [Everyday life with illness—Patients' cultural perspectives on healthcare practice in Greenland]. *Inussuk—Arktisk forskningsjournal*, 1, 1–221. Retrieved from: http://naalakkersuisut.gl/~/media/Nanoq/Files/Attached%20Files/Forskning/Inussuk/DK%20og%20ENG/Inussuk%201%202015.pdf.

Aagaard, T. (2017). Patient involvement in healthcare professional practice – a question about knowledge. *International Journal of Circumpolar Health*, 76(1): 1403258

Abomics. (2017). GeneRx—Pharmacogenomic database. Retrieved from http://www.abomics.fi/en/solutions/generx-en.

Abyad, A. (2001). Health care for older persons: A country profile—Lebanon. *Journal of American Geriatrics Society*, 49, 1366–1370.

Adambekov, S., Kaiyrlykyzy, A., Igissinov, N., & Linkov, F. (2016). Health challenges in Kazakhstan and Central Asia. *Journal of Epidemiology and Community Health*, 70(1), 104–108.

Adams, H. P., Jr. & Biller, J. (2014). Future of subspecialty training in vascular neurology. *Stroke*, 45(12), 3730–3733.

Agency for Healthcare Research and Quality. (2017). TeamSTEPPS® 2.0. Retrieved from https://www.ahrq.gov/teamstepps/instructor/index.html.

Agins, B. D., Young, M. T., Ellis, W. C., Burke, G. R., & Rotunno, F. F. (1995). A statewide program to evaluate the quality of care provided to persons with HIV infection. *Joint Commission Journal on Quality Improvement*, 21(9), 439–456.

Ailisto, H. (2017). Finland will be a winner in the artificial intelligence disruption [blog post], May 23. Retrieved from https://vttblog.com/2017/05/23/finland-will-be-a-winner-in-the-artificial-intelligence-disruption/.

Akala, F. A. & El-Saharty, S. (2006). Public-health challenges in the Middle East and North Africa. *Lancet*, 367(9515), 961–964.

Akdag, R. (2008). *HTP Progress Report*. Ankara: Ministry of Health.

Al-Abbadi, I. (2009). Health care equity issues in Middle East. *Policy Analysis*, July/August, 8–9.

Al-Ajlouni, M. (2010). Human resources for health: Jordan country profile. Retrieved from http://apps.who.int/medicinedocs/documents/s17239e/s17239e.pdf.

Alakeson, V. (2013). The individual as service integrator. *Journal of Integrated Care*, 21(4), 188–197.

Alakeson, V., Boardman, J., Boland, C., Crimlisk, H., Harrison, C., Iliffe, S., et al. (2016). Debating personal health budgets. *BJPsych Bulletin*, 40(1), 34–37.

Al-Awadhi, T., Ramadan, E., Choudri, B. S., & Charabi, Y. (2016). Growth of coastal population: Likely exposure to sea level rise and associated storm surge flooding in the Sultanate of Oman. *Journal of Environmental Management & Tourism*, 7(2), 340–344.

Al-balushi, M. S., Ahmed, M. S., Islam, M. M., & Khan, M. H. (2016). Contraceptive method choices among women in Oman: A multilevel analysis. *Journal of Data Science*, 14(1), 117–132.

Albiger, B., Glasner, C., Struelens, M. J., Grundmann, H., & Monnet, D. L., the European Survey of Carbapenemase-Producing Enterobacteriaceae (EuSCAPE) working group. (2015). Carbapenemase-producing Enterobacteriaceae in Europe: Assessment by national experts from 38 countries. *Euro Surveillance*, 20(45), 30062.

Alhinai, S. (2015). Healthy villages: An experience from Oman. Retrieved from http://hehp.kums.ac.ir/kums_content/media/image/2015/05/58002_orig.pdf.

Al-Lawati, J. A., Mabry, R., & Mohammed, A. J. (2008). Addressing the threat of chronic diseases in Oman. *Preventing Chronic Disease*, 5(3), A99.

Al-Shammari, S. A., Al Mazrou, Y., Jarahhal, J. S., Al Ansary, L., El-Shabrawy, A. M., & Bamgboye, E. A. (2000). Appraisal of clinical, psychological and environmental health of the elderly in Saudi Arabia: A household survey. *International Journal of Aging and Human Development*, 50(1), 43–60.

Al-Sinawi, H., Al-Alawi, M., Al-Lawati, R., Al-Harrasi, A., Al-Shafaee, M., & Al-Adawi, S. (2012). Emerging burden of frail young and elderly persons in Oman: For whom the bell tolls? *Sultan Qaboos University Medical Journal*, 12(2), 169–176.

Al-Surimi, K. (2017). Improvement of basic health services in Yemen: A successful donor-driven improvement initiative. In J. Braithwaite, R. Mannion, Y. Matsuyama, P. Shekelle, S. Whittaker, & S. Al-Adawi (Eds.), *Health Systems Improvement Across the Globe: Success Stories from 60 Countries* (pp. 361–371). Boca Raton, FL: Taylor & Francis.

Amalberti, R., Nicklin, W., & Braithwaite, J. (2016). Preparing national health systems to cope with the impending tsunami of ageing and its associated complexities: Towards more sustainable health care. *International Journal for Quality in Health Care*, 28(3), 412–414.

American Public Health Association. (2017). What is public health? Retrieved from https://www.apha.org/what-is-public-health.

Amilasan, Al-S. T., Ujiie, M., Suzuki, M., Salva, E., Belo, M. C. P., Koizumi, N., et al. (2012). Outbreak of leptospirosis after flood, the Philippines, 2009. *Emerging Infectious Diseases*, 18(1), 91–94.

Anderson, E. & Panov, L. (2015). Family doctor's offices' outpatient consultations data in the e-health system. Tervise Arengo Instituut. Retrieved from https://intra.tai.ee//images/prints/documents/149872700952_Family_doctors_offices_outpatient_consultations_data_in_the_e-health_system_2015.pdf.

Antoun, J., Phillips, F., & Johnson, T. (2011). Post-Soviet transition: Improving health services delivery and management. *Mount Sinai Journal of Medicine*, 78(3), 436–448.

Anwar, F. & Shamim, A. (2011). Barriers in adoption of health information technology in developing societies. *International Journal of Advanced Computer Science and Applications*, 2(8), 40–45.

Aponte-Moreno, M. & Lattig, L. (2012). Chavez: Rhetoric made in Havana. *World Policy Journal*. Retrieved from http://www.worldpolicy.org/journal/spring2012/ch%C3%A1vez-rhetoric-made-havana.

App Annie. (2016). Mobile app forecast: The path to 100 billion. Retrieved from http://www.mobuzz.org/wp-content/uploads/2016/11/app-annie-02-2016-forecast-en.pdf.

Arce, H. (2010). *El sistema de salud: De dónde viene y hacia dónde va* [*The Health System: Where It Comes from and Where It Goes to*]. Buenos Aires, Argentina: Editorial Prometeo Libros.

Arce, H., García Elorrio, E., & Rodríguez, V. (2017). Successful initiatives in quality and patient safety in Argentina. In J. Braithwaite, R. Mannion, Y. Matsuyama, P. Shekelle, S. Whittaker, & S. Al-Adawi (Eds.), *Health Systems Improvement Across the Globe: Success Stories from 60 Countries* (pp. 3–9). Boca Raton, FL: Taylor & Francis.

Aronson, S. J. & Rehm, H. L. (2015). Building the foundation for genomics in precision medicine. *Nature, 526,* 336–342.

Arrighi, E., Blancafort, S., Jovell, A. J., & Navarro, M. D. (2015). Quality of cancer care in Spain: Recommendations of a patients' jury. *European Journal of Cancer Care, 24*(3), 387–394.

Article 15a B-VG. (2016). Agreement in accordance with Article 15a of the federal constitutional law on organization and finance of the health-care system, valid 2017–2020. Retrieved from https://www.parlament.gv.at/PAKT/VHG/XXV/I/I_01340/fname_572713.pdf.

Asaba, S. (2016). Babyl launches digital healthcare system. *New Times,* July 29. Retrieved from http://www.newtimes.co.rw/section/article/2016-07-29/202166/.

Ascencio, C. (2017). Cuestionados médicos integrales comunitarios formados en Venezuela trabajan en la salud pública chilena. Retrieved from http://www.economiaynegocios.cl.

Ashley, E. A. (2015). The precision medicine initiative. A new national effort. *JAMA: The Journal of the American Medical Association, 313*(21), 2119–2120.

Aspria, M., de Mul, M., Adams, S., & Bal, R. (2016). Of blooming flowers and multiple sockets: Infrastructure integration and the sociotechnical imaginary. *Science, Technology & Society, 29*(3), 68–87.

Association of Southeast Asian Nations. (n.d.). About ASEAN. Retrieved from http://asean.org/asean/about-asean/.

Australian Institute of Health and Welfare. (2017). *Australia's Health 2016.* Canberra: Commonwealth Government of Australia.

Azhar, S., Hassali, M. A., Ibrahim, M. I. M., Ahmad, M., Masood, I., & Shafie, A. A. (2009). The role of pharmacists in developing countries: The current scenario in Pakistan. *Human Resources for Health, 7*(1), 54.

Azzopardi-Muscat, N. (1999). *Health Care Systems in Transition: Malta.* Copenhagen, Denmark: European Observatory on Health Systems, 1–84.

Azzopardi-Muscat, N., Calleja, N., Calleja, A., & Cylus, J. (2014). Malta: Health system review. *Health Systems in Transition, 19*(1), 1–137.

Bachelet, M. (2013). Chile de Todos. Programa de Gobierno de Michelle Bachelet 2014–2018 [Chile of all. Government program of Michelle Bachelet 2014–2018]. Retrieved from http://www.onar.gob.cl/wp-content/upLoads/2014/05/ProgramaMB.pdf.

Baird, S., McIntosh, C., & Özler, B. (2011). Cash or condition? Evidence from a cash transfer experiment. *Quarterly Journal of Economics, 126*(4), 1709–1753.

Balabanova, D., Roberts, B., Richardson, E., Haerpfer, C., & McKee, M. (2012). Health care reform in the former Soviet Union: Beyond the transition. *Health Services Research, 47*(2), 840–864.

Barker, P., Reid, A., & Schall, M. (2016). A framework for scaling up health interventions: Lessons from large-scale improvement initiatives in Africa. *Implementation Science, 11,* 12.

Bärnighausen, T. & Bloom, D. E. (2009). Designing financial-incentive programmes for return of medical service in underserved areas: Seven management functions. *Human Resources for Health, 7,* 52.

Barr, N. (2012). *The Economics of the Welfare State* (5th ed.). Oxford, UK: Oxford University Press.

Batalden, P. B. & Davidoff, F. (2007). What is "quality improvement" and how can it transform healthcare? *Quality and Safety in Health Care,* 16, 2–3.

Batenburg, R., Kroneman, M., & Sagan, A. (2015). The impact of the crisis on the health system and health in the Netherlands. In A. Maresso, P. Mladovsky, & S. Thomson (Eds.), *Economic Crisis, Health Systems and Health in Europe* (pp. 247–282). Copenhagen, Denmark: WHO Regional Office for Europe/European Observatory on Health Systems and Policies.

Beaglehole, R., Bonita, R., Horton, R., Adams, C., Alleyne, G., Asaria, P., et al. (2011). Priority actions for the non-communicable disease crisis. *Lancet,* 377(9775), 1438–1447.

Beard, J. R., Officer, A., de Cavalho, I. S., Sadana, R., Pot, A. M., Michel, J.-P., et al. (2016). The world report on aging and health: A policy framework for healthy aging. *Lancet,* 387(10033), 2145–2154.

Behrman, J. R., Calderon, M. C., Mitchell, O., Vasquez, J., & Bravo, D. (2011). First-round impacts of the 2008 Chilean pension system reform. PARC Working Paper Series. University of Pennsylvania Scholarly Commons. Retrieved from http://repository.upenn.edu/cgi/viewcontent.cgi?article=1032&context=parc_working_papers.

Bengoa, R. (2016). Systems, not structures: Changing health & social care. Expert Panel Report. Retrieved from https://www.health-ni.gov.uk/sites/default/files/publications/health/expert-panel-full-report.pdf.

Beratarrechea, A., Lee, A. G., Willner, J. M., Jahangir, E., Ciapponi, A., & Rubinstein, A. (2014). The impact of mobile health interventions on chronic disease outcomes in developing countries: A systematic review. *Telemedicine and e-Health,* 20(1), 75–82.

Bermúdez, I. (2016). Las obras sociales cubren cerca del 70% de la atención médica. Retrieved from http://carlosfelice.com.ar/blog/2016/08/las-obras-sociales-cubren-cerca-del-70-de-la-atencion-medica/.

Bernstein, S. J., McGlynn, E. A., Siu, A. L., Roth, C. P., Sherwood, M. J., Keesey, J. W., et al. (1993). The appropriateness of hysterectomy: A comparison of care in seven health plans. *JAMA: The Journal of the American Medical Association,* 269(18), 2398–2402.

Berwick, D. (2013). *A Promise to Learn—A Commitment to Act: Improving the Safety of Patients in England.* London, UK: Department of Health.

Berwick, D. (2016). Era 3 for medicine and health care. *JAMA: The Journal of the American Medical Association,* 315(13), 1329–1330.

Berwick, D. M., Feeley, D., & Loehrer, S. (2015). Change from the inside out: Health care leaders taking the helm. *JAMA: The Journal of the American Medical Association,* 313(17), 1707–1708.

Berwick, D. M., Nolan, T. W., & Whittington, J. (2008). The triple aim: Care, health, and cost. *Health Affairs,* 27(3), 759–769.

Best, A., Greenhalgh, T., Lewis, S., Saul, J. E., Carroll, S., & Bitz, J. (2012). Large-system transformation in health care: A realist review. *Milbank Quarterly,* 90(3), 421–456.

Betrán, A. P., Ye, J., Moller, A.-B., Zhang, J., Gülmezoglu, A. M., & Torloni, M. R. (2016). The increasing trend in caesarean section rates: Global, regional and national estimates: 1990–2014. *PLoS One,* 11(2), e0148343.

Biosca, O. & Brown, H. (2015). Boosting health insurance coverage in developing countries: Do conditional cash transfer programmes matter in Mexico? *Health Policy Plan*, 30(2), 155–162.

Bjerregaard, P. (2005). Sundhedsforskningens historie i Grønland [History of health research in Greenland]. In K. Thisted (Ed.), *Grønlandsforskning—historie og perspektiver* [*Greenlandic Research—History and Perspectives*]. København: Det Grønlandske Selskab.

Black, N. (2013). Patient reported outcome measures could help transform healthcare. *BMJ*, 346, f167.

Blancafort, S., Fernández-Maldonado, L., Gabriele, J., Pineda, E., & Salvà, A. (2013). Training patients in chronic disease self-management: Results of the implementation of the "Stanford model." *Presented at the 5th National Congress of Chronic Patients Health Care*, Barcelona, Spain, April 11–13.

Blancou, J., Chomel, B. B., Belotto, A., & Meslin, F. X. (2005). Emerging or re-emerging bacterial zoonosis: Factors of emergence, surveillance and control. *Veterinary Research*, 36, 507–522.

Bolormaa, T., Natsagdorj, T., Tumurbat, B., Bujin, T., Bulganchimeg, B., Soyoltuya, B., et al. (2007). Mongolia health system review. *Health Systems in Transition*, 9(4), 1–151.

Bousmah, M.-a.-Q., Ventelou, B., & Abu-Zaineh, M. (2016). Medicine and democracy: The importance of institutional quality in the relationship between health expenditure and health outcomes in the MENA region. *Health Policy*, 120(8), 928–935.

Boyle, P. & Levin, B. (2008). World Cancer Report 2008. Lyon, France: International Agency for Research on Cancer, World Health Organization. Retrieved from http://www.iarc.fr/en/publications/pdfs-online/wcr/2008/wcr_2008.pdf

Braithwaite, J. (2014). The medical miracles delusion. *Journal of the Royal Society of Medicine*, 107(3), 92–93.

Braithwaite, J., Greenfield, D., Westbrook, J., Pawsey, M., Westbrook, M., Gibberd, R., et al. (2010). Health service accreditation as a predictor of clinical and organisational performance: A blinded, random, stratified study. *Quality and Safety in Healthcare*, 19(1), 14–21.

Braithwaite, J., Lazarus, L., Vining, R., & Soar, J. (1995). Hospitals: To the next millennium. *International Journal of Health Planning and Management*, 10(2), 87–98.

Braithwaite, J., Mannion, R., Matsuyama, Y., Shekelle, P., Whittaker, S., & Al-Adawi, S. (Eds.). (2017a). *Health Systems Improvement Across the Globe: Success Stories from 60 Countries*. Boca Raton, FL: Taylor & Francis Group.

Braithwaite, J., Mannion, R., Matsuyama, Y., Shekelle, P., Whittaker, S., Al-Adawi, S., et al. (2017b). Accomplishing reform: Successful case studies drawn from the health systems of 60 countries. *International Journal for Quality in Health Care*, 29(6), 880–886.

Braithwaite, J., Hillman, K., Pain, C., & Hughes, C. (2017c). Two decades of evolving a patient safety system: An Australian story of reducing harm to deteriorating patients. In J. Braithwaite, R. Mannion, Y. Matsuyama, P. Shekelle, S. Whittaker, & S. Al-Adawi (Eds.), *Health Systems Improvement Across the Globe: Success Stories from 60 Countries* (pp. 373–378). Boca Raton, FL: Taylor & Francis Group.

Braithwaite, J., Matsuyama, Y., Johnson, J., & Mannion, R. (Eds.). (2015). *Healthcare Reform, Quality and Safety: Perspectives, Participants, Partnerships and Prospects in 30 Countries*. Farnham, UK: Ashgate.

Braithwaite, J., Matsuyama, Y., Mannion, R., Johnson, J., Bates, D. W., & Hughes, C. (2016). How to do better health reform: A snapshot of change and improvement initiatives in the health systems of 30 countries. *International Journal for Quality in Health Care,* 28(6), 843–846.

Brett, J., Staniszewska, S., Mockford, C., Herron-Marx, S., Hughes, J., Tysall, C., et al. (2014). A systematic review of the impact of patient and public involvement on service users, researchers and communities. *Patient,* 7(4), 387–395.

Brocchetto, M. (2017). Venezuela asks UN for help as medicine shortages grow severe. *CNN.* Retrieved from http://edition.cnn.com.

Brook, R. H. (2017). Should the definition of health include a measure of tolerance? *JAMA: The Journal of the American Medical Association,* 317(6), 585–586.

Brook, R. H., Chassin, M. R., Fink, A., Solomon, D. H., Kosecoff, J., & Park, R. E. (1986). A method for the detailed assessment of the appropriateness of medical technologies. *International Journal of Technology Assessment in Healthcare,* 2(1), 53–63.

Brook, R. H., Ware, J. E., Davies-Avery, A., Stewart, A. L., Donald, C. A., Rogers, W. H., et al. (1979). Overview of adult health status measures fielded in Rand's Health Insurance Study. *Medical Care,* 17(7), 1–131.

Brook, R. H., Ware, J. E., & Rogers, W. (1983). Does free care improve adults' health? Results from a randomized controlled trial. *New England Journal of Medicine,* 309(23), 1426–1434.

Brotherton, S. E., Rockey, P. H., & Etzel, S. I. (2005). US graduate medical education, 2004–2005: Trends in primary care specialties. *JAMA: The Journal of the American Medical Association,* 294(9), 1075–1082.

BuddeComm. (2016). Papua New Guinea—telecoms, mobile and broadband—statistics and analyses. Retrieved from https://www.budde.com.au/research/Papua-New-Guinea-telecoms-mobile-and-broadband-statistics-and-analyses.

Burjorjee, R. & Al-Adawi, S. (1992a). The role of the health workers in the healthcare delivery system of the country, Sultanate of Oman. *Bahrain Medical Bulletin,* 14, 79–81.

Burjorjee, R. & Al-Adawi, S. (1992b). The Sultanate of Oman: An experiment in community care. *Psychiatric Bulletin,* 16, 646–648.

Caldas de Almeida, J. M., Mateus, P., & Tomé, G. (2015). Joint action on mental health and well-being: Towards community-based and socially inclusive mental health. *Joint Action on Mental Health and Well-Being.* Retrieved from https://ec.europa.eu/health/sites/health/files/mental_health/docs/2017_towardsmhcare_en.pdf.

Callaghan, M., Ford, N., & Schneider, H. (2010). A systematic review of task-shifting for HIV treatment and care in Africa. *Human Resources for Health,* 8(1), 8.

Canadian Institute for Health Information. (2011). Health care use at the end of life in Atlantic Canada. Retrieved from https://secure.cihi.ca/free_products/end_of_life_2011_en.pdf.

Cancer Research UK. (2017). Women's cancer rates rising faster than men's. Retrieved from http://www.cancerresearchuk.org/about-us/cancer-news/press-release/2017-02-03-womens-cancer-rates-rising-faster-than-mens.

CancerIndex. (2012). Malta: Cancer statistics. Retrieved from http://www.cancerindex.org/Malta#sthash.XxpCmXyZ.dpuf.

Canterbury Health System. (2017). Canterbury District Health: Improvement plan. Retrieved from http://nsfl.health.govt.nz/system/files/documents/pages/cant_0.pdf.

Cardona-Morrell, M. & Hillman, K. (2015). Development of a tool for defining and identifying the dying in hospital: Criteria for screening and triaging to appropriate alternative care. *BMJ Supportive and Palliative Care*, 5, 78–90.

Cardona-Morrell, M., Kim, J. C. H., Turner, R. M., Anstey, M., Mitchell, I. A., & Hillman, K. (2016). Non-beneficial treatments at the end of life: A systematic review on extent of the problem. *International Journal for Quality in Health Care*, 28(4), 456–469.

Carrol, P. (2006). *Nursing Leadership and Management: A Practical Guide*. Clifton Park, NY: Delmar Cengage Learning.

Carson-Stevens, A., Edwards, A., Panesar, S., Parry, G., Rees, P., Sheikh, A., & Donaldson, L. (2015). Reducing the burden of iatrogenic harm in children. *Lancet*, 385(9978), 1593.

Carson-Stevens, A. P. (2017). Generating learning from patient safety incident reports from general practice. Doctoral dissertation, School of Medicine, Cardiff University.

Casassus, B. (2017). Macron's vision for the French health system. *Lancet*, 389(10082), 1871–1872.

Castro Lopes, S., Nove, A., ten Hoope-Bender, P., de Bernis, L., Bokosi, M., Moyo, N. T., & Homer, C. S. E. (2016). A descriptive analysis of midwifery education, regulation and association in 73 countries: The baseline for a post-2015 pathway. *Human Resources for Health*, 14(1), 34–37.

Center for Health Development. (2016). *Health Indicators, 2016*. Ulaanbaatar, Mongolia: Center for Health Development.

Centers for Disease Control and Prevention. (2001). *Campaign to Prevent Antimicrobial Resistance in Healthcare Settings: Why a Campaign?* Atlanta, GA: Centers for Disease Control and Prevention.

Centers for Disease Control and Prevention. (2013a). Antibiotic resistance threats in the United States, 2013. Retrieved from https://www.cdc.gov/drugresistance/pdf/ar-threats-2013-508.pdf.

Centers for Disease Control and Prevention. (2013b). CDC in Cambodia: Fact sheet. Retrieved from https://www.cdc.gov/globalhealth/countries/cambodia/pdf/cambodia.pdf.

Centers for Disease Control and Prevention. (2013c). CDC in Indonesia: Fact sheet. Retrieved from https://www.cdc.gov/globalhealth/countries/indonesia/pdf/indonesia.pdf.

Chan, J. C., So, W.-Y., Yeung, C.-Y., Ko, G. T., Lau, I.-T., Tsang, M.-W., et al. (2009). Effects of structured versus usual care on renal endpoint in type 2 diabetes: The SURE Study: A randomized multicenter translational study. *Diabetes Care*, 32(6), 977–982.

Chang, I. & Pratt, R. W. (2012). Neurohospitalists: An emerging subspecialty. *Current Neurology and Neuroscience Reports*, 12(4), 481–488.

Chase, D. (2013). The 7 habits of highly patient centric providers. *Forbes*, February 8. Retrieved from https://www.forbes.com/sites/davechase/2013/02/18/the-7-habits-of-highly-patient-centric-providers/#265ff6b76704.

Chassin, M. R., Brook, R. H., Park, R. E., Keesey, J., Fink, A., Kosecoff, J., et al. (1986). Variations in the use of medical and surgical services by the Medicare population. *New England Journal of Medicine*, 314(5), 285–290.

Chen, J., Ou, L., Flabouris, A., Hillman, K., Bellomo, R., & Parr, M. (2016). Impact of a standardized rapid response system on outcomes in a large health jurisdiction. *Resuscitation*, 107, 47–56.

Chen, J., Ou, L., & Hollis, S. J. (2013). A systematic review of the impact of routine collection of patient reported outcomes measures on patients, providers and health organisations in an oncologic setting. *BMC Health Service Research*, 13, 211. doi:10.1186/1472-6963-13-211.

Chen, L., Evans, T., Anand, S., Boufford, J. I., Brown, H., Chowdhury, M., et al. (2004). Human resources for health: Overcoming the crisis. *Lancet*, 364(9449), 1984–1990.

陳育群 [Chen, Y.-C.] and 李偉強 [Lee, W.-C.], (2016). 醫療大數據：健保資料庫之臨床應用與研究. [Examples of data analysis and application using Taiwan's National Health Insurance Database in healthcare]. 台灣醫學 [*Journal of the Formosan Medical Association*], 20(6): 602–608. doi: 10.6320/fjm.2016.20(6).6

Chernew, M. & Bush, J. (2016). Let efficient providers prosper. *NEJM Catalyst*, December 6. Retrieved from https://catalyst.nejm.org/health-care-payment-efficient-providers/.

Chernew, M. & Newhouse, J. P. (2008). What does the RAND Health Insurance Experiment tell us about the impact of patient cost sharing on health outcomes? *American Journal of Managed Care*, 14(7), 412–414.

Chimeddagva, D. (2011). The role of institutional design and organizational practice, for health financing performance in Mongolia. OASIS Mongolia Report. Geneva, Switzerland: World Health Organization.

Chisholm, B. (1954). Outline for a study group on world health and the survival of the human race. *World Health Organization*. Retrieved from whqlibdoc.who.int/hist/../ChisholmBrock_1953_Compilation.pdf.

Chowdary, S. (2017). Medical tourist arrivals in India up 25%. *Business Standard*. Retrieved from http://www.business-standard.com/article/companies/medical-tourist-arrivals-in-india-up-25-117041900577_1.html.

Cibulskis, R. E. & Hiawalyer, G. (2002). Information systems for health sector monitoring in Papua New Guinea. *Bulletin of the World Health Organization*, 80(9), 752–758.

Cicchetti, A., Fiore, A., Coretti, S., Iacopino, V., Marchetti, M., & Mennini, F. S. (2016). Coverage under evidence development: The experience of the Italian Health Policy Forum. Working Paper.

Clark, D., Armstrong, M., Allan, A., Graham, F., Carnon, A., & Isles, C. (2014). Imminence of death among a national cohort of hospital inpatients. *Palliative Medicine*, 28(6), 474–479.

Clyne, K. (2012). Fuad Khan: Human resource issues hurting health sector. *Trinidad and Tobago Guardian*. Retrieved from http://m.guardian.co.tt/news/2012-11-19/fuad-khan-human-resource-issues-hurting-health-sector.

Coddington, D. C., Ackerman, F. K., Jr. & Moore, K. D. (2001). Integrated health care systems: Major issues and lessons learned. *Healthcare Leadership & Management Report*, 9(1), 1–9.

Coiera, E. (2015). *Guide to Health Informatics* (3rd ed.). Boca Raton, FL: CRC Press.

Coiera, E. & Hovenga, E. J. (2007). Building a sustainable health system. *IMIA Yearbook of Medical Informatics*, 1, 11–18.

Coker, R., Rushton, J., Mounier-Jack, S., Karimuribo, E., Lutumba, P., Kambarage, D., et al. (2011). Towards a conceptual framework to support one-health research for policy on emerging zoonoses. *Lancet Infectious Diseases*, 11(4), 326–331.

Commonwealth Fund. (2016). 2016 Commonwealth Fund international health policy survey of adults. Retrieved from http://www.commonwealthfund.org/interactives-and-data/surveys/international-health-policy-surveys/2016/2016-international-survey.

Communications Authority of Kenya. (2015). First quarter of the financial year 2014/15. Quarterly Sector Statistics Report. Retrieved from http://www.ca.go.ke/index.php/component/content/category/99-research-statistics.

Conner, M. & Norman, P. (2005). *Predicting Health Behaviour.* Maidenhead, UK: McGraw-Hill Education.

Cook, D. A., Hatala, R., Brydges, R., Zendejas, B., Szostek, J. H., Wang, A. T., et al. (2011). Technology-enhanced simulation for health professions education: A systematic review and meta-analysis. *JAMA: The Journal of the American Medical Association, 306*(9), 978–988.

Cooper, A., Edwards, A., Williams, H., Evans, H. P., Avery, A., Hibbert, P., et al. (2017). Sources of unsafe primary care for older adults: A mixed-methods analysis of patient safety incident reports. *Age and Ageing, 46*(5), 833–839.

Coulter, A. & Magee, H. (2003). *The European Patient of the Future.* Maidenhead, PA: Open University Press.

Council of Shopping Centers. (2016). Activity report of Council of Shopping Centers—Turkey. Retrieved from http://www.ayd.org.tr/tr/pdfs/web_faaliyet_raporu_2015.pdf.

Cumming, J. (2011). Integrated care in New Zealand. *International Journal of Integrated Care, 11*, e138.

Cumming, J. & Mays, N. (2011). New Zealand's primary health care strategy: Early effects of the new financing and payment system for general practice and future challenges. *Health Economics, Policy and Law, 6*(1), 1–21.

Cumming, J., Mays, N., & Gribben, B. (2008). Reforming primary health care: Is New Zealand's primary health care strategy achieving its early goals? *Australia and New Zealand Health Policy, 5,* 24.

Cumming, J., McDonald, J., Barr, C., Martin, G., Gerring, Z., & Daubé, J. (2014). New Zealand health system review. *Health Systems in Transition, 3*(4), 1–244.

Cunningham, A. A. (2005). A walk on the wild side—Emerging wildlife diseases. *BMJ, 331,* 1214–1215.

Cuschieri, S., Vassallo, J., Calleja, N., Camilleri, R., Borg, A., Bonnici, G., et al. (2016). Prevalence of obesity in Malta. *Obesity Science and Practice, 2*(4), 466–470.

Czypionka, T., Röhrling, G., & Mayer, S. (2017). The relationship between outpatient department utilisation and non-hospital ambulatory care in Austria. *European Journal of Public Health, 27*(1), 20–25.

Dahl-Petersen, I. K., Larsen, C. V. L., Nielsen, N. O., Jørgensen, M. E., & Bjerregaard, P. (2014). *Befolkningsundersøgelsen i Grønland 2014—levevilkår, livsstil og helbred [Population Study in Greenland 2014—Living Conditions, Lifestyle and Health].* København: Statens Institut for Folkesundhed og Naalakkersuisut/Government of Greenland.

DANBIO. (2015). National clinical quality database for treatment of patients with rheumatology. Annual Report 2014. *Regions Clinical Quality Improvement Program (RKKP).* Retrieved from https://danbio-online.dk/formidling/dokumentmappe/danbios-arsrapport-2014.

Daniels, N. (2000). Accountability for reasonableness. *BMJ, 321*(7272), 1300–1301.

Daniels, N. & Sabin, J. E. (2008). *Setting Limits Fairly—Learning to Share Resources for Health.* Oxford, UK: Oxford University Press.

Danish Regions. (2015a). Pressure on healthcare. Why hospital costs are rising—and how to keep them down. Oxford, UK: Retrieved from http://klartekst.dk/wp-content/uploads/2015/11/pres-p%C3%A5-sundhedsv%C3%A6senet.pdf.

Danish Regions. (2015b). The citizens healthcare system. Retrieved from http://www.regioner.dk/media/1269/plan-bs.pdf.

Davey, P., Marwick, C. A., Scott, C. L., Charani, E., McNeil, K., Brown, E., et al. (2017). Interventions to improve antibiotic prescribing practices for hospital inpatients. *Cochrane Database of Systematic Reviews*, 2, CD003543.

Davis, K., Stremikis, K., Squires, D., & Schoen, C. (2014). Mirror, mirror on the wall, 2014 update: How the U.S. health care system compares internationally. Retrieved from http://www.commonwealthfund.org/publications/fund-reports/2014/jun/mirror-mirror.

DeFulio, A. & Silverman, K. (2012). The use of incentives to reinforce medication adherence. *Preventive Medicine*, 55 Suppl., S86–S94.

Degos, L., Romaneix, F., Michel, P., & Bacou, J. (2008). Can France keep its patients happy? *BMJ*, 336, 254.

Deilkås, E. T., Ingebrigtsen, T., & Ringard, Å. (2015). Norway. In J. Braithwaite, Y. Matsuyama, J. Johnson, & R. Mannion (Eds.), *Healthcare Reform, Quality and Safety: Perspectives, Participants, Partnerships and Prospects in 30 Countries* (pp. 261–271). Farnham, UK: Ashgate.

De Leeuw, E. (2001). Global and local (glocal) health: The WHO healthy cities programme. *Global Change and Human Health*, 2(1), 34–45.

Delgado, P. & Azpurua, L. (2017). Mision Barrio Adentro: Universal health coverage efforts in Venezuela. In J. Braithwaite, R. Mannion, Y. Matsuyama, P. Shekelle, S. Whittaker, & S. Al-Adawi (Eds.), *Health Systems Improvement Across the Globe: Success Stories from 60 Countries* (pp. 63–67). Boca Raton, FL: Taylor & Francis Group.

Delnoij, D., Klazinga, N., & Glasgow, I. K. (2002). Integrated care in an international perspective. *International Journal of Integrated Care*, 2(1), 1–4.

Deloitte. (2010). *Grønlands Sundhedsvæsen—Udfordringer for fremtiden. Økonomi- og strukturanalyse af det grønlandske sundhedsvæsen [Greenland's Healthcare System—Challenges for the Future—Economic and Structural Analysis of the Greenlandic Healthcare System]*. København, Denmark: Deloitte Business Consulting A/S.

Department of Health [England]. (2008). *High Quality Care for All: Final Darzi Report*. London, UK: The Stationery Office.

Department of Health [England]. (2010). *Equity and Excellence*. London, UK: The Stationery Office.

Department of Health [Northern Ireland]. (2016). Health and wellbeing 2026: Delivering together. Retrieved from https://www.health-ni.gov.uk/sites/default/files/publications/health/health-and-wellbeing-2026-delivering-together.pdf.

Department of Health [United Kingdom]. (2000). *An Organisation with a Memory: Report of an Expert Group on Learning from Adverse Events in the NHS*. Norwich, UK: Stationery Office.

Department of Health Information and Research [Malta]. (2014). Smoking attributable deaths. Retrieved from https://health.gov.mt/en/dhir/Documents/dhir_fact_sheet_smoking_attributable_deaths.pdf.

Department of Health Information and Research [Malta]. (2017). Deaths: National mortality register. Ministry for Health, Malta. Retrieved from https://deputyprimeminister.gov.mt/en/dhir/Pages/Registries/deaths.aspx.

Department of Health, Social Care and Public Safety [Northern Ireland]. (2011). Quality 2020: A 10-year strategy to protect and improve quality in health and social care in Northern Ireland. Retrieved from https://www.healthni.gov.uk/sites/default/files/publications/dhssps/q2020-strategy.pdf.

Department of Health, Social Care and Public Safety [Northern Ireland]. (2014). Supporting leadership for quality improvement and safety: An Attributes Framework for Health and Social Care. Retrieved from http:// nipecportfolio.hscni.net/compro/attributes/doc%20Quality2020Attributes%20 FrameworkJul14.pdf.

Department of Quality and Accreditation in Health. (2015). *Announcements of Accreditation System of Health (AASH), 2013–2015.* Ankara: Ministry of Health [Turkey].

Department of Statistics [Jordan]. (2016). The 2015 Jordan population and housing census. Retrieved from http://census.dos.gov.jo/wpcontent/uploads/ sites/2/2016/02/Census_results_2016.pdf.

Department of Statistics [Jordan] & ICF International. (2012). 2012 Demographic and Health Survey (DHS)/Jordan Population and Family Health Survey. Retrieved from https://dhsprogram.com/pubs/pdf/FR282/FR282.pdf.

DeSilva, M., Samele, C., Saxena, S., Patel, V., & Darzi, A. (2014). Policy actions to achieve integrated community-based mental health services. *Health Affairs,* 33(9), 1595–1602.

Dhillon, P. K., Jeemon, P., Arora, N. K., Mathur, P., Maskey, M., Sukirna, R. D., et al. (2012). Status of epidemiology in the WHO South-East Asia region: Burden of disease, determinants of health and epidemiological research, workforce and training capacity. *International Journal of Epidemiology,* 41(3), 847–860.

Diamond, C. C., Mostashari, F., & Shirky, C. (2009). Collecting and sharing data for population health: A new paradigm. *Health Affairs,* 28, 454–466.

Dini, L., Sarganas, G., Heintze, C., & Braun, V. (2012). Home visit delegation in primary care: Acceptability to general practitioners in the state of Mecklenburg-Western Pomerania, Germany. *Deutsches Arzteblatt International,* 109(46), 795–801.

Dodani, S. & LaPorte, R. E. (2005). Brain drain from developing countries: How can brain drain be converted into wisdom gain? *Journal of the Royal Society of Medicine,* 98(11), 487–491.

Dolor, R. J. & Schulman, K. A. (2013). Financial incentives in primary care practice: The struggle to achieve population health goals. *JAMA: The Journal of the American Medical Association,* 310(10), 1031–1032.

Donabedian, A. (1978). The quality of medical care. *Science,* 200, 856–864.

Donaldson, L., Rutter, P., & Henderson, M. (2014). The right time, the right place. Retrieved from https://www.health-ni.gov.uk/sites/default/files/publications/ dhssps/donaldsonreport270115_0.pdf.

Dorjdagva, J. (2017). *Socioeconomic-related Health Inequalities and Health Care Utilization in Mongolia.* Kuopio, Finland: University of Eastern Finland.

Dorjdagva, J., Batbaatar, E., Dorjsuren, B., Svensson, M., & Kauhanen, J. (2016). Catastrophic health expenditure and impoverishment in Mongolia. *International Journal for Equity in Health,* 15, 105.

Doupi, P. & Ruotsalainen, P. (2004). eHealth in Finland: Present status and future trends. *International Journal of Circumpolar Health,* 63(4), 322–327.

Doyle, D. (2007). Palliative medicine in Britain. *Omega,* 56(1), 77–88.

Dreier, O. (2008). *Psychotherapy in Everyday Life.* Cambridge, UK: Cambridge University Press.

Drummond, P., Thakoor, V. J., & Yu, S. (2014). Africa rising: Harnessing the demographic dividend. *Institute for Health Metrics and Evaluation: Global Burden of Disease.* Retrieved from http://www.healthdata.org/gbd.

Dubai Health Authority. (2016). Dubai health strategy 2016–2021. Retrieved from https://www.dha.gov.ae/Documents/Dubai_Health_Strategy_2016-2021_En.pdf.

Dubai Health Authority. (2017). Salama electronic medical record system now live in 5 DHA facilities. Retrieved from https://www.dha.gov.ae/en/DHANews/Pages/DHANews619602078-09-04-2017.aspx.

Duffy, S. (2015). Will personal health budgets destroy the NHS? Socialist Health Association. Retrieved from https://www.sochealth.co.uk/2015/09/02/will-personal-health-budgets-destroy-the-nhs/.

East African Community. (2014). Institutionalizing One Health approach in the East African Community to prevent and control zoonotic diseases and other events of public health concern. Presented at the 9th Ordinary Meeting of the EAC Sectoral Council of Ministers of Health. Retrieved from http://publications.universalhealth2030.org/uploads/eac_sectoral_council_of_ministers_of_health_17th_april_2014__3_.pdf.

Eijkenaar, F. (2013). Key issues in the design of pay-for-performance. *European Journal of Health Economics*, 14, 117–131.

El-Khoury, G. (2016). Labour force and unemployment in Arab countries: Selected indicators. *Contemporary Arab Affairs*, 9(3), 487–491.

Elmslie, K. (2012). *Against the Growing Burden of Disease*. Ottawa: Public Health Agency of Canada.

Emery, D. J., Shojania, K. G., Forster, A. J., Mojaverian, N., & Feasby, T. E. (2013). Overuse of magnetic resonance imaging. *JAMA Internal Medicine*, 173(9), 823–825.

EMPATHiE Consortium. (2014). EMPATHiE—Empowering patients in the management of chronic diseases. Retrieved from http://ec.europa.eu/health//sites/health/files/patient_safety/docs/empathie_frep_en.pdf.

EMRO-WHO (Eastern Mediterranean Regional Office, World Health Organization). (2017). National health strategy for Yemen 2011–2025. Retrieved from http://www.emro.who.int/yem/programmes/health-systems-development.html.

England, K. (2013). Annual mortality report—National mortality registry. Retrieved from https://health.gov.mt/en/dhir/Documents/annual_mortality_report_2013.pdf.

Ennis, L. & Wykes, T. (2013). Impact of patient involvement in mental health research: Longitudinal study. *British Journal of Psychiatry*, 203(5), 381–386.

Epstein, R. M. & Street, R. L. (2011). The values and value of patient-centered care. *Annals of Family Medicine*, 9(2), 100–103.

European Centre for Disease Prevention and Control. (2014). 2014 surveillance of antimicrobial consumption in Europe 2011. Retrieved from http://ecdc.europa.eu/en/publications/Publications/antimicrobial-consumption-europe-surveil-lance-2011.pdf.

European Centre for Disease Prevention and Control. (2016). *Rapid Risk Assessment: Carbapenem-resistant Enterobacteriaceae*. Stockholm: European Centre for Disease Prevention and Control.

European Centre for Disease Prevention and Control. (2017). Summary of the latest data on antibiotic consumption in the European Union. ESAC-Net surveillance data. Retrieved from https://ecdc.europa.eu/sites/portal/files/documents/Final_2017_EAAD_ESAC-Net_Summary-edited%20-%20FINALwith%20erratum.pdf.

European Commission. (2003). EU Council recommendation on cancer screening. Retrieved from https://ec.europa.eu/jrc/sites/jrcsh/files/20151020-21-breast-cancer-standards-hubel_en.pdf.

European Commission. (2012). Eurobarometer qualitative study patient involvement aggregate report. Retrieved from https://ec.europa.eu/eip/ageing/library/eurobarometer-qualitative-study-patient-involvement-healthcare_en.

European Commission. (2013). Use of '-omics' technologies in the development of personalised medicine. Commission Staff Working Document 436. Brussels: European Commission.

European Commission. (2015). Pilot project on promoting self-management for chronic diseases in the EU (PRO-STEP). Retrieved from http://www.eu-patient.eu/whatwedo/Projects/prostep/.

European Council. (2016). Council conclusions on the next steps under a One Health approach to combat antimicrobial resistance. Retrieved from http://www.consilium.europa.eu/press-releases-pdf/2016/6/47244642809_en.pdf.

European Observatory on Health Care Systems. (2002). Health care in Central Asia: Policy brief. Retrieved from http://www.euro.who.int/__data/assets/pdf_file/0009/108828/Carbrief120202.pdf?ua=1.

European Social Policy Network. (2016). In-depth reform of the healthcare system in Finland. Retrieved from ec.europa.eu/social/BlobServlet?docId=15887.

European Union. (2011). Directive 2011/24/EU of the European Parliament and of the Council of 9 March 2011 on the application of patients' rights in cross-border healthcare. *Official Journal of the European Union, L* 88, 45.

Eurostat. (2016). Eurostat statistics explained: Unmet health care needs statistics. Retrieved from http://ec.europa.eu/eurostat/statistics-explained/index.php/Unmet_health_care_needs_statistics.

Evans, D. B., Hsu, J., & Boerma, T. (2013). Universal health coverage and universal access. *Bulletin of the World Health Organization,* 91(8), 546–546A.

Evans, P. & Wurster, T. (1999). *Blown to Bits: How the New Economics of Information Transforms Strategy.* Cambridge, MA: Harvard Business School Press.

Evidence-Based Medicine Working Group. (1992). Evidence-based medicine. A new approach to teaching the practice of medicine. *JAMA: The Journal of the American Medical Association,* 268(17), 2420.

Fagan, E. B., Gibbons, C., Finnegan, S. C., Petterson, S., Peterson, L. E., Phillips, R. L., Jr., et al. (2015). Family medicine graduate proximity to their site of training: Policy options for improving the distribution of primary care access. *Family Medicine,* 47(2), 124–130.

Falagas, M. E., Tansarli, G. S., Karageorgopoulos, D. E., & Vardakas, K. Z. (2014). Deaths attributable to carbapenem-resistant Enterobacteriaceae infections. *Emerging Infectious Diseases,* 20(7), 1170–1175.

Federal Ministry of Health [Nigeria]. (2014a). National Health Bill, 2014. Retrieved from http://nigeriahealthwatch.com/wp-content/uploads/bsk-pdf-manager/746_894_NATIONAL_HEALTH_BILL_%28HARMONISED__SENATE_&_HOUSE%29_15TH_JULY_2014.PDF.

Federal Ministry of Health [Nigeria]. (2014b). FG inaugurates committee on Nigerian public health training initiative and the national steering committee on the Nigerian health workforce registry. Retrieved from http://health.gov.ng/index.php/about-us/9-uncategorised/176-fg-inaugurates-committee-on-nigerian-public-health-training-initiative-and-the-national-steering-committee-on-the-nigerian-health-workforce-registry.

Feero, W. G. (2017). Introducing "genomics and precision health." *JAMA: The Journal of the American Medical Association*, 317(18), 1842–1843.

Feero, W. G., Guttmacher, A. E., & Collins, F. S. (2010). Genomic medicine—An updated primer. *New England Journal of Medicine*, 362(21), 2001–2011.

Fegeler, U., Jäger-Roman, E., Martin, R., & Nentwich, H. J. (2014). Ambulante allgemeinpädiatrische Grundversorgung [Outpatient general pediatric primary care]. *Monatsschrift Kinderheilkunde*, 162(12), 1117–1130.

Fernald, L. C., Gertler, P. J., & Neufeld, L. M. (2009). 10-year effect of Oportunidades, Mexico's conditional cash transfer programme, on child growth, cognition, language, and behaviour: A longitudinal follow-up study. *Lancet*, 374(9706), 1997–2005.

Fernald, L. C. H., Gertler, P. J., & Hou, X. (2008). Cash component of conditional cash transfer program is associated with higher body mass index and blood pressure in adults. *Journal of Nutrition*, 138(11), 2250–2257.

Fernandes, P. A., Silva, M. G., Cruz, A. P., & Paiva, J. A. (2016). Prevenção e controlo de infeções e de resistência aos antimicrobianos em números—2015 [Prevention and control of infections and resistance to antimicrobials in numbers—2015]. *Direção Geral da Saúde*. Retrieved from https://www.sns.gov.pt/noticias/2016/03/16/prevencao-e-controlo-de-infecoes-e-de-resistencia-aos-antimicrobianos-em-numeros/.

Fernández-Maldonado, L., Blancafort Alias, S., Gabriele, G., & Pineda, E. (2013). Acceso, uso y percepción de la información sanitaria por parte de la población española: resultados del estudio de alfabetización sanitaria. Autocuidado y responsabilización de los pacientes [Access, use and perception of health information by the Spanish population: results of the health literacy study. Self-care and responsibility of patients]. *V Congreso Nacional de Atención Sanitaria al Paciente Crónico (Barcelona)*, F-032. Presented at the 5th National Congress of Chronic Patients Health Care, Barcelona, April 11–13.

Fidler, A., Bredenkamp, C., & Schlippert, S. (2009). Innovations in health services delivery from transition economies in Eastern Europe and Central Asia. *Health Affairs*, 28(4), 1011–1021.

Filippini, M., Ortiz, L. G., & Masiero, G. (2013). Assessing the impact of national antibiotic campaigns in Europe. *European Journal of Health Economics*, 14, 587–599.

Financial Times. (2016). Chile Pension Reform Comes Under World Spotlight, September 12. Retrieved from https://www.ft.com/content/b9293586-7680-11e6-bf48-b372cdb1043a.

Finnish Government. (2015). Finland, a land of solutions. Prime Minister Juha Sipilä's program. Retrieved from http://valtioneuvosto.fi/documents/10184/1427398/Ratkaisujen+Suomi_EN_YHDISTETTY_netti.pdf/8d2e1a66-e24a-4073-8303-ee3127fbfcac and Homepage of the Health, Social care and regional government reform progress: http://alueuudistus.fi/en/frontpage.

Fins, J. J. (2015). The expert-generalist: A contradiction whose time has come. *Academic Medicine*, 90(8), 1010–1014.

Fisher, J. F. (2016). Permanent resident. *Medical Education Online*, 21(1), 31160.

Flanagan, P. (2015). Expenditure in PNG's 2016 budget—A detailed analysis. DevPolicy Blog: Development Policy Centre. Retrieved from http://devpolicy.org/expenditure-in-pngs-2016-budget-a-detailed-analysis-20151202/.

Flores, M. (2014). Ministros de Salud en Venezuela, 1936–2017. Retrieved from http://maiquiflores.over-blog.es/article-ministros-de-salud-de-venezuela-1936-2014-122169402.html.

Food and Agriculture Organization of the United Nations, International Fund for Agricultural Development, & World Food Programme. (2015). Meeting the 2015 international hunger targets: Taking stock of uneven progress. The state of food insecurity in the world. Retrieved from http://www.fao.org/3/a-i4646e.pdf.

Food and Drug Administration. (2009). Guidance for industry: Patient-reported outcome measures: Use in medical product development to support labeling claims. Retrieved from http://www.fda.gov/downloads/drugs/guidances/ucm193282.pdf.

Førde, R. & Pedersen, R. (2011). Clinical ethics committees in Norway—What do they do, and does it make a difference? *Cambridge Quarterly of Healthcare Ethics*, 20, 389–395.

Førde, R. & Ruud Hansen, T. W. (2014). Do organizational and clinical ethics in a hospital setting need different venues? *HEC Forum*, 26(2), 147–158.

Forder, J., Jones, K., Glendinning, C., Caiels, J., Welch, E., Baxter, K., et al. (2012). *Evaluation of the Personal Health Budget Pilot Programme*. London, UK: Department of Health.

Forse, R. A., Bramble, J., & McQuillan, R. (2011). Team training can improve operating room performance. *Surgery*, 150(4), 771–778.

Fox, S. (2011). 80% of Internet users look for health information online. Pew Research Center's Internet & American Life Project. Retrieved from http://www.pewinternet.org/files/old-media/Files/Reports/2011/PIP_Health_Topics.pdf.

Frankel, A. S., Leonard, M. W., & Denham, C. R. (2006). Fair and just culture, team behavior, and leadership engagement: The tools to achieve high reliability. *Health Services Research*, 41(4p2), 1690–1709.

Fraser, B. & Willer, H. (2016). Venezuela: Aid needed to ease health crisis. *Lancet*, 338, 947–949.

Free, C., Phillips, G., Felix, L., Galli, L., Patel, V., & Edwards, P. (2010). The effectiveness of M-health technologies for improving health and health services: A systematic review protocol. *BMC Research Notes*, 3(1), 250.

Freedman, D. H. (2017). A reality check for IBM's AI ambitions. *MIT Technology Review*. Retrieved from https://www.technologyreview.com/s/607965/a-reality-check-for-ibms-ai-ambitions/.

Freil, M. & Knudsen, J. L. (2009). User involvement in healthcare. *Danish Medical Bulletin*, 171(20), 1663–1666.

Fröschl, B. & Antony, K. (2017). Evaluation des projekts PHC—primärversorgungszentrum medizin mariahilf, kurzbericht zum 1. Evaluierungsjahr, gesundheit österreich forschungs- und planungs Gmbh [Evaluation of the project PHC—Primary care center medicine Mariahilf, short report to the first evaluation year. Health Austria research and planning GmbH]. Retrieved from http://www.medizinmariahilf.at/wp-content/uploads/2017/06/kurzbericht_evaluierung_phc_mm_1_zb.pdf.

Gabriele, G., Navarro, M. D., & Jovell, A. J. (2012). Rheumatoid arthritis: A vision of the present and look into the future. Results of a multidisciplinary qualitative study. Presented at Societal Impact of Pain (SIP) Symposium, Copenhagen, Denmark.

Gadsby, E. (2013). *Personal Budgets and Health: A Review of the Evidence*. Canterbury, UK: PRUComm Centre for Health Services Studies, University of Kent.

General Directorate of Health Research [Turkey]. (2016). *Health Statistics Yearbook—2015*. Ankara: Ministry of Health.

Genomics England. (n.d.). Homepage. Retrieved from https://www.genomicsengland.co.uk/the-100000-genomes-project/.

Gertler, P., Martinez, S., & Rubio-Codina, M. (2012). Investing cash transfers to raise long term living standards. *American Economic Journal: Applied Economics*, 4(1), 164–192.

GRUG [Gesundheitsreformumsetzungsgesetz]. (2017). *131 Bundesgesetz [131 Federal Law]*. Wien: Bundesgesetzblatt.

Gibbs, P. J. (2014). The evolution of One Health: A decade of progress and challenges for the future. Retrieved from http://veterinaryrecord.bmj.com/content/174/4/85.

Gillam, S. (2013). Public health practice in primary care. In C. Guest, W. Ricciardi, I. Kawachi, & I. Lain (Eds.), *Oxford Handbook of Public Health Practice*. Oxford, UK: Oxford University Press.

GLA:D. (2016). What is GLA:D? Retrieved from https://www.glaid.dk/training.html.

Global Health Workforce Alliance & WHO (World Health Organization). (2013). A universal truth: No health without a workforce. Retrieved from http://www.who.int/workforcealliance/knowledge/resources/hrhreport2013/en/.

Global Health Workforce Alliance & WHO (World Health Organization). (2014). Health workforce 2030: A global strategy on human resources for health, 2014. Retrieved from http://www.who.int/hrh/documents/strategy_brochure9-20-14.pdf.

Goldman, D. & Smith, J. P. (2011). The increasing value of education to health. *Social Science & Medicine*, 72(10), 1728–1737.

Goldsbury, D. E., O'Connell, D. L., Girgis, A., Wilkinson, A., Phillips, J. L., Davidson, P. M., & Ingham, J. M. (2015). Acute hospital-based services used by adults during the last year of life in New South Wales, Australia: A population-based retrospective cohort study. *BMC Health Services Research*, 15, 537.

Gönenç, R., Hofmarcher, M. M., & Wörgötter, A. (2011). Reforming Austria's highly regarded but costly health system. OECD Economics Department. Working Paper No. 895. Paris, France: OECD Publishing.

Gostin, L. O., Abou-Taleb, H., Roache, S. A., & Alwan, A. (2017). Legal priorities for prevention of non-communicable diseases: Innovations from WHO's Eastern Mediterranean region. *Public Health*, 144, 4–12.

Government of Denmark and Danish Regions. (2015). *Agreement on the Regions Economy for 2016*. Copenhagen, Denmark: Government of Denmark and Danish Regions.

Government of Denmark and Danish Regions. (2016). *Agreement on the Regions Economy for 2017*. Copenhagen, Denmark: Government of Denmark and Danish Regions.

Government of Pakistan. (2016). Pakistan Health Council Research Act, 2016. Retrieved from http://www.na.gov.pk/uploads/documents/1459926224_610.pdf.

Grant Thornton. (2009). Transforming the Middle East's healthcare model. Healthcare guide 2009. Retrieved from http://www.mynewsdesk.com/se/_grant_thornton_/documents/transforming-the-middle-easts-healthcare-model-6569.

Grech, K., Podesta, M., Calleja, A., & Calleja, N. (2015). *Performance of the Maltese Health System*. Valletta, Malta: Ministry for Energy and Health.

Greer, S. L., Hervey, T. K., Mackenbach, J. P., & McKee, M. (2013). Health law and policy in the European Union. *Lancet*, 381(9872), 1135–1144.

Groene, O. & Sunol, R. (2015). Patient involvement in quality management: Rationale and current status. *Journal of Health Organization and Management*, 29(5), 556–569.

Groenewegen, P. P., Dourgnon, P., Gress, S., Jurgutis, A., & Willems, S. (2013). Strengthening weak primary care systems: Steps towards stronger primary care in selected Western and Eastern European countries. *Health Policy*, 113(1–2), 170–179.

Guo, Q. (2004). The current status, challenges and solutions of general practitioner education in China. *Chinese General Practice*, 7(5), 291–297.

Guttmacher, A. E. & Collins, F. S. (2002). Genomic medicine—A primer. *New England Journal of Medicine*, 347(19), 1512–1520.

Haac, B. E., Gallaher, J. R., Mabedi, C., & Charles, A. G. (2016). Task-shifting: Use of laypersons for acquisition of vital signs data for clinical decision-making in the emergency room after traumatic injury. *Journal of the American College of Surgeons*, 223(4), S70–S71.

Hafeez, A., Mohamud, B. K., Shiekh, M. R., Shah, S. A. I., & Jooma, R. (2011). Lady health workers programme in Pakistan: Challenges, achievements and the way forward. *Journal of the Pakistan Medical Association*, 61(3), 210.

Hall, J. J. & Taylor, R. (2003). Health for all beyond 2000: The demise of the Alma Ata Declaration and primary health care in developing countries. *Medical Journal of Australia*, 178(1), 17–20.

Hanlon, C., Wondimagegn, D., & Alem, A. (2010). Lessons learned in developing community mental health care in Africa. *World Psychiatry: Official Journal of the World Psychiatric Association*, 9(3), 185–189.

Harris, C. (2016). Which is the most obese country in the EU? *Euronews*. Retrieved from http://www.euronews.com/2016/10/20/which-is-the-most-obese-country-in-the-eu.

Harrison, S. & Mort, M. (1998). Which champions, which people? Public and user involvement in health care as a technology of legitimation. *Social Policy & Administration*, 32(1), 60–70.

Hart, J. T. (1971). The inverse care law. *Lancet*, 297(7696), 405–412.

Hauptverband. (2015). Umfrageergebnisse: Bevölkerungsstudie—Gesundheit 2015 [Population survey—Health 2015]. Retrieved from http://www.hauptverband. at/portal27/hvbportal/content?contentid=10007.758811&viewmode=content.

Haynes, R. B., McKibbon, A., Fitzgerald, D., Guyatt, G. H., Walker, C. J., & Sackett, D. L. (1986). How to keep up with medical literature: 1. Why try to keep up and how to get started. *Annals of Internal Medicine*, 105(1), 149–153.

Hayward-Jones, J. (2016). The future of Papua New Guinea: Old challenges for new leaders. Retrieved from www.lowyinstitute.org.

Health Consumer Powerhouse. (2017). *Euro Health Consumer Index 2016*. France: Health Consumer Powerhouse Ltd.

Heerden, A. V., Tomlinson, M., & Swartz, L. (2012). Point of care in your pocket: A research agenda for the field of m-health. *Bulletin of the World Health Organization*, 90(5), 393–394.

Hefford, M., Cumming, J., Finlayson, M., Raymont, A., Love, T., & van Essen, E. (2010). Practice nurse cost benefit analysis: Report to the Ministry of Health. Retrieved from https://www.health.govt.nz/system/files/documents/publications/practise-nurse-cost-benefit-analysis.pdf.

Heuer, A. J. (2004). Hospital accreditation and patient satisfaction: Testing the relationship. *Journal for Healthcare Quality*, 26, 46–51.

Hildebrandt, H., Schulte, T., & Stunder, B. (2012). Triple aim in Kinzigtal, Germany: Improving population health, integrating health care and reducing costs of care—Lessons for the UK. *Journal of Integrated Care*, 20(4), 205–222.

Hillman, K. (1999). The changing role of acute-care hospitals. *Medical Journal of Australia*, 170(7), 325–328.

Hillman, K. (2009). *Vital Signs: Stories from Intensive Care*. Sydney, Australia: UNSW Press.

Hillman, K. (2017). *A Good Life to the End*. Crows Nest, Australia: Allen & Unwin.

Hillman, K., Chen, J., Cretikos, M., Bellomo, R., Brown, D., Doig, G., et al. (2005). Introduction of the medical emergency team (MET) system: A cluster randomised controlled trial. *Lancet*, 365(9477), 2091–2097.

Hjøllund, N. H., Larsen, L. P., Biering, K., Johnson, S. P., Riiskjær, E., & Schougaard, L. M. (2014). Use of patient-reported outcome (PRO) measures at group and patient levels: Experiences from the generic integrated PRO system, WestChronic. *Interactive Journal of Medical Research*, 3(1), 1–13.

HM Treasury. (2016). *Autumn Statement CM9362*. London, UK: The Stationery Office.

Hoffman, K. R., Loong, B., & van Haren, F. (2016). Very old patients urgently referred to the intensive care unit: Long-term outcomes for admitted and declined patients. *Critical Care and Resuscitation*, 18, 157–164.

Hoffmann, K., George, A., Dorner, T. E., Katharina Süß, K., Schäfer, W. L. A., & Maier, M. (2015). Primary health care teams put to the test a cross-sectional study from Austria within the QUALICOPC project. *BMC Family Practice*, 16(1), 168.

Hofmarcher, M. M. (2010). Ambulatory care reforms fail to face the facts? *Health Policy Monitor*. Retrieved from http://hpm.org/en/Surveys/GOEG_-_Austria/15/Ambulatory_care_reforms_fail_to_face_the_facts_.html.

Hofmarcher, M. M. (2013). Austria health system review. *Health Systems in Transition*, 15(7), 1–291.

Hofmarcher, M. M. (2014). The Austrian health reform 2013 is promising but requires continuous political ambition. *Health Policy*, 118(1), 8–13.

Hofmarcher, M. M., Simon, J., & Haidinger, G. (2017). Stroke-units in Austria: Incubators for improved health outcomes. In J. Braithwaite, R. Mannion, Y. Matsuyama, P. Shekelle, S. Whittaker, & S. Al-Adawi (Eds.), *Health Systems Improvement Across the Globe: Success Stories from 60 Countries* (pp. 127–132). Boca Raton, FL: Taylor & Francis Group.

Hollnagel, E. (2014). *Safety-I and Safety-II: The Past and Future of Safety Management*. Farnham, UK: Ashgate.

Homøe, P., Bjarnsholt, T., Wessmann, M., Sørensen, H. C. F., & Johansen, H. K. (2009). Morphological evidence of biofilm formation in Greenlanders with chronic supprative otitis media. *European Archives of Oto-Rhino-Laryngology*, 266(10), 1533–1538.

Hong Kong Hospital Authority. (2015). *Patient Experience and Satisfaction Survey on Specialist Outpatient Service 2014*. Hong Kong: Hong Kong Hospital Authority.

Hong Kong Hospital Authority. (2016). *Patient Experience and Satisfaction Survey on Inpatient Service 2015*. Hong Kong: Hong Kong Hospital Authority.

Hong Kong Hospital Authority. (2017a). *Patient Experience and Satisfaction Survey on Accident & Emergency Service 2016*. Hong Kong: Hong Kong Hospital Authority.

Hong Kong Hospital Authority. (2017b). *Strategic Plan 2017–2022: An Overview*. Hong Kong: Hong Kong Hospital Authority.

Hong Kong Hospital Authority. (n.d.). Vision, mission and values. Retrieved from http://www.ha.org.hk/visitor/ha_visitor_index.asp?Content_ID=10009&Lang=ENG&Dimension=100&Parent_ID=10004.

Hongoro, C. & McPake, B. (2004). How to bridge the gap in human resources for health. *Lancet*, 364(9443), 1451–1456.

Hounsgaard, L., Jensen, A. B., Wilche, J. P., & Dolmer, I. (2013). The nature of nursing practice in rural and remote areas of Greenland. *International Journal of Circumpolar Health*, 72(1), 20964.

Hounsgaard, L., Pedersen, B. D., & Wagner, L. (2011). The daily living for informal caregivers with a partner with Parkinson's disease—An interview study of women's experiences of care decisions and self-management. *Journal of Nursing and Healthcare of Chronic Illness*, 3(4), 504–512.

House, J. S. (2016). Social determinants and disparities in health: Their crucifixion, resurrection, and ultimate triumph (?) in health policy. *Journal of Health Politics, Policy and Law*, 41(4), 599–626.

Hsiao, W. C. (1995). Abnormal economics in the health sector. In P. Berman (Ed.), *Health Sector Reform in Developing Countries: Making Health Development Sustainable*. Boston, MA: Harvard University Press.

Huang, P. & Yi, L. (2015). Regional medical combination based on classified medicine, *Chinese Health Quality Management*, 22(4), 102–104.

Huber, M., Knottnerus, A., Green, L., van der Horst, H., Jadad, A., Kromhout, D., et al. (2011). How should we define health? *BMJ*, 343, d4163.

Hudon, C., Fortin, M., Haggerty, J. L., Lambert, M., & Poitras, M. (2011). Measuring patients' perception of patient-centered care: A systematic review of tools for family medicine. *Annual of Family Medicine*, 9, 155–154.

Human Resource Services Group. (2017). A proven way to improve organizational performance. Retrieved from http://www.hrsg.ca/why-competencies.

Human Rights Watch. (2016). Venezuela's humanitarian crisis: Severe medical and food shortages, inadequate and repressive government response. Retrieved from https://www.hrw.org/sites/default/files/report_pdf/venezuela1016_brochure_web_0.pdf.

Ieder(in). (2017). *Klem in zorgsysteem. Signaalrapport april* [*Clamp in Care System. Signal Report April*]. Utrecht: Ieder(in).

ILO Social Security Department. (2008). Social health protection: An ILO strategy towards universal access to health care. Social Security Policy Briefings, Vol. 1. Geneva, Switzerland: International Labour Organization.

Institute for Health Metrics and Evaluation (2015). GBD compare. Retrieved from http://vizhub.healthdata.org/gbd-compare/.

Institute for Health Metrics and Evaluation. (2017). GBD compare. Retrieved from http://vizhub.healthdata.org/gbd-compare/.

Institute for Health Metrics and Evaluation. (n.d.). Global burden of disease. Retrieved from http://www.healthdata.org/gbd.

Institute for Healthcare Improvement. (2003). The breakthrough series: IHI's collaborative model for achieving breakthrough improvement. IHI Innovation Series White Paper. Boston, MA: Institute for Healthcare Improvement.

Institute for the Future. (2003). *Health and Health Care 2010: The Forecast, the Challenge* (2nd ed.). Princeton, NJ: Jossey-Bass.

Institute of Medicine. (2001). Crossing the quality chasm: A new health system for the 21st century. *BMJ*, 323, 1192.

Institutos de Salud Previsional. (2016). El rol de las ISAPRE en la seguridad social. Hacia los principios de la seguridad social [The role of the ISAPRE in social security. Toward the principles of social security]. *Isapres de Chile*. Retrieved from http://www.isapre.cl/index.php/reformaalasisapres/31-comunicados-de-prensa/76-el-rol-de-las-isapres-en-la-seguridad-social.

International Alliance of Patients' Organizations. (2012). Patient-centred healthcare indicators review. London, UK: IAPO. Retrieved from http://iapo.org.uk/sites/default/files/files/IAPO%20Patient-Centred%20Healthcare%20Indicators%20Review.pdf.

International Alliance of Patients' Organizations. (2016). Declaration on patient-centered healthcare. London, UK: IAPO. Retrieved from https://www.iapo.org.uk/sites/default/files/files/IAPO_declaration_ENG_2016.pdf.

International Monetary Fund. (2017). World Economic Outlook Database. Retrieved from http://www.imf.org/external/pubs/ft/weo/2017/01/weodata/index.aspx.

International Society for Quality in Health Care. (2016). White paper: Health systems and their sustainability: Dealing with the impending pressures of ageing, chronic and complex conditions, technology and resource constraints. Dublin, UK: International Society for Quality in Health Care.

Internews. (2013). Lost: Syrian refugees and the information gap. Retrieved from https://www.internews.org/resource/lost-syrian-refugees-and-information-gap.

Iturriaga, M. V. & Valdivia, L. (2017). Chile: Constructing symbolic capital: A case study from Chile. In J. Braithwaite, R. Mannion, Y. Matsuyama, P. Shekelle, S. Whittaker, & S. Al-Adawi (Eds.), *Health Systems Improvement Across the Globe: Success Stories from 60 Countries* (pp. 25–31). Boca Raton, FL: Taylor & Francis.

Jaafar, S., Noh, K. M., Muttalib, K. A., Othman, N. H., & Healy, J. (2013). Malaysia health system review. *Health Systems in Transition*, 3(1), 1–103.

Jakovljevic, M. M., Arsenijevic, J., Pavlova, M., Verhaeghe, N., Laaser, U., & Groot, W. (2017). Within the triangle of healthcare legacies: Comparing the performance of South-Eastern European health systems. *Journal of Medical Economics*, 20(5), 483–492.

Jing, L. (2014). Study on the models and development status of regional longitudinal medical alliance in China. *Medicine and Society*, 27(5), 35–39.

Johns Hopkins University Bloomberg School of Public Health & Médicins du Monde. (2015). Syrian refugee and affected host population health access survey in Lebanon. Retrieved from https://data.unhcr.org/syrianrefugees/download.php?id=9550.

Johnson, A. B. (1990). *Out of Bedlam: The Truth about Deinstitutionalization*. New York, NY: Basic Books.

Jolly, P., Erikson, C., & Garrison, G. (2013). U.S. graduate medical education and physician specialty choice. *Academic Medicine*, 88(4), 468–474.

Jones, D., Bagshaw, S. M., Barrett, J., Bellomo, R., Bhatia, G., Bucknall, T. K., et al. (2012). The role of the medical emergency team in end-of-life care: A multicentre prospective observational study. *Critical Care Medicine*, 40(1), 98–103.

Jones, R. (2008). Hugo Chavez's health-care programme misses its goals. *Lancet*, 371(9629), 1988.

Jonsson, E. & Banta, D. (1999). Management of health technologies: An international view. *BMJ*, 319(7220), 1293.

Joumard, I., André, C., & Nicq, C. (2010). Health care systems: Efficiency and institutions. OECD Economics Department. Working Paper 769. Paris, France: OECD Publishing.

Jovell, A. J., Navarro Rubio, M. D., Fernandez Maldonado, L., & Blancafort, S. (2006). Involvement of the patient: The new role of patients in the health system. *Atencion Primaria*, 38(4), 234–237.

Joyner, M. J. & Paneth, N. (2015). Seven questions for personalized medicine. *JAMA: The Journal of the American Medical Association*, 314(10), 999–1000.

Kana, M. A., Doctor, H. V., Peleteiro, B., Lunet, N., & Barros, H. (2015). Maternal and child health interventions in Nigeria: A systematic review of published studies from 1990 to 2014. *BMC Public Health*, 15, 334.

Kantar Emor. (2016). Eesti elanike hinnangud tervisele ja arstiabile, Eesti Haigekassa ja Sotsiaalministeerium [Estimates of Estonian residents for health and medical care, Estonian Health Insurance Fund and Ministry of Social Affairs]. Retrieved from https://www.sm.ee/sites/default/files/content-editors/Ministeerium_kontaktid/Uuringu_ja_analuusid/Tervisevaldkond/arstiabi_uuringu_aruanne_2016_kantar_emor.pdf.

Kassenärztliche Bundesvereinigung. (2013). Die neue Bedarfsplanung Grundlagen, Instrumente und regionale Möglichkeiten [The new requirements planning basics, instruments and regional opportunities]. Retrieved from http://www.kbv.de/html/bedarfsplanung.

Kayral, İ. (2014). Perceived service quality in healthcare organizations and a research in Ankara by hospital type. *Journal of Ankara Studies*, 2(1), 22–34.

Kelley, A. S., Deb, P., Du W., Aldridge Carlson, M. D., & Morrison R. S. (2013). Hospice enrolment saves money for Medicare and improves care quality across a number of different lengths-of-stay. *Health Affairs*, 32(3), 552–561.

Keränen, T. (2017). Pocket-fitting chat is displacing video consultations. *Finnish Medical Journal (Suomen Lääkärilehti)*, 35, 1830–1833.

Khatri, S. (2012). As population ages, Qatar tackles elder care. *Doha News*. Retrieved from http://dohanews.co/post/13445483480/as-population-ages-qatar-tackles-elder-care.

Khoury, M. J. & Galea, S. (2016). Will precision medicine improve population health? *JAMA*, 316(13), 1357–1358.

Kieny, M.-P., Evans, D. B., Schmets, G., & Kadandale, S. (2014). Health-system resilience: Reflections on the Ebola crisis in western Africa. *Bulletin of the World Health Organization*, 92(12), 850–850.

Kim, S. K. & Park, M. (2017). Effectiveness of person-centered care on people with dementia: A systematic review and meta-analysis. *Clinical Interventions in Aging*, 12, 381–397.

King, A. (2000). *The New Zealand Health Strategy*. Wellington, New Zealand: Ministry of Health.

King, A. (2001). *The Primary Health Care Strategy*. Wellington, New Zealand: Ministry of Health.

Klynveld Peat Marwick Goerdeler. (2014). The global economic impact of antimicrobial resistance. Retrieved from https://www.kpmg.com/UK/en/IssuesAndInsights/ArticlesPublications/Documents/PDF/Issues%20and%20Insights/amr-report-final.pdf.

Knaul, F. M., Arreola-Ornelas, H., Méndez-Carniado, O., Bryson-Cahn, C., Barofsky, J., Maguire, R., et al. (2006). Evidence is good for your health system: Policy reform to remedy catastrophic and impoverishing health spending in Mexico. *Lancet*, 368(9549), 1828–1841.

Knaul, F. M., González-Pier, E., Gómez-Dantés, O., García-Junco, D., Arreola-Ornelas, H., et al. (2012). The quest for universal health coverage: Achieving social protection for all in Mexico. *Lancet*, 380(9849), 1259–1279.

Kodner, D. L. & Kyriacou, C. K. (2000). Fully integrated care for frail elderly: Two American models. *International Journal of Integrated Care*, 1, e08.

Kommuneqarfik Sermersooq. (2016). Sektorplan 2. fase [Sector plan 2. Phase]. Retrieved from http://www.sermersooq2028.gl/download/Sektorplan_aeldreomraadet_dk.pdf.

Koornneef, E., Robben, P., Al Seiari, M., & Al Siksek, Z. (2012). Health system reform in the Emirate of Abu Dhabi, United Arab Emirates. *Health Policy*, 108(2–3), 115–121.

Koplan, J. P., Bond, T. C., Merson, M. H., Reddy, K. S., Rodriguez, M. H., Sewankambo, N. K., & Wasserheit, J. N. (2009). Towards a common definition of global health. *Lancet*, 373(9679), 1993–1995.

Koppel, A., Meiesaar, K., Valtonen, H., Metsa, A., & Lember, M. (2003). Evaluation of primary health care reform in Estonia. *Social Science & Medicine*, 56(12), 2461–2466.

Kringos, D. S., Boerma, W., van der Zee, J., & Groenewegen, P. (2013). Europe's strong primary care systems are linked to better population health but also higher health spending. *Health Affairs*, 32(4), 686–694.

Kritzer, B. (2008). Chile's next generation pension reform. *Social Security Bulletin*, 68(2). Retrieved from https://www.ssa.gov/policy/docs/ssb/v68n2/v68n2p69.html.

Kroneman, M., Boerma, W., van den Berg, M., Groenewegen, P., de Jong, J., & van Ginneken, E. (2016). Netherlands Health system review. *Health Systems in Transition*, 18(2), 1–239.

Kroneman, M. & Maarse, H. (2015). The reform of long-term care in 2015. Retrieved from http://www.hspm.org/countries/netherlands25062012/countrypage.aspx.

Kruk, M. E., Myers, M., Varpilah, S. T., & Dahn, B. T. (2015). What is a resilient health system? Lessons from Ebola. *Lancet*, 385(9980), 1910–1912.

Krzyzanowska, M., Kaplan, R., & Sullivan, R. (2011). How may clinical research improve healthcare outcomes? *Annals of Oncology*. 22(Suppl. 7), vii10–vii15.

Kunnamo, I. (2015). How to build an ideal healthcare information system? In *WONCA Europe World Book of Family Medicine—European Edition, 2015*. Retrieved from http://www.woncaeurope.org/sites/default/files/World%20Book%202015.pdf.

Lagarde, M., Haines, A., & Palmer, N. (2009). The impact of conditional cash transfers on health outcomes and use of health services in low and middle income countries. *Cochrane Database of Systematic Reviews*, 7(4), CD008137.

Lagomarsino, G., Garabrant, A., Adyas, A., Muga, R., & Otoo, N. (2012). Moving towards universal health coverage: Health insurance reforms in nine developing countries in Africa and Asia. *Lancet*, 380(9845), 933–943.

Lai, T., Habicht, T., Kahur, K., Reinap, M., Kiivet, R., & van Ginneken, E. (2013). Estonia. Health system review. *Health Systems in Transition*, 15(6), 1–196.

Lander, E. S., Linton, L. M., Birren, B., Nusbaum, C., Zody, M. C., Baldwin, J., et al. (2001). Initial sequencing and analysis of the human genome. *Nature*, 409(6822), 860–921.

Langley, G. L., Nolan, K. M., Nolan, T. W., Norman, C. L., & Provost, L. P. (2009). *The Improvement Guide: A Practical Approach to Enhancing Organisational Performance* (2nd ed.). San Francisco, CA: Jossey-Bass Publishers.

Larsson, S., Lawyer, P., Garellick, G., Lindahl, B., & Lundström, M. (2012). Use Of 13 disease registries in 5 countries demonstrates the potential to use outcome data to improve health care's value. *Health Affairs*, 31(1), 220–227.

Last, J. M. (Ed.). (1995). *A Dictionary of Epidemiology*. Vol. 3. Toronto, Canada: Oxford University Press.

Leape, L. L., Park, R. E., Solomon, D. H., Chassin, M. R., Kosecoff, J., & Brook, R. H. (1990). Does inappropriate use explain small-area variations in the use of health-care services. *JAMA: The Journal of the American Medical Association*, 263(5), 669–672.

Learning Healthcare Project. (2015). Professor Charles Friedman interview, 2015. Retrieved from http://www.learninghealthcareproject.org/section/evidence/25/50/professor-charles-friedman-interview.

Leatherman, S., Ferris, T. G., Berwick, D., Omaswa, F., & Crisp, N. (2010). The role of quality improvement in strengthening health systems in developing countries. *International Journal for Quality in Health Care*, 22(4), 237–243.

Lehmann, U., Van Damme, W., Barten, F., & Sanders, D. (2009). Task shifting: The answer to the human resources crisis in Africa? *Human Resources for Health*, 7(1), 49.

Lember, M. (2002). A policy of introducing a new contract and funding system of general practice in Estonia. *International Journal of Health Planning and Management*, 17(1), 41–53.

Lemmens, L. C., Hanneke, W. D., Lette, M., & Baan, C. A. (2017). Populatiegerichte aanpak voor verbinding van preventie, zorg en welzijn: De beweging in beeld [Population-oriented approach to prevention, care and well-being: The movement in view]. *Nederlands Tijdschrift voor Geneeskunde*, 161, D849.

Levi, M. (2017). Generalism in modern subspecializing medicine. *European Journal of Internal Medicine*, 39, 36–38.

Liddy, C., Rowan, M. S., Afkham, A., Maranger, J., & Keely, E. (2013). Building access to specialist care through e-consultation. *Open Medicine*, 7(1), e1–e8.

Liu, W. (2016). Medical alliance: How far is it from the integrated stratified healthcare system? *China Hospital CEO*, 17, 40–43.

López, J. (1997). Latin American health care reforms at the crossroads: An introduction. *Eurohealth*, 3(3), 21–23.

Lopreiato, J., Downing, D., Gammon, W., Lioce, L., Sittner, B., Slot, V., et al. (2016). Healthcare simulation dictionary. Retrieved from http://www.ssih.org/dictionary.

Lorig, K., Sobel, D. S., Stewart, A. L., Brown, B. W., Bandura, A., Ritter, P., et al. (1999). Evidence suggesting that a chronic disease self-management program can improve health status while reducing hospitalization: A randomized trial. *Medical Care*, 37(1), 5–14.

Loring, B. J., Friel, S., Kitau, R., Matheson, D., Ake, I., Kitur, U., et al. (2013). Reducing community health inequity: The potential role for mHealth in Papua New Guinea. *Journal of Community Informatics*, 9(2). Retrieved from http://ci-journal.net/index.php/ciej/article/view/838/1002.

Love, T. & Blick, G. (2014). Primary care funding—A discussion paper. Retrieved from http://www.srgexpert.com/wp-content/uploads/2015/08/GPNZ_funding_paper-20_November.pdf.

Lovelock, K., Martin, G., Cumming, J., & Gauld, R. (2014). The evaluation of the better, sooner, more convenient business cases in Midcentral and the West Coast District Health Boards. Retrieved from http://www.otago.ac.nz/healthsystems/otago089473.pdf.

Lozano, R., Soliz, P., Gakidou, E., Abbott-Klafter, J., Feehan, D. M., Vidal, C., et al. (2007). Benchmarking of performance of Mexican states with effective coverage. *Salud Publica de Mexico*, 49(1), S53–S69.

Lu, C., Chin, B., Lewandowski, J. L., Basinga, P., Hirschhorn, L. R., Hill, K., et al. (2012). Towards universal health coverage: An evaluation of Rwanda Mutuelles in its first eight years. *PLoS One*, 7(6), e39282.

Lubetkin, E. I., Sofaer, S., Gold, M. R., Berger, M. L., Murray, J. F., & Teutsch, S. M. (2003). Aligning quality for populations and patients: Do we know which way to go? *American Journal of Public Health*, 93(3), 406–411.

Luepke, K. H., Suda, K. J., Boucher, H., Russo, R. L., Bonney, M. W., Hunt, T. D., et al. (2017). Past, present and future of antibacterial economics: Increasing bacterial resistance, limited antibiotic pipeline, and societal implications. *Pharmacotherapy*, 37(1), 71–84.

Lutz, W., Sanderson, W., & Scherbov, S. (2008). The coming acceleration of global population ageing. *Nature*, 451(7179), 716–719.

Lyon, A. R., Stirman, S. W., Kerns, S. E. U., & Bruns, E. J. (2011). Developing the mental health workforce: Review and application of training approaches from multiple disciplines. *Administration and Policy in Mental Health*, 38(4), 238–253.

Maben, J., Al-Thowini, K., West, E., & Rafferty, A. M. (2010). Uneven development: Comparing the indigenous health care workforce in Saudi Arabia, Bahrain and Oman. *International Journal of Nursing Studies*, 47(3), 392–396.

Macdonald, J. (2008). *Blended Learning and Online Tutoring: Planning Learner Support and Activity Design* (2nd ed.). Aldershot, UK: Gower.

Mackenbach, J. P., Karanikolos, M., & McKee, M. (2013). The unequal health of Europeans: Successes and failures of policies. *Lancet*, 381(9872), 1125–1134.

Macrae, C. (2016). The problem with incident reporting. *BMJ Quality & Safety*, 25(2), 71–75.

Mahler, H. (2016). The meaning of "health for all by the year 2000." *American Journal of Public Health*, 106(1), 36–38.

Makary, M. & Daniel, M. (2016). Medical error—The third leading cause of death in the US. *The BMJ*, 353, i2139.

Marmot, M. (2004). Status syndrome: How your social standing directly affects your health and life expectancy. *Significance*, 1(4), 150–154.

Marshall, W. E. & Garrick, N. W. (2010). Effect of street network design on walking and biking. *Transportation Research Record*, 2198, 103–115.

Marshall, W. E., Piatkowski, D. P., & Garrick, N. W. (2014). Community design, street networks, and public health. *Journal of Transport & Health*, 1(4), 326–340.

Martin, J. C., Avant, R. F., Bowman, M. A., Bucholtz, J. R., Dickinson, J. R., Evans, K., et al. (2004). The future of family medicine: A collaborative project of the family medicine community. *Annals of Family Medicine*, 2(Suppl. 1), S3–S32.

Martineau, T. & Buchan, J. (2000). Human resources and the success of health sector reform. *Human Resources for Health Development Journal*, 4(3), 174–183.

Mathura, R. (2013). Doctors sell soul of medicine [Letter to the editor]. *Newsday*, June 16. Retrieved from http://archives.newsday.co.tt/letters/0,179235.html.

Matiz Cortes, S. (2015). Los cuestionados médicos 'express' venezolanos que migraron a Colombia. *El Espectador*, September 17. Retrieved from http://www.elespectador.com.

Mattick, J. S., Dziadek, M. A., Terrill, B. N., Kaplan, W., Spigelman, A. D., Bowling, F. G., et al. (2014). The impact of genomics on the future of medicine and health. *Medical Journal of Australia*, 201(1), 17–20.

Mayer, M. L. & Skinner, A. C. (2004). Too many, too few, too concentrated? A review of the pediatric subspecialty workforce literature. *Archives of Pediatrics & Adolescent Medicine*, 158(12), 1158–1165.

Mays, N. (2013). Evaluating the Labour Government's English NHS health system reforms: The 2008 Darzi reforms. *Journal of Health Services Research & Policy*, 18(Suppl. 2), 1–10.

McCartney, M. (2017). Don't rush into precision medicine. *The BMJ*, 356, j1168.

McClarty, K. & Gaertner, M. (2015). Measuring mastery: Best practices for assessment in competency-based education. Retrieved from https://www.luminafoundation.org/files/resources/measuring-mastery.pdf.

McClellan, M., Kent, J., Beales, S., Macdonnell, M., Thoumi, A., Shuttleworth, B., et al. (2013). Accountable care: Focusing accountability on the outcomes that matter. Doha, Qatar: Accountable Care Working Group. Retrieved from http://www.wish-qatar.org/app/media/384.

McConnell, H., Haile-Mariam, T., & Rangarajan, S. (2004). The world health channel: An innovation for health and development. *World Hospitals & Health Services*, 40(4), 36–39.

McDermott, A., Steel, D., McKee, L., Hamel, L., & Flood, P. C. (2015). Scotland 'bold and brave'? Conditions for creating a coherent national healthcare quality strategy. In S. Boch Waldorff, A. Reff Pederson, L. Fitzgerald, & E. Ferlie (Eds.), *Managing Change: From Health Policy to Practice* (pp. 189–205). London, UK: Palgrave Macmillan.

McGlynn, E. A., Asch, S. M., Adams, J., Keesey, J., Hicks, J., DeCristofaro, A., et al. (2003). The quality of healthcare delivered to adults in the United States. *New England Journal of Medicine*, 348(26), 2635–2645.

McGlynn, E. A., Naylor, C. D., Anderson, G. M., Leape, L. L., Park, R. E., Hilborne, L. H., et al. (1994). Comparison of the appropriateness of coronary angiography and coronary artery bypass graft surgery between Canada and New York State. *JAMA: The Journal of the American Medical Association*, 272(12), 934–940.

McMillan, S. S., Kendall, E., Sav, A., King, M. A., Whitty, J. A., Kelly F., et al. (2013). Patient-centered approaches to health care: A systematic review of randomized controlled trials. *Medical Care Research and Review*, 70(6), 567–596.

Meadows, K. A. (2011). Patient-Reported outcome measures: An overview. *British Journal of Community Nursing*, 16(3),146–151.

Mediuutiset. (2017). "It's still in its infancy"—IBM's artificial intelligence Watson not nearly as good as Finnish oncologists [in Finnish]. Retrieved from http://www.mediuutiset.fi/uutisarkisto/se-on-lapsen-kengissa-ibm-n-tekoaly-watson-ei-parjaakaan-syopalaakarille-6677181.

Micheletti, L., Preti, M., Bogliatto, F., & Lynch, P. J. (2002). Vulvology. A proposal for a multidisciplinary subspecialty. *Journal of Reproductive Medicine*, 47(9), 715–717.

Middleton, L. & Cumming, J. (2016). At risk individuals model of care: An evaluation. Retrieved from https://www.victoria.ac.nz/health/centres/health-services-research-centre/our-publications/reports/hsrc-ari-report.pdf.

Minister of Health [New Zealand]. (2016a). New Zealand health strategy future direction: All New Zealanders live well, stay well, get well. Retrieved from http://www.health.govt.nz/system/files/documents/publications/new-zealand-health-strategy-future-direction-apr16_1.pdf.

Minister of Health [New Zealand]. (2016b). New Zealand health strategy: Roadmap of actions 2016. Retrieved from http://www.health.govt.nz/new-zealand-health-system/new-zealand-health-strategy-roadmap-actions-2016.

Ministério da Saúde [Brazil]. (2013). Portaria n° 529, de 1° de abril de 2013, que instituiu o Programa Nacional de Segurança do Paciente (PNSP). Retrieved from http://bvsms.saude.gov.br/bvs/saudelegis/gm/2013/prt0529_01_04_2013.html.

Ministério da Saúde [Brazil]. (2013). Fundação Oswaldo Cruz, & Agência Nacional de Vigilância Sanitária. Documento de referência para o programa nacional de Segurança do paciente. Retrieved from http://proqualis.net/sites/proqualis.net/files/documento_referencia_programa_nacional_seguranca.pdf.

Ministério da Saúde [Portugal]. (2013a). Despacho n° 15423/2013, D.R. n° 229, Parte C, Série II de 2013b. Retrieved from http://legislacaoportuguesa.com/despacho-n-o-154232013-d-r-n-o-229-parte-c-serie-ii-de-2013-11-26/.

Ministério da Saúde [Portugal]. (2013b). Despacho n° 2902/2013 de 22 de fevereiro, publicado no Diário da República, 2ª Série, n° 38, de 22 de fevereiro de 2013a. Lisbon, Portugal: Ministério da Saúde.

Ministério da Saúde [Portugal]. (2016). Despacho n° 3844-A/2016—Diário da República n° 52/2016. 1° Suplemento, Série II de 2016-03-15. Retrieved from https://dre.pt/home/-/dre/73865548/details/maximized?p_auth=Dm4lj3la&serie=II.

Ministerio del Poder Popular para Educación Universitaria, Ciencia y Tecnología [Venezuela]. (2010). Ley Orgánica de Ciencia y Tecnología e Innovación. Retrieved from https://www.mppeuct.gob.ve/sites/default/files/descargables/ley_organica_de_ciencia_tecnologia_e_innovacion_2010.pdf.

Ministerio del Poder Popular para la Comunicación y la Información [Venezuela]. (2016). Universidad de las Ciencias de la Salud proyecta graduar 60 mil médicos integrales. Retrieved from http://minci.gob.ve/2016/10/universidad-de-las-ciencias-de-la-salud-proyecta-graduar-60-mil-medicos-integrales/.

Ministerio de Salud [Argentina]. (2016). Generación de una cultura evaluativa en el Sector Salud: principales resultados de la agenda de estudios y evaluación del Plan Nacer/Programa SUMAR. Buenos Aires: Author. Retrieved from http://www.msal.gob.ar/sumar/images/stories/pdf/generacion-cultura-evaluativa-sector-salud.pdf.

Ministerio de Salud [Argentina]. (n.d.). Sistema Integrado de Información Sanitaria Argentina. Retrieved from https://www.msal.gob.ar.

Ministerio de Sanidad, Servicios Sociales e Igualdad [Spain]. (2012). Documento de trabajo: Red Escuelas de Salud para ciudadanos. Retrieved from http://www.escuelas.msssi.gob.es/conocenos/laRed/docs/Documento_trabajo_Red_Escuelas.pdf.

Ministry for Energy and Health [Malta]. (2014). The National Health Systems Strategy 2014–2020: Securing our health systems for future generations. Retrieved from https://health.gov.mt/en/CMO/Documents/alert_nhss_eng.pdf.

Ministry for Energy and Health [Malta]. (2016). Patient's charter. Retrieved from https://socialdialogue.gov.mt/en/Public_Consultations/MEH-HEALTH/Documents/Patient%20Charter%20(in%20English).pdf.

Ministry for Health [Malta]. (2017). Malta national health screening programs. Retrieved from https://health.gov.mt/en/phc/nbs/Pages/Home.aspx.

Ministry for Health, the Elderly and Community Care [Malta]. (2011). The national cancer plan: 2011–2015. Retrieved from https://health.gov.mt/en/CMO/Documents/the_national_cancer_plan.pdf.

Ministry for Health, the Elderly and Community Care [Malta]. (2017). The national cancer plan: 2017–2021. Retrieved from https://deputyprimeminister.gov.mt/en/CMO/Documents/NationalCancerPlan2017.pdf.

Ministry of Employment and the Economy [Finland]. (2012). Health technology and pharmaceutical research as the cornerstone of growth in Finland. Retrieved from https://docslide.com.br/healthcare/getpersonalized-health-technology-and-pharmaceutical-research-as-the-cornerstone-of-growth-in-finland-mikko-alkio.html.

Ministry of Employment and the Economy [Finland]. (2014). Health sector growth strategy for research and innovation activities. Report 16/2014. Retrieved from https://tem.fi/documents/1410877/3437254/Health+Sector+Growth+Strategy+for+Research+and+Innovation+Activities+26052014.

Ministry of Employment and the Economy [Finland]. (2016). Innovating together: Health sector growth strategy for research and innovation activities roadmap for 2016–2018. Report 6/2016. Retrieved from https://julkaisut.valtioneuvosto.fi/bitstream/handle/10024/75145/MEE_guidelines_8_2016_Health_sector_growth_strategy_17062016_web.pdf?sequence=1.

Ministry of Health [Austria]. (2014). "Das Team rund um den Hausarzt." Konzept zur multiprofessionellen und interdisziplinären Primärversorgung in Österreich. Retrieved from http://www.bmgf.gv.at/cms/home/attachments/1/2/6/CH1443/CMS1404305722379/primaerversorgung.pdf.

Ministry of Health [Denmark]. (2017). Healthcare in Denmark, an overview. Retrieved from http://www.sum.dk/Aktuelt/Publikationer/~/media/Filer%20-%20Publikationer_i_pdf/2016/Healthcare-in-dk-16-dec/Healthcare-english-V16-dec.ashx.

Ministry of Health [Italy]. (2016). National HTA programme for medical devices. Strategic document. Internal document. Rome: Ministry of Health.

Ministry of Health [Jordan]. (2012). Ministry of health strategic plan 2013–2017. Retrieved from https://jordankmportal.com/resources/ministry-of-health-strategic-plan-2013-2017.

Ministry of Health [Malaysia]. (2013). Perinatal care manual. Retrieved from http://fh.moh.gov.my/v3/index.php/component/jdownloads/send/18-sektor-kesihatan ibu/224-perinatal-care-manual-3rd-edition-2013?option=com_jdownloads on 25 February 2017.

Ministry of Health [New Zealand]. (2016). System level measures framework. Retrieved from http://www.health.govt.nz/new-zealand-health-system/system-level-measures-framework.

Ministry of Health [New Zealand]. (2017). New Zealand health survey: Annual update of key findings 2015/16. Wellington: Ministry of Health. Retrieved from https://minhealthnz.shinyapps.io/nz-health-survey-2015-16-annual-update/.

Ministry of Health [Oman]. (2015). Annual health report. Retrieved from http://ghdx.healthdata.org/organizations/ministry-health-oman.

Ministry of Health [Turkey]. (2003). *Transformation in Health (HTP)*. Ankara, Turkey: Ministry of Health.

Ministry of Health and Care Services [Norway]. (2015). På ramme alvor: Alvorlighet og prioritering. Rapport fra arbeidsgruppe nedsatt av Helse- og omsorgsdepartementet 2015 [Report on severity of illness and priority setting in Norway]. Oslo, Norway: Ministry of Health and Care Services.

Ministry of Health and Care Services [Norway]. (2016). Verdier i pasientens helsetjeneste. Melding om prioritering. Meld. St. 34 [Values in the health care services—On priority setting. White Paper No. 34]. Oslo, Norway: Departementenes sikkerhets og serviceorganisasjon.

Ministry of Health, Labor and Welfare [Japan]. (2015). A comprehensive strategy for the promotion of dementia measures: Towards a community friendly to the elderly with dementia, etc. A New Orange Plan. Retrieved from http://www.mhlw.go.jp/stf/houdou/0000072246.html.

Ministry of Health and Prevention [UAE]. (2017). World class healthcare. Retrieved from https://www.vision2021.ae/en/national-priority-areas/world-class-healthcare.

Ministry of Health and Social Services [Namibia]. (2015). The Namibia AIDS response progress report 2015. Retrieved from http://www.unaids.org/sites/default/files/country/documents/NAM_narrative_report_2015.pdf.

Ministry of National Education [Turkey]. (2016). National education statistics, 2016. Retrieved from http://www.meb.gov.tr/index.php.

Ministry of National Planning and Economic Policy [Chile]. (2000). Propuesta de Políticas para la seguridad social en Chile: Componente salud. Unidad de Estudios Prospectivos, Ministerio de Planificación y Coordinación. Santiago de Chile: Ministry of National Planning and Economic Policy.

Ministry of Public Health [Guyana]. (2017). Mental Health Unit Expenditure Records. Integrated Financial Management and Accountability System (IFMAS) Reports. Georgetown, Guyana: Ministry of Finance.

Ministry of Public Health [Lebanon]. (2007). The MOH strategic plan 2007. No. DG/HK0701001. Beirut, Lebanon: Ministry of Public Health.

Ministry of Public Health [Lebanon]. (2017). National e-health program. Retrieved from http://www.moph.gov.lb/en/Pages/6/2651/national-e-health-program.

Ministry of Public Health [Qatar]. (2017). National cancer strategy: National health strategy. Retrieved from http://www.nhsq.info/strategy-goals-and-projects/national-cancer-strategy/national-cancer-strategy-home.

Ministry of Public Health and Population [Yemen]. (2000). Health sector reform in the republic of Yemen. Retrieved from http://www.mophp-ye.org/docs/HSR_Strategy.pdf.

Ministry of Social Affairs and Health [Finland]. (2013). Peer review: eHealth strategy and action plan of Finland in a European context. Reports and Memorandums of the Ministry of Social Affairs and Health 2013:11. Retrieved from http://urn.fi/.

Ministry of Social Affairs and Health [Finland]. (2015). Improving health through the use of genomic data: Finland's genome strategy. Working group proposal. Retrieved from http://urn.fi/URN:ISBN:978-952-00-3598-3.

Ministry of Social Affairs and Health [Finland]. (2016). Information to support well-being and service renewal: eHealth and eSocial strategy 2020. Retrieved from http://julkaisut.valtioneuvosto.fi/bitstream/handle/10024/74459/URN_ISBN_978-952-00-3575-4.pdf.

Ministry of Social Affairs and Health [Finland]. (n.d.). Key projects at MSAH. Retrieved from http://stm.fi/en/key-projects

Mitchell, C., Moraia, L. B., & Kaye, J. (2014). Health database: Restore public trust in care.data project. *Nature*, 508(7497), 458–458.

Mokdad, A. H., Forouzanfar, M. H., Daoud, F., El Bcheraoui, C., Moradi-Lakeh, M., Khalil, I., et al. (2016). Health in times of uncertainty in the eastern Mediterranean region, 1990–2013: A systematic analysis for the global burden of disease study 2013. *Lancet Global Health*, 4(10), e704–e713.

Mørland, B., Ringard, Å., & Røttingen, J. A. (2010). Supporting tough decisions in Norway: A healthcare system approach. *International Journal of Technology Assessment in Health Care*, 26(4), 398–404.

Mosadeghrad, A. M. (2012a). Towards a theory of quality management: an integration of strategic management, quality management and project management. *International Journal of Modelling in Operations Management*, 2(1), 89–118.

Mosadeghrad, A. M. (2012b). Implementing strategic collaborative quality management in healthcare sector. *International Journal of Strategic Change Management*, 4(3/4), 203–228.

Mosadeghrad, A. M. (2013). Verification of a quality management theory: Using a Delphi study. *International Journal of Health Policy and Management*, 1(4), 261–271.

Mosadeghrad, A. M. (2016). Comments on Iran hospital accreditation system. *Iranian Journal of Public Health*, 45(6), 837–842.

Mosadeghrad, A. M. (2017). Iran's Health Transformation Plan. In J. Braithwaite, R. Mannion, Y. Matsuyama, P. Shekelle, S. Whittaker, & S. Al-Adawi (Eds.), *Health Systems Improvement Across the Globe: Success Stories from 60 Countries* (pp. 309–316). Boca Raton, FL: Taylor & Francis Group.

Mosadeghrad, A. M., Akbari-sari, A., & Yousefinezhadi, T. (2017a). Evaluation of hospital accreditation method in Iran. *Tehran University Medical Journal*, 75(4), 288–298.

Mosadeghrad, A. M., Akbari-sari, A., & Yousefinezhadi, T. (2017b). Evaluation of hospital accreditation standards. *Razi Journal of Medical Sciences*, 23(153), 50–61.

Mosadeghrad, A. M. & Ferlie, E. (2016). Total quality management in healthcare. In A. Ortenblad, C. Abrahamson Lofstrom, & R. Sheaff (Eds.), *Management Innovations for Healthcare Organizations: Adopt, Abandon or Adapt?* (pp. 378–396). London, UK: Routledge.

Mullan, F. (1984). Community-oriented primary care: Epidemiology's role in the future of primary care. *Public Health Reports*, 99(5), 442–445.

Mullan, F. & Epstein, L. (2002). Community-oriented primary care: New relevance in a changing world. *American Journal of Public Health*, 92(11), 1748–1755.

Nandraj, S., Khot, A., Menon, S., & Brugha, R. (2001). A stakeholder approach towards hospital accreditation in India. *Health Policy Plan*, 16(2), 70–79.

Nasim, M. (2011). Short communication—Medical education needs to change in Pakistan. *Journal of the Pakistan Medical Association*, 61(8), 808–811.

National Accreditation Board for Hospitals and Healthcare Providers. (2017). About us. Retrieved from http://www.nabh.co/introduction.aspx#.

National Council for Priority Setting. (2015). Hvilket ansvar hard et offentlig helsevesen for pasienter med egenbetalte legemidler? [What is the responsibility for a public health care system for patients who buy their own medicines?]. Retrieved from http://www.prioritering.no/saker/hvilket-ansvar-har-det-offentlige-helsevesen-for-pasienter-med-egenbetalte-legemidler.

National Department of Health [Papua New Guinea]. (2010). *National Health Plan 2011–2020. Volume 1: Policies and Strategies*. Port Moresby, Papua New Guinea: Government of Papua New Guinea.

National Department of Health [Papua New Guinea]. (2016a). *Independent Review of the PNG NHIS for Rural Primary Health Services Delivery*. Port Moresby, Papua New Guinea: Government of Papua New Guinea.ef

National Department of Health [Papua New Guinea]. (2016b). *Minutes of the 3rd Meeting of the National e-health Steering Committee*. Port Moresby, Papua New Guinea: Government of Papua New Guinea.

National Department of Health [Papua New Guinea] & WHO. (2015). *Developing an eHealth Strategy for Papua New Guinea: Final Report*. Port Moresby, Papua New Guinea: Government of Papua New Guinea.

National Department of Health [South Africa]. (2004). National Health Act. *Government Gazette*, 469(26595), 2–94. Retrieved from https://www.gov.za/sites/www.gov.za/files/a61-03.pdf.

National Department of Health [South Africa]. (2013). National Health Amendment Act. *Government Gazette*, 577, 2–35. Retrieved from http://www.ohsc.org.za/dev/images/NationalHealthAmendmentAct12of2013.pdf.

National Department of Health [South Africa]. (2017). *National Health Insurance for South Africa: Towards Universal Health Coverage*. Pretoria: Government Printing Works.

National Health and Family Planning Commission [China]. (2013). 2013 China health statistics. Retrieved from http://www.nhfpc.gov.cn/htmlfiles/zwgkzt/ptjnj/year2013/index2013.html.

National Health and Family Planning Commission [China]. (2015). Medical institutions report. Retrieved from http://www.nhfpc.gov.cn/mohwsbwstjxxzx/s7967/201604/c647068ab129448e95f205e18dbe1a18.shtml.

National Health and Family Planning Commission [China]. (2016a). Medical institutions report. Retrieved from http://www.moh.gov.cn/mohwsbwstjxxzx/s7967/201702/0a644a51bfc347ccab43fb1766aa5089.shtml.

National Health and Family Planning Commission [China]. (2016b). Medical service report. Retrieved from http://www.moh.gov.cn/mohwsbwstjxxzx/s7967/201702/79b6d9e3bf9e40e6a8efa1328b80ada9.shtml.

National Health and Family Planning Commission [China]. (2016c). Health china 2030 planning outline. Retrieved from http://www.nhfpc.gov.cn/zwgk/jdjd/201610/a2325a1198694bd6ba42d6e47567daa8.shtml.

National Health and Family Planning Commission [China]. (2016d). Notification on establishing pilot cities for hierarchy diagnosis and treatment. Retrieved from http://www.moh.gov.cn/yzygj/s3593g/201608/eed94f9b48e441929f-35f9e721064c01.shtml.

National Health Insurance Administration, Ministry of Health and Welfare, [Taiwan]. (2015a). Value-added innovative application: NHI information now accessible. Retrieved from http://www.nhi.gov.tw/english/HotNewsEnglish/HotnewsEnglish_Detail.aspx?News_ID=101&menu=1&menu_id=291.

National Health Insurance Administration, Ministry of Health and Welfare, [Taiwan]. (2015b). New NHI mobile app brings NHI services at your fingertips. Retrieved from https://www.nhi.gov.tw/english/News_Content.aspx?n=996D1B4B5DC48343&sms=F0EAFEB716DE7FFA&s=3F2D8EF2E5AEC431.

National Institute of Population and Social Security Research [Japan]. (2012). Population projections for Japan (January 2012): 2011 to 2060. Retrieved from http://www.ipss.go.jp/site-ad/index_english/esuikei/ppfj2012.pdf.

National Institute of Population and Social Security Research [Japan]. (2013). Household projections for Japan 2010–2035. Retrieved from http://www.ipss.go.jp/pp-ajsetai/e/hhprj2013/t-page_e.asp.e

National Institutes of Health. (2017). All of us research program. Retrieved from https://allofus.nih.gov/.

National Primary Health Care Development Agency [Nigeria]. (2016). Management guidelines for primary health care under one roof. Abuja: National Primary Health Care Development Agency. Retrieved from https://niftng.com/wp-content/uploads/2016/04/PHCUOR-Scorecard-3-Narrative-Report-final.pdf

National Statistics Office [Malta]. (2014). Malta in figures 2014. Retrieved from https://nso.gov.mt/en/publicatons/Publications_by_Unit/Documents/D2_External_Cooperation_and_Communication/Malta_in_Figures_2014.pdf.

Neily, J., Mills, P. D., Eldridge, N., Dunn, E. J., Samples, C., Turner, J. R., et al. (2009). Incorrect surgical procedures within and outside of the operating room. *Archives of Surgery*, 144(11), 1028–1034.

Nelson, E. C., Hvidtfeldt, H., Reid, R., Grossman, D., Lindblad, S., Mastanduno, M., et al. (2012). Using patient-reported information to improve health outcomes and healthcare value: Case studies from Dartmouth, Karolinska and group health. Lebanon, NH: The Dartmouth Institute for Health Policy and Clinical Practice, Center for Population Health. Retrieved from https://kiedit.ki.se/sites/default/files/using_patientreported_information_to_improve_health_outcomes_and_health_care_value.pdf.

Nemec, J. & Kolisnichenko, N. (2006). Market-based health care reforms in Central and Eastern Europe: Lessons after ten years of change. *International Review of Administrative Sciences*, 72(1), 11–26.

Newhouse, J. P. & Garber, A. M. (2013). Geographic variation in medicare services. *New England Journal of Medicine*, 368(16), 1465–1468.

Newhouse, J. P. & the Insurance Experiment Group. (1993). *Free for All? Lessons from the RAND Health Insurance Experiment*. Cambridge, MA: Harvard University Press.

Newton, D. A. & Grayson, M. S. (2003). Trends in career choice by US medical school graduates. *JAMA: The Journal of the American Medical Association*, 290(9), 1179–1182.

New Zealand Government. (2016). Budget at a glance. Retrieved from http://www.treasury.govt.nz/budget/2016/at-a-glance/b16-at-a-glance.pdf.

Nigenda, G. (2013). Servicio social en medicina en Mexico: Una reforma urgente y posible. *Salud Pública de México*, 55(5), 519–527.

Nigenda, G., Wirtz, V. J., González-Robledo, L. M., & Reich, M. R. (2015). Evaluating the implementation of Mexico's health reform: The case of seguro popular. *Health Systems & Reform*, 1(3), 217–228.

Nilsson, E. & Lindblom, H. (2015). Overview of PROM and PREM in the national quality registries. Retrieved from http://rcso.se/patientmedverkan/prom/.

Ninomiya, T. (2015). *The Study on Projections of the Elderly Population with Dementia in Japan 2015*. Tokyo: Ministry of Health, Labor and Welfare.

Nolte, E. & McKee, M. (2008). Measuring the health of nations: 23 updating an earlier analysis. *Health Affairs*, 27(1), 58–71.

Norheim, O. F. (2003). Norway. In C. Ham & G. Robert (Eds.), *Reasonable Rationing: International Experience of Priority Setting in Health Care* (pp. 94–114). Maidenhead, PA: Open University Press.

Norheim, O. F. (2008). Clinical priority setting. *BMJ*, 337, a1846.

Northern Ireland Practice & Education Council for Nursing and Midwifery. (2014). The attributes competency assessment tool. Retrieved from https://nipecport-folio.hscni.net/compro/ReadOnly/attributes/default.asp.

NTN24 (2017). 180 Médicos Cubanos escapan de Venezuela y esperan en Bogotá visa de EEUU. Nuestra Tele Noticias, April 17. Retrieved from http://www.ntn24america.com/noticia/180-medicos-cubanos-esperan-en-bogota-visa-de-ee-uu-tras-desertar-venezuela-138741.

Nuño-Solinis, R., Rodríguez, C., & Piñera, K. (2013). Panorama de las iniciativas de educación para el autocuidado en España. *Gac Sanit*, 27, 332–337.

Nyatanyi, T., Wilkes, M., McDermott, H., Nzietchueng, S., Gafarasi, I., Mudakikwa, A., et al. (2017). Implementing One Health as an integrated approach to health in Rwanda. *BMJ Global Health*, 2(1), e000121.

Observatorio Nacional de Salud. (2016). Barrio Adentro no puede considerarse un sistema de salud. Retrieved from http://www.ovsalud.org.

OECD (Organisation for Economic Co-operation and Development). (2013). *Cancer Care: Assuring Quality to Improve Survival*. Paris, France: OECD Publishing.

OECD (Organisation for Economic Co-operation and Development). (2015a). *Health at a Glance 2015—OECD Indicators*. Paris, France: OECD Publishing.

OECD (Organisation for Economic Co-operation and Development). (2015b). Reforming health and long-term care to ensure fiscal sustainability and improve service delivery. Retrieved from https://www.oecd.org/policy-briefs/slovenia-reforming-health-and-long-term-care.pdf.

OECD (Organisation for Economic Co-operation and Development). (2016a). *OECD Reviews of Health Care Quality: United Kingdom 2016: Raising Standards*. Paris, France: OECD Publishing.

OECD (Organisation for Economic Co-operation and Development). (2016b). Health policy in Chile: OECD health policy overview. Retrieved from http://www.oecd.org/chile/Health-Policy-in-Chile-February-2016.pdf.

OECD (Organisation for Economic Co-operation and Development). (2016c). Health at a glance: Europe 2016. Retrieved from http://www.oecd.org/health/health-at-a-glance-europe-23056088.htm.

OECD (Organisation for Economic Co-operation and Development). (2016d). Reviews of health systems: Mexico. Retrieved from https://www.oecd.org/health/health-systems/OECD-Reviews-of-Health-Systems-Mexico-2016-Assessment-and-recommendations-English.pdf.

Office of Health Standards Compliance [South Africa]. (2014). Overview on complaints management and the work of the ombud. Retrieved from http://www.ohsc.org.za/index.php/overview-on-complaints-management.

Office of Health Standards Compliance [South Africa]. (2017). Strategic plan 2015/16–2019/20. Retrieved from http://ohsc.org.za/images/publications/OHSC%20STRAT%20PLAN%202017%20web.pdf.

Okello, A. L., Gibbs, E. P., Vandersmissen, A., & Welburn, S. C. (2011). One Health and the neglected zoonoses: Turning rhetoric into reality. *Veterinary Record*, 169(11), 281–285.

Olesen, I. (2016). 'Det grønlandske sundhedsdilemma'—En kvalitativ undersøgelse om sundhedsopfattelse og selvvurderet helbred i en grønlandsk kontekst ['The Greenlandic health dilemma'—A qualitative study about health understandings and self-estimated health in a Greenlandic context]. *Tikiusaaq*, 24(3), 15–17.

Oleszczyk, M., Svab, I., Seifert, B., Krzton-Krolewiecka, A., & Windak, A. (2012). Family medicine in post-communist Europe needs a boost. Exploring the position of family medicine in healthcare systems of Central and Eastern Europe and Russia. *BMC Family Practice*, 13(1), 15.

Oman Information Center. (n.d.). Primary healthcare in Sultanate of Oman. Retrieved from http://www.omaninfo.com/health/primary-health-care-sultanate-oman.asp.

O'Neill, J. (2016). Tackling drug-resistant infections globally: Final report and recommendations. Retrieved from https://amr-review.org/sites/default/files/160518_Final%20paper_with%20cover.pdf.

Organización Panamericana de la Salud. (2001). Recursos Humanos en Salud en Argentina—2001. Retrieved from http://publicaciones.ops.org.ar/publicaciones/coleccionOPS/pub/pub53.pdf.

Ottersen, T., Førde, R., Kakad, M., Kjellevold, A., Melberg, H. O., Moen, A., Ringard, Å., & Norheim O. F. (2016). A new proposal for priority setting in Norway: Open and fair. *Health Policy*, 120(3), 246–251.

Øvretveit, J. & Keel, G. (2014). *Full Evidence Review of the Swedish Rheumatology Quality Register and Care and Learning System*. Stockholm, Sweden: Medical Management Center, Karolinska Institutet.

Øvretveit, J., Keller, C., Hvitfeldt Forsberg, H., Essén, A., Lindblad, S., & Brommels, M. (2013). Continuous innovation: The development and use of the Swedish rheumatology register to improve the quality of arthritis care. *International Journal for Quality in Health Care*, 25(2), 118–124.

Øvretveit, J., Nelson, E., & James, B. (2016). Building a learning health system using clinical registers: A non-technical introduction. *Journal of Health Organization and Management*, 30(7), 1105–1118.

Oxford Business Group. (2015). PNG's regulators improve ICT performance. Country report: New Guinea. Retrieved from https://oxfordbusinessgroup.com/overview/pngs-regulators-improve-ict-performance.

Oxford Policy Management. (2016). Improving access to maternal health care in Sheikhupura district, Pakistan. Retrieved from https://medium.com/@OPMglobal/improving-access-to-maternal-health-care-in-sheikhupura-district-pakistan-d94c216a97aa.

Pagès-Puigdemont, N., Mangues, M. A., Masip, M., Gabriele, G., Fernández, L., Blancafort, S., & Tuneu, L. (2016). Patients' perspective of medication adherence in chronic conditions: A qualitative study. *Advances in Therapy*, 33(10), 1740–1754.

Pathmanathan, I. & Liljestrand, J. (2002). Investing in maternal health: Learning from Malaysia and Sri Lanka. Washington DC: World Bank Publications. Retrieved from http://documents.worldbank.org/curated/en/367761468760748311/pdf/2 59010REPLACEM10082135362401PUBLIC1.pdf.

Pedersen, M. L. (2012). Diabetes mellitus in Greenland: Prevalence, organisation and quality in the management of type 2 diabetes mellitus. Effect of a diabetes health care project. *Danish Medical Journal*, 59(2), 1–21.

Pedersen, R. & Førde, R. (2005). Hva gjør de kliniske etikkomiteene? [What does the clinical ethics committee do?]. *Tidsskrift for den Norske Lægeforening*, 125, 3127–3129.

Peilin, L. (2015). *Society of China Analysis and Forecast (2016)*. Beijing, China: Social Sciences Academic Press.

Pelzang, R. (2010). Time to learn: Understanding patient-centred care. *British Journal of Nursing*, 19(14), 912–917.

Phommasack, B., Jiraphongsa, C., Ko Oo, M., Bond, K. C., Phaholyothin, N., Suphanchaimat, R., et al. (2013). Mekong basin disease surveillance (MBDS): A trust-based network. *Emerging Health Threats Journal*, 6, 19944.

Pieber, K., Stamm, T. A., Hoffmann, K., & Dorner, T. E. (2015). Synergistic effect of pain and deficits in ADL towards general practitioner visits. *Family Practice*, 32(4), 426–430.

Põlluste, K., Kalda, R., & Lember, M. (2017). Primary healthcare reform as a promoter of quality in the Estonian healthcare system. In J. Braithwaite, R. Mannion, Y. Matsuyama, P. Shekelle, S. Whittaker, & S. Al-Adawi (Eds.), *Health Systems Improvement Across the Globe: Success Stories from 60 Countries*. Boca Raton, FL: Taylor & Francis.

Pons, C. (2017). Venezuela 2016 inflation hits 800 percent, GDP shrinks 19 percent: Document. *Reuters*, January 21. Retrieved from www.reuters.com.

Population Reference Bureau. (2017). 2016 world population data sheet. Retrieved from http://www.prb.org/pdf16/prb-wpds2016-web-2016.pdf.

Portuguese Alliance for the Preservation of the Antibiotic. (2011). Memorandum, November 18. Retrieved from https://www2.arsalgarve.min-saude.pt/portal/sites/default/files//images/centrodocs/dia_europeu_antibioticos_2013/memorando_APAPA_dia_antibiotico_2013.pdf.

Poushter, J. (2016). Smartphone ownership and Internet usage continues to climb in emerging economies, but advanced economies still have higher rates of technology use. *Pew Research Center*. Retrieved from http://www.pewglobal.org/2016/02/22/smartphone-ownership-and-internet-usage-continues-to-climb-in-emerging-economies/.

Powell, M., Greener, I., Szmigin, I., Doheny, S., & Mills, N. (2010). Broadening the focus of public service consumerism. *Public Management Review*, 12(3), 323–339.

Presidential Commission. (2010). Informe final: Comisión presidencial de salud. Retrieved from http://www.minsal.cl/portal/url/item/96c1350fbf1a856ce040 01011f015405.pdf.

Presidential Commission. (2014). Informe final: Comisión asesora presidencial para el estudio y propuesta de un nuevo régimen jurídico para el sistema de salud privado. Retrieved from https://www.researchgate.net/publication/303255601_INFORME_FINAL_COMISION_ASESORA_PRESIDENCIAL_PARA_EL_ESTUDIO_Y_PROPUESTA_DE_UN_NUEVO_REGIMEN_JURIDICO_PARA_EL_SISTEMA_DE_SALUD_PRIVADO_2014.

Price, C. (2015). Revealed: NHS funding splashed on holidays, games consoles and summer houses. *Pulse*, September 1. Retrieved from http://www.pulsetoday.co.uk/news/commissioning-news/revealed-nhs-funding-splashed-on-holidays-games-consoles-and-summer houses/20010960.article#.VeasfpZIgYA.

Prioriteringsutvalget. (2014). Åpent og rettferdig—prioriteringer i helsetjenesten [Open and fair—Priority setting in the health service]. Official Norwegian Reports 2014:12. Oslo: Departementenes sikkerhets- og serviceorganisasjon.

Program PRO. (2016). Use of PRO-data in quality improvement in Danish healthcare: Recommendations and knowledge base. Retrieved from https://danskepatienter.dk/files/media/Publikationer%20-%20Egne/B_ViBIS/A_Rapporter%20og%20unders%C3%B8gelser/program_pro-rapport.pdf.

PROMcenter. (2015). The national quality registries. Retrieved from http://rcso.se/patientmedverkan/promsv/prom-ikvalitetsregister/.

Provan, K. G. & Kenis, P. (2007). Modes of network governance: Structure, management, and effectiveness. *Journal of Public Administration Research and Theory*, 18, 229–252.

Quintero, G. A. (2014). Medical education and the healthcare system: Why does the curriculum need to be reformed? *BMC medicine*, 12(1), 213.

Qureshi, Q. A., Nawaz, A., Khan, I. U., Waseem, M., & Muhammad, F. (2014). E-Readiness: A crucial factor for successful implementation of e-health projects in developing countries like Pakistan. *Public Policy and Administration Research*, 4(8), 97–103.

Ramadan, M. (2012). Preliminary meeting on "Is there a place for IAGG in the Middle East?" September 22. Doha, Qatar: International Association for Gerontology and Geriatrics.

Ramones, M. (2016). Alrededor de 15 mil médicos se han ido de Venezuela en los últimos años: FMV. *Panorama*, June 6. Retrieved from http://www.panorama.com.ve.

Ranchod, S., Erasmus, D., Abraham, M., Bloch, J., Chigiji, K., & Dreyer, K. (2016). Effective health financing models in SADC: Three case studies. Retrieved from https://docgo.org/effective-health-financing-models-in-sadc-three-case-studies.

Ranganathan, M. & Lagarde, M. (2012). Promoting healthy behaviours and improving health outcomes in low and middle income countries: A review of the impact of conditional cash transfer programmes. *Preventive Medicine*, 55, S95–S105.

Rapport, F., Clay-Williams, R., Churruca, K., Shih, P., Hogden, A., & Braithwaite, J. (2017). The struggle of translating science into action: Foundational concepts of implementation science. *Journal of Evaluation in Clinical Practice*. Advance online publication. doi: 10.1111/jep.12741.

Ravichandran, J. & Ravindran, J. (2014). Lessons from the confidential enquiry into maternal deaths, Malaysia. *International Journal of Obstetrics and Gynaecology*, 121(Suppl. 4), 47–52.

Ravindran, J., Shamsuddin, K., & Selvaraju, S. (2003). Did we do it right? An evaluation of the colour coding system for antenatal care in Malaysia. *Medical Journal of Malaysia*, 58(1), 37–53.

Raymont, A., Cumming, J., & Gribben, B. (2013). Evaluation of the primary health care strategy: Changes in fees and consultation rates between 2001 and 2007. Retrieved from http://www.victoria.ac.nz/sog/researchcentres/health-services-research-centre/publications/reports.

Rechel, B., Ahmedov, M., Akkazieva, B., Katsaga, A., Khodjamurodov, G., & McKee, M. (2012). Lessons from two decades of health reform in Central Asia. *Health Policy and Planning*, 27(4), 281–287.

Rechel, B. & McKee, M. (2009). Health reform in central and eastern Europe and the former Soviet Union. *Lancet*, 374(9696), 1186–1195.

Rechel, B., Roberts, B., Richardson, E., Shishkin, S., Shkolnikov, V. M., Leon, D. A., et al. (2013). Health and health systems in the Commonwealth of Independent States. *Lancet*, 381(9872), 1145–1155.

Redgrave, L. S., Sutton, S. B., Webber, M. A., & Piddock, L. (2014). Fluoroquinolone resistance: Mechanisms, impact on bacteria, and role in evolutionary success. *Trends in Microbiology*, 22, 438–45.

Reichard, S. (1996). Ideology drives health care reforms in Chile. *Journal of Public Health and Policy*, 17(1), 80–98.

Reichman, M. (2007). Optimizing referrals and consults with a standardized process. *Family Practice Management*, 14(10), 38–42.

Republic of Rwanda. (2013). Economic development and poverty reduction strategy II: 2013–2018. Retrieved from http://www.rdb.rw/uploads/tx_sbdownloader/EDPRS_2_Main_Document.pdf.

Republic of Rwanda Ministry of Health. (2014). One Health strategic plan 2014–2018. Retrieved from http://www.rbc.gov.rw/IMG/pdf/one_health.pdf.

Richards, T. (2017). Tessa Richards: Better together in Central and Eastern Europe. *BMJ Opinion*, June 14. Retrieved from http://blogs.bmj.com/bmj/2017/06/14/tessa-richards-better-together-in-central-and-eastern-europe/.

Ricketts, T. C., Adamson, W. T., Fraher, E. P., Knapton, A., Geiger, J. D., Abdullah, F., & Klein, M. D. (2017). Future supply of pediatric surgeons: Analytical study of the current and projected supply of pediatric surgeons in the context of a rapidly changing process for specialty and subspecialty training. *Annals of Surgery*, 265(3), 609–615.

Rigsby, D. (2011). The anatomy of poor health care [Letter to the editor]. *Trinidad Express*, March 20. Retrieved from http://www.trinidadexpress.com/letters/The_anatomy_of_poor_health_care-118342819.html.

Ringard, Å., Mørland, B., & Røttingen, J. A. (2010). Åpne prosesser for prioritering [Open and fair processes for priority setting]. *Tidsskrift for den Norske Lægeforening*, 130(22), 2264–2266.

Ringard, Å., Sagan, A., Saunes, I. S., & Lindahl, A. K. (2013). Norway: Health systems review. *Health Systems in Transition*, 15(8), 1–164.

Rodriguez, J. (2016). Community-based mental health model: Conceptual framework and experiences. Presented at the Mental Health Situation in Guyana, October 17, Georgetown, Guyana.

Rodriguez-Manas, L. & Fried, L. P. (2015). Frailty in the clinical scenario. *Lancet*, 385(9968), e7–e9.

Roemer, M. I. (1991). *National Health Systems of the World, Vol. I, The Countries*. New York, NY: Oxford University Press.

Ross, C. & Swetlitz, I. (2017). IBM pitched its Watson supercomputer as a revolution in cancer care. It's nowhere close. *STAT News*, September 5. Retrieved from https://www.statnews.com/2017/09/05/watson-ibm-cancer/.

Rotheram-Borus, M. J., Le Roux, I. M., Tomlinson, M., Mbewu, N., Comulada, W. S., Le Roux, K., et al. (2011). Philani Plus (+): A mentor mother community health worker home visiting program to improve maternal and infants' outcomes. *Prevention Science*, 12(4), 372–388.

Rubinstein, A. (2017). Informe del Ministerio de Salud a marzo 2017. *Jornada FeMeBA*, June 8.

Runciman, W., Hibbert, P., Thomson, R., Van Der Schaaf, T., Sherman, H., & Lewalle, P. (2009). Towards an international classification for patient safety: Key concepts and terms. *International Journal for Quality in Health Care*, 21(1), 18–26.

Saam, M., Huttner, B., & Harbarth, S. (2017). Evaluation of antibiotic awareness campaigns. Retrieved from http://www.who.int/selection_medicines/committees/expert/21/applications/antibacterials-ccps_rev/en/.

Saar Poll. (2016). E-konsultatsiooni barjäärianalüüsi läbiviimine. Retrieved from https://www.haigekassa.ee/sites/default/files/TTL/e_konsultatsioon/e-konsultatsiooni_barjaarianaluus_aruanne_final_pdf.pdf.

Sabel, C. F. & Zeitlin, J. (2012). Experimentalist governance. In D. Levi-Faur (Ed.), *The Oxford Handbook of Governance*. Oxford, UK: Oxford University Press.

Sacanella, E., Pérez-Castejón, J. M., Nicolás, J. M., Masanés, F., Navarro, M., Castro, P., & López-Soto, A. (2001). Functional status and quality of life 12 months after discharge from a medical ICU in healthy elderly patients: A prospective observational study. *Critical Care*, 15, R105.

Sack, C., Scherag, A., Lutkes, P., Gunther, W., Jockel, K. H., & Holtmann, G. (2011). Is there an association between hospital accreditation and patient satisfaction with hospital care? A survey of 37,000 patients treated by 73 hospitals. *International Journal for Quality in Health Care*, 23(3), 278–283.

Sakunphanit, T. (2006). *Universal Health Care Coverage through Pluralistic Approaches: Experience from Thailand*. Bangkok: ILO Subregional Office for East Asia.

Salas, E., Diaz Granados, D., Klein, C., Burke, C. S., Stagl, K. C., Goodwin, G. F., & Halpin, S. M. (2008). Does team training improve team performance? A meta-analysis. *Human Factors*, 50(6), 903–933.

Saleh, S., Khodor, R., Alameddine, M., & Baroud, M. (2016). Readiness of healthcare providers for eHealth: The case from primary healthcare centers in Lebanon. *BMC Health Services Research*, 16(1), 644.

Sanabria, T. & Orta M. (2012). The MANIAPURE Program—Lessons learned from a rural experience: Two decades delivering primary healthcare through telemedicine. *Telemedicine and e-Health*, 18, 1–5.

Sanyahumbi, A. S., Sable, C. A., Karlsten, M., Hosseinipour, M. C., Kazembe, P. N., Minard, C. G., & Penny, D. J. (2017). Task shifting to clinical officer-led echocardiography screening for detecting rheumatic heart disease in Malawi, Africa. *Cardiology in the Young*, 27(6), 1133–1139.

Schwandt, T. A. (2005). The centrality of practice to evaluation. *American Journal of Evaluation*, 26(1), 95–105.

Scottish Government. (2016). Health and social care delivery plan. Edinburgh: Scottish Government. Retrieved from http://www.gov.scot/Resource/0051/00511950.pdf.

Scottish Health Council. (2017). Gathering views and experience of maternity and neonatal services. Glasgow: Scottish Health Council. Retrieved from http://www.scottishhealthcouncil.org/publications/gathering_public_views/maternity_and_neonatal_review.aspx#.Wngy3Pll_IV.

Scott-Samuel, A. (2015). Personal health budgets in England: Mood music or death knell for the National Health Service? *International Journal of Health Services*, 45(1), 73–86.

Secretaría de Gobernación. (2016). Diario Oficial de Federación: Presupuesto de Egresos de la Federación para el Ejercicio Fiscal 2016. Retrieved from http://www.dof.gob.mx/nota_detalle.php?codigo=5463187&fecha=30/11/2016.

Seifert, B., Svab, I., Madis, T., Kersnik, J., Windak, A., Steflova, A., et al. (2008). Perspectives of family medicine in Central and Eastern Europe. *Family Practice*, 25(2), 113–118.

Semrau, M., Lempp, H., Keynejad, R., Evans-Lacko, S., Mugisha, J., Raja, S., et al. (2016). Service user and caregiver involvement in mental health system strengthening in low- and middle-income countries: Systematic review. *BMC Health Services Research*, 16, 79.

Senge, P. (1990). *The Fifth Discipline: The Art and Practice of the Learning Organization*. New York, NY: Doubleday.

Sensor Market Research. (2014). Einstellung zum derzeitigen Primärversorgungs-System: Gruppendiskussion mit MedizinstudentInnen [Study commissioned by the Main Association of Social Security Institutions]. Retrieved from http://www.hauptverband.at/cdscontent/load?contentid=10008.603049&version=1414058399.

Seppänen, A. (2016). Finnish. Virtuaa¬lik¬li¬nikka muuttaa sairaalatyötä [Virtual clinic changes hospital work]. *Finnish Medical Journal (Suomen Lääkärilehti)*, 16, 1118–1121.

Shah, R. (2015). Medical system a mess. *Trinidad and Tobago News Blog*, January 11. Retrieved from http://www.trinidadandtobagonews.com/blog/?p=8556.

Shahebrahimi, S. (2016). Evaluation of hospitals' performance in Tehran province using data mining and data envelopment analysis. Unpublished thesis, Iran University of Science and Technology, Tehran.

Shahraz, S., Forouzanfar, M., Sepanlou, S., Dicker, D., Naghavi, P., Pourmalek, F., et al. (2014). Population health and burden of disease profile of Iran among 20 countries in the region: From Afghanistan to Qatar and Lebanon. *Archive of Iranian Medicine*, 17(5), 336–342.

Shead, S. (2017). Investors backed an AI startup that puts a doctor on your smartphone with $60 million. *Business Insider*, April 25. Retrieved from http://www.businessinsider.com/babylon-raises-60-million-for-its-doctor-in-your-pocket-app-2017-4?r=UK&IR=T.

Sheikh, A., Panesar, S. S., Larizgoitia, I., Bates, D. W., & Donaldson, L. J. (2013). Safer primary care for all: A global imperative. *Lancet Global Health*, 1(4), e182–e183.

Sheppard, C. (2010). The development of a competency model for head nurses at the Port-of-Spain General Hospital. Unpublished thesis, University of Trinidad and Tobago.

Sherbourne, C. D., Edelen, M. O., Zhou, A., Bird, C., Duan, N., & Wells, K. B. (2008). How a therapy-based quality improvement intervention for depression affected life events and psychological well-being over time: A 9-year longitudinal analysis. *Medical Care*, 46(1), 78–84.

Shi, L. (2012). The impact of primary care: A focused review. *Scientifica*, 2012, 432892. doi:10.6064/2012/432892.

Shojania, K. G. & Thomas, E. J. (2013). Trends in adverse events over time: Why are we not improving? *BMJ Quality & Safety*, 22, 273–277.

Siebolds, M., Münzel, B., Müller, R., Häußermann, S., Paul, M., & Kahl, C. (2016). [Comprehensive implementation of interprofessional quality circles regarding early prevention of childhood disadvantage in Baden Wurttemberg (Germany)]. *Bundesgesundheitsblatt Gesundheitsforschung Gesundheitsschutz*, 59(10), 1310–1314.

Siemieniuk, R. A. C., Harris, I. A., Agoritsas, T., Poolman, R. W., Brignardello-Petersen, R., Van de Velde, S., et al. (2017). Arthroscopic surgery for degenerative knee arthritis and meniscal tears: A clinical practice guideline. *The BMJ*, 357, j1982.

Simmons, R., Powell, M., & Greener, I. (Eds.). (2009). *The Consumer in Public Services*. Bristol: Policy Press.

Sitra. (n.d.). Well-being data. Retrieved from https://www.sitra.fi/en/topics/well-being-data/.

Skatte-og Velfærdskommissionen [Tax and Welfare Commission]. (2010). Hvordan sikres vækst og velfærd i Grønland? Baggrundsrapport [How to secure growth and welfare in Greenland? Background report]. Retrieved from http://naalakkersuisut.gl/~/media/Nanoq/Files/Attached%20Files/Finans/DK/Skatte%20og%20velfaerdskommissionen/Baggrundsrapporten.pdf.

Sloan, F. A. (1970). Lifetime earnings and physicians' choice of specialty. *ILR Review*, 24(1), 47–56.

Smith, J. (2009). *Critical Analysis of the Primary Health Care Strategy and Framing of Issues for the Next Phase*. Wellington, New Zealand: Ministry of Health.

Smith, J. P. (2007). The impact of socioeconomic status on health over the life-course. *Journal of Human Resources*, 42(4), 739–764.

Snowdon, A. & Alessi, C. (2016). Visibility: The new value proposition for health systems. Ontario, Canada: World Health Innovation Network. Retrieved from https://issuu.com/worldhealthinnovationnetwork/docs/full_paper_-_win_visibility_thought/1?e=25657717/39177387.

Snowdon, A., Schnarr, K., & Alessi, C. (2014). It's all about me: The personalization of health systems. *Ivey International Centre for Health Innovation*. Retrieved from http://sites.ivey.ca/healthinnovation/files/2014/02/Its-All-About-Me-The-Personalization-of-Health-Systems.pdf.

Sørensen, K., Van den Broucke, S., Fullam, J., Doyle, G., Pelikan, J., Slonska, Z., et al. (2012). Health literacy and public health: A systematic review and integration of definitions and models. *BMC Public Health*, 12, 80.

Sousa, A., Scheffler, R. M., Nyoni, J., & Boerma, T. (2013). A comprehensive health labour market framework for universal health coverage. *Bulletin of the World Health Organization*, 91, 892–894. Retrieved from http://www.who.int/bulletin/volumes/91/11/13-118927.pdf.

South East Asia Infectious Disease Clinical Research Network. (2013). Effect of double dose oseltamivir on clinical and virological outcomes in children and adults admitted to hospital with severe influenza: Double blind randomised controlled trial. *BMJ*, 346. doi: https://doi.org/10.1136/bmj.f3039.

South East Asia Infectious Disease Clinical Research Network. (n.d.a). Profile. Retrieved from http://www.seaicrn.org/Infobox.aspx?pageID=1.

South East Asia Infectious Disease Clinical Research Network. (n.d.b). Research. Retrieved from http://www.seaicrn.org/info.aspx?pageID=102.

Spindler, E., Shattuck, D., & Menstell, E. (2016). Population, plateaus, and geopolitics: What's up with the fertility stall in Jordan? Washington DC: Georgetown University, Institute for Reproductive Health. Retrieved from http://irh.org/blog/population-plateaus-geopolitics-whats-fertility-stall-jordan/.

Sprangers, M. A. G., Cull, A., Bjordal, K., Grønvold, M., & Aaronson, N. K. (1993). The European Organization for Research and Treatment of Cancer Approach to Quality of Life Assessment: guidelines for developing questionnaire modules. *Quality of Life Research*, 2(4), 287–295.

Staines, A., Bezzola, P., & Albisetti, P. (2017). Switzerland's use of breakthrough collaboratives to improve patient safety. In J. Braithwaite, R. Mannion, Y. Matsuyama, P. Shekelle, S. Whittaker, & S. Al-Adawi (Eds.), *Health Systems Improvement Across the Globe: Success Stories from 60 Countries*. Boca Raton, FL: Taylor & Francis.

Starfield, B. (1994). Is primary care essential? *Lancet*, 344(8930), 1129–1133.

Starfield, B. (2004). Summing up: Primary health care reform in contemporary health care systems. In R. Wilson, S. E. D. Shortt, & J. Dorland (Eds.), *Implementing Primary Care Reform: Barriers and Facilitators*. Montréal, Canada: McGill-Queen's University Press.

Starfield, B. & Shi, L. (2002). Policy relevant determinants of health: An international perspective. *Health Policy*, 60(3), 201–218.

Starfield, B., Shi, L., & Macinko, J. (2005). Contribution of primary care to health systems and health. *Milbank Quarterly*, 83(3), 457–502.

State Council [China]. (2015). Guide to advance the integrated stratified healthcare system. Retrieved from http://www.nhfpc.gov.cn/yzygj/s3593g/201509/c30041e1016a427f9477774c9e864eb4.shtml.

Statistisches Bundesamt. (2017). Koordinierte Bevölkerungsvorausberechnung nach Bundesländern. Retrieved from https://service.destatis.de/laenderpyramiden.

Stevanovic, V. & Fujisawa, R. (2011). Performance of systems of cancer care in OECD countries: Exploration of the relation between resources, process quality, governance and survival in patients with breast, cervical, colorectal and lung cancers. Presented at HCQI Expert Group Meeting, Paris, France, May 27. Retrieved from http://www.oecd.org/els/health-systems/48098832.pdf.

Stevens, G., Dias, R. H., Thomas, K. J. A., Rivera, J. A., Carvalho, N., Barquera, S., et al. (2008). Characterizing the epidemiological transition in Mexico: National and subnational burden of diseases, injuries, and risk factors. *PLoS Medicine*, 5(6), 0900–0910.

Stevenson Rowan, M., Hogg, W., & Huston, P. (2007). Integrating public health and primary care. *Health Policy*, 3(1), e160–e181.

Stoddard, J. J., Cull, W. L., Jewett, E. A., Brotherton, S. E., Mulvey, H. J., & Alden, E. R. (2000). Providing pediatric subspecialty care: A workforce analysis. *Pediatrics*, 106(6), 1325–1333.

Sudat, S. E., Franco, A., Pressman, A. R., Rosenfeld, K., Gornet, E., & Stewart, W. (2017). Impact of home-based, patient-centered support for people with advanced illness in an open health system: A retrospective claims analysis of health expenditures, utilization, and quality of care at end of life. *Palliative Medicine*, 32(2), 485–492.

Suñer, R. & Santiñà, M. (Eds.). (2016). *Health Literacy: Standards and Recommendations for Health Professionals*. Girona: Documenta Universitaria.

Svab, I., Pavlic, D. R., Radic, S., & Vainiomaki, P. (2004). General practice east of Eden: An overview of general practice in Eastern Europe. *Croatian Medical Journal*, 45(5), 537–542.

Swain, R. (2017). Assessment centres. Retrieved from https://www.prospects.ac.uk/careers-advice/interview-tips/assessment-centres.

Swedish Agency for Growth Policy Analysis. (2013). China's healthcare system: Overview and quality improvement. Retrieved from https://www.tillvaxtanalys.se/download/18.5d9caa4d14d0347533bcf93a/1430910410539/direct_response_2013_03.pdf.

Taylor, D. H., Jr., Ostermann, J., Van Houtven, C. H., Tulsky, J. A., & Steinhauser, K. (2007). What length of hospice use maximizes reduction in medical expenditures near death in the US Medicare program? *Social Science & Medicine*, 65, 1466–1478.

Tekes. (2016). Finland and IBM partner to develop personalized healthcare and spark economic growth with Watson. Retrieved from https://www.tekes.fi/en/whats-going-on/news-2016/finland-and-ibm-partner-to-develop-personalized-healthcare-and-spark-economic-growth-with-watson/.

Temperton, J. (2016). NHS care.data scheme closed after years of controversy. *Wired*, July 6. Retrieved from http://www.wired.co.uk/article/care-data-nhs-england-closed.

The Economist. (2017). How Chávez and Maduro have impoverished Venezuela, April 6. Retrieved from https://www.economist.com/news/finance-and-economics/21720289-over-past-year-74-venezuelans-lost-average-87kg-weight-how.

Thomas, D. R. (2006). A general inductive approach for analyzing qualitative evaluation data. *American Journal of Evaluation*, 27(2), 237–246.

Thompson, A. & Steel, D. (2015). Scotland. In J. Braithwaite, Y. Matsuyama, J. Johnson, & R. Mannion (Eds.), *Healthcare Reform, Quality and Safety: Perspectives, Participants, Partnerships and Prospects in 30 Countries* (pp. 261–272). Farnham, UK: Ashgate.

Thompson, A. & Steel, D. (2017). Partnership and collaboration as the hallmark of Scottish healthcare improvement. In J. Braithwaite, R. Mannion, Y. Matsuyama, P. Shekelle, S. Whittaker, & S. Al-Adawi (Eds.), *Health Systems Improvement Across the Globe: Success Stories from 60 Countries* (pp. 243–248). Boca Raton, FL: Taylor & Francis.

Thornicroft, G., Deb, T., & Henderson, C. (2016). Community mental health care worldwide: Current status and further developments. *World Psychiatry*, 15(3), 276–286.

Thyen, U. (2010). *Bundesgesundheitsblatt Gesundheitsforschung Gesundheitsschutz* [Early childhood intervention—Relationship management with risks], 53(11), 1117–1118.

Tinetti, M. E. & Fried, T. (2004). The end of the disease era. *American Journal of Medicine*, 116(3), 179–185.

Topol, E. (2017). The future of medicine is in your hands. Lecture at Arizona State University, W. P. Carey School of Business. Retrieved from https://vimeo.com/wpcoas/review/211748247/ea6ec69fc9.

Tragakes, E., Brigis, G., Karaskevica, J., Rurane, A., Stuburs, A., & Zusmane, E. (2008). Latvia: Health system review. *Health Systems in Transition*, 10(2), 1–251.

Trinidad Express. (2014). Manager defends nurses against claims, December 21. Retrieved from http://www.trinidadexpress.com/news/Manager-defends-nurses-against-claims-286526481.html.

Truitt, F. E., Pina, B. J., Person-Rennell, N. H., & Angstman, K. B. (2013). Outcomes for collaborative care versus routine care in the management of postpartum depression. *Quality in Primary Care*, 21(3), 171–177.

Tsilaajav, T., Nanzad, O., & Ichinnorov, E. (2015). *Analysis of Catastrophic Health Payments and Benefit Incidence of Government Spending for Health in Mongolia, 2015*. Geneva, Switzerland: World Health Organization.

Tsilaajav, T., Ser-Od, E., Baasai, B., Byamba, G., & Shagdarsuren, O. (2013). Mongolia health system review. *Health Systems in Transition*, 3(2), 1–162.

Tsouros, A. D. (1995). The WHO Healthy Cities Project: State of the art and future plans. *Health Promotion International*, 10(2),133–141.

Turkish Great National Assembly Plan. (2016). Turkish Great National Assembly Plan and Budget Commission. *Journal of Minutes*, November 14. Retrieved from https://www.tbmm.gov.tr/develop/owa/komisyon_tutanaklari.goruntule?pTutanakId=1780.

Tyson, A. (2107). *The Decision*. Sydney, Australia: Film Buff Productions.

Ulikpan, A., Mirzoev, T., Jimenez, E., Malik, A., & Hill, P. S. (2014). Central Asian Post-Soviet health systems in transition: Has different aid engagement produced different outcomes? *Global Health Action, 7,* 24978.

UNAIDS (Joint United Nations Programme on HIV and AIDS). (1983). The Denver principles. Retrieved from http://data.unaids.org/pub/externaldocument/2007/gipa1983denverprinciples_en.pdf.

UNAIDS (Joint United Nations Programme on HIV and AIDS). (2007). UNAIDS policy brief: The greater involvement of people living with HIV (GIPA). Retrieved from http://data.unaids.org/pub/briefingnote/2007/jc1299_policy_brief_gipa.pdf.

United Nations. (2004). Aging in the Arab countries: Regional variation, policies and programs. Economic and Social Commission for Western Asia. Retrieved from http://www.monitoringris.org/documents/strat_reg/escwa_ageing_arab_countries.pdf.

United Nations. (2017a).Goal 3: Ensure healthy lives and promote well-being for all at all ages. Retrieved from http://www.un.org/sustainabledevelopment/health/.

United Nations. (2017b). Millennium development goals and beyond 2015. Retrieved from http://www.un.org/millenniumgoals/.

United Nations. (2017c). UN data: United Arab Emirates. Retrieved from http://data.un.org/CountryProfile.aspx?crName=United%20Arab%20Emirates.

United Nations, Department of Economic and Social Affairs, Population Division. (2015). World population prospects: The 2015 revision, key findings and advance tables. Working Paper No. ESA/P/WP.241. New York, NY: United Nations.

United Nations Development Programme. (2014). Human development report. Retrieved from http://hdr.undp.org/en/2015-report.

United Nations Development Programme. (2015). Informe sobre Desarrollo Humano en Chile 2015. Los tiempos de la politización [Human development report in Chile 2015. The times of politicization. Human Development Program]. Programa de Desarrollo Humano. Retrieved from http://hdr.undp.org/sites/default/files/informe_2015.pdf.

United Nations Development Programme. (2017a). DESIGUALES Orígenes, cambios y desafíos de la brecha social en Chile [DESIGUALES Origins, changes and challenges of the social gap in Chile]. Programa de las Naciones Unidas Para el Desarrollo. Retrieved from http://www.cl.undp.org/content/chile/es/home/presscenter/pressreleases/2017/06/14/estudio-del-pnud-revela-seis-nudos-de-reproducci-n-y-cambio-de-la-desigualdad-en-chile.html.

United Nations Development Programme. (2017b). Human development report: Oman. Retrieved from http://hdr.undp.org/en/countries/profiles/OMN.

United Nations Development Programme. (2017c). *Lebanon Stabilization and Recovery Programme 2017.* Beirut, Lebanon: United Nations Development Programme.

Unit for Evaluation and User Involvement. (2016). The national survey of patient experiences. Retrieved from http://patientoplevelser.dk/sites/patientoplevelser.dk/files/dokumenter/filer/LUP/2015/lup_national_rapport_2015.pdf.

Universidad Católica Argentina. (2016). Tiempo de Balance; deudas sociales pendientes al final del Bicentenario. ODSA-UCA. Barómetro de Deuda Social. Serie del Bicentenario. Buenos Aires: Universidad Católica Argentina.

UWI Campus News. (2009). UWI launches master's in advanced nursing, April 16. Retrieved from http://sta.uwi.edu/news/releases/release.asp?id=265.

van den Berg, N., Beyer, A., & Hoffmann, W. (2017). Projektbericht delegation ärztlicher aufgaben an nichtärztliche gesundheitsberufe in der pädiatrischen versorgung - Bedarfsentwicklung und konzeptentwicklung. Retrieved from http://www2.medizin.uni-greifswald.de/icm/index.php?id=326.

van den Berg, N., Beyer, A., Stentzel, U., & Hoffmann, W. (2016). Versorgungsepidemiologische analyse der kinder- und jugendmedizin in Deutschland (Expertise im auftrag des DAKJ e.V. - 2016). Retrieved from http://www2.medizin.uni-greifswald.de/icm/index.php?id=326.

van den Berg, N., Fiss, T., Meinke, C., Heymann, R., Scriba, S., & Hoffmann, W. (2009). GP-support by means of AGnES-practice assistants and the use of telecare devices in a sparsely populated region in northern Germany—Proof of concept. *BMC Family Practice, 10*, 44.

van den Berg, N., Meinke, C., Matzke, M., Heymann, R., Flessa, S., & Hoffmann, W. (2010). Delegation of GP-home visits to qualified practice assistants: assessment of economic effects in an ambulatory healthcare centre. *BMC Health Services Research, 10*, 155.

van den Berg, N., Seidlitz, G., Meinke-Franze, C., Pieper, C., Lode, H., & Hoffmann, W. (2014). Auswirkungen des demografischen Wandels auf die kinder- und jugendmedizinische Versorgung in der Region Ostvorpommern. In F. Dünkel, M. Herbst, & T. Schlegel (Eds.), *Think rural! Dynamiken des wandels in peripheren ländlichen räumen und ihre implikationen für die daseinsvorsorge* (pp. 65–72). Weisbaden, Germany: Springer.

van Dijk, H. M., Cramm, J. M., & Nieboer, A. P. (2013). The experiences of neighbour, volunteer and professional support-givers in supporting community dwelling older people. *Health & Social Care in the Community, 21*(2), 150–158.

van Ginneken, E. & Kroneman, M. (2015). Long-term care reform in the Netherlands: Too large to handle? *Eurohealth, 21*(3), 47–50.

Van Lerberghe, W. (2008). *The World Health Report 2008: Primary Health Care: Now More than Ever.* Geneva, Switzerland: WHO.

Van Lerberghe, W., Ammar, W., el Rashidi, R., Sales, A., & Mechbal, A. (1997a). Reform follows failure: I. Unregulated private care in Lebanon. *Health Policy and Planning, 12*(4), 296–311.

Van Lerberghe, W., Ammar, W., el Rashidi, R., Awar, M., Sales, A., & Mechbal A. (1997b). Reform follows failure: II. Pressure for change in the Lebanese health sector. *Health Policy and Planning, 12*(4), 312–319.

Van Lerberghe, W., Matthews, Z., Achadi, E., Ancona, C., Campgell, J., Channon, A., et al. (2014). Country experience with strengthening of health systems and deployment of midwives in countries with high maternal mortality. *Lancet, 384*(9949), 1215–1225.

Venter, J. C., Adams, M. D., Myers, E. W., Li, P. W., Mural, R. J., Sutton, G. G., et al. (2001). The sequence of the human genome. *Science, 291*(5507), 1304–1351.

Verdecchia, A., Francisci, S., Brenner, H., Gatta, G., Micheli, A., Mangone, L., et al. (2007). Recent cancer survival in Europe: A 2000–02 period analysis of EUROCARE-4 data. *Lancet Oncology, 8*(9), 784–796.

Verver, D., Merten, H., Robben, P., & Wagner, C. (2017). Perspectives on the risks for older adults living independently. *British Journal of Community Nursing, 22*(7), 338–345.

Victora, C. G., Hanson, K., Bryce, J., & Vaughan, J. P. (2004). Achieving universal coverage with health interventions. *Lancet, 364*(9444), 1541–1548.

Vimalananda, V. G., Gupte, G., Seraj, S. M., Orlander, J., Berlowitz, D., Fincke, B. G., et al. (2015). Electronic consultations (e-consults) to improve access to specialty care: A systematic review and narrative synthesis. *Journal of Telemedicine and Telecare, 6*, 323–330.

Vincent, C. & Amalberti, R. (2016). *Safer Healthcare: Strategies for the Real World.* New York, NY: Springer.

Vision 2021. (n.d.). UAE Vision 2021. Retrieved from https://www.vision2021.ae/en/our-vision.

Vital Wave Consulting. (2009). *mHealth for Development: The Opportunity of Mobile Technology for Healthcare in the Developing World.* Washington DC: UN Foundation–Vodafone Foundation Partnership.

Warren, D. W., Jarvis, A., LeBlanc, L., Gravel, J., CTAS National Working Group, Canadian Association of Emergency Physicians, et al. (2008). Revisions to the Canadian Triage and Acuity Scale paediatric guidelines (PaedCTAS). *CJEM, 10*(3), 224–243.

Watts, J. (2016). 'Like doctors in a war': Inside Venezuela's healthcare crisis, October 19. *The Guardian.* Retrieved from https://www.theguardian.com.

Webber, M., Treacy, S., Carr, S., Clark, M., & Parker, G. (2014). The effectiveness of personal budgets for people with mental health problems: A systematic review. *Journal of Mental Health, 23*(3), 146–155.

Weckmann, M. T., Freund, K., Camden, B., & Broderick, A. (2012). Medical manuscripts impact of hospice enrollment on cost and length of stay of a terminal admission. *American Journal of Hospice and Palliative Care, 30*(6), 576–578. doi:10.1177/1049909112459368.

Wehrens, R., Oldenhof, L., Verweij, L., Francke, A., & Bal, R. (2017). *Experimenteel sturen in netwerken: een evaluatie van proces en structuur van het Nationaal Programma Ouderenzorg.* Rotterdam, the Netherlands: iBMG & Nivel.

Weiland, K. (2007). The puzzle of policy diffusion. In K. Weiland (Ed.), *Bounded Rationality and Policy Diffusion: Social Sector Reform in Latin America* (pp. 1–28). Princeton, NJ: Princeton University Press.

Wells, K. B., Sturm, R., Sherbourne, C. D., & Meredith, L. S. (1996). *Caring for Depression.* Cambridge, MA: Harvard University Press.

Wertheim, H. F. L., Puthavathana, P., Nghiem, N. M., van Doorn, H. R., Nguyen, T. V., Pham, H. V., et al. (2010). Laboratory capacity building in Asia for infectious disease research: Experiences from the South East Asia Infectious Disease Clinical Research Network (SEAICRN). *PLOS Medicine, 7*(4), e1000231.

WHO (World Health Organization). (1978). Primary health care: Report of the International Conference on Primary Health Care. Retrieved from http://www.who.int/publications/almaata_declaration_en.pdf.

WHO (World Health Organization). (1995). *Building a Healthy City: A Practitioner's Guide.* Geneva, Switzerland: WHO.

WHO (World Health Organization). (2000). *The World Health Report 2000—Health Systems: Improving Performance.* Geneva, Switzerland: WHO.

WHO (World Health Organization). (2002). *World Health Report 2002—Reducing Risks, Promoting Healthy Life.* Geneva, Switzerland: WHO.

WHO (World Health Organization). (2003). *Primary Health Care: A Framework for Future Strategic Directions.* Geneva, Switzerland: WHO.

WHO (World Health Organization). (2005). Joint external evaluation tool: International Health Regulations. Retrieved from http://apps.who.int/iris/bitstream/10665/204368/1/9789241510172_eng.pdf.

WHO (World Health Organization). (2006a). *Health System Profile—Yemen.* Geneva, Switzerland: WHO.

WHO (World Health Organization). (2006b). World health report 2006—Working together for health. Retrieved from http://www.who.int/whr/2006/whr06_en.pdf?ua=1.

WHO (World Health Organization). (2007). Provider payments and cost-containment. Lessons from OECD countries. Retrieved from http://www.who.int/health_financing/documents/pb_e_07_2-provider_payments.pdf.

WHO (World Health Organization). (2008a). First global conference on task shifting. Retrieved from http://www.who.int/mediacentre/events/meetings/task_shifting/en/.

WHO (World Health Organization). (2008b). The global burden of disease: 2004 update. Retrieved from http://www.who.int/entity/healthinfo/global_burden_disease/GBD_report_2004update_full.pdf?ua=1.

WHO (World Health Organization). (2010). Communicable diseases in the South-East Asia region of the World Health Organization: Towards a more effective response. *Bulletin of the World Health Organization.* Retrieved from http://www.who.int/bulletin/volumes/88/3/09-065540/en/.

WHO (World Health Organization). (2011). Global status report on non-communicable diseases. Retrieved from http://whqlibdoc.who.int/publications/2011/9789240686458_eng.pdf.

WHO (World Health Organization). (2012a). Dementia—A public health priority. Retrieved from http://www.who.int/mental_health/publications/dementia_report_2012/en/.

WHO (World Health Organization). (2012b). Severe hand, foot and mouth disease killed Cambodian children. Retrieved from http://www.wpro.who.int/mediacentre/releases/2012/20120713/en/.

WHO (World Health Organization). (2013). *Investing in Mental Health: Evidence for Action.* Geneva, Switzerland: WHO.

WHO (World Health Organization). (2014). Nigeria in the frontline—Council approves first national health workforce registry in Africa. Retrieved from http://www.afro.who.int/news/nigeria-frontline-council-approves-first-national-health-workforce-registry-africa.

WHO (World Health Organization). (2015a). Global Health Observatory data, antenatal care: Situation. Retrieved from http://www.who.int/gho/countries/mys.pdf?ua=1 on 18.2.2017.

WHO (World Health Organization). (2015b). *Mental Health Atlas 2014.* Geneva, Switzerland: WHO.

WHO (World Health Organization). (2015c). National health accounts. Retrieved from http://www.who.int/nha/en/.

WHO (World Health Organization). (2015d). Palliative care fact sheet no. 402. Retrieved from http://www.who.int/mediacentre/factsheets/fs402/en/.

WHO (World Health Organization). (2015e). Progress towards the Health 2020 targets: European health report 2015. Retrieved from http://www.euro.who.int/en/data-and-evidence/european-health-report/european-health-report-2015/chapter-2-progress-towards-the-health-2020-targets.

WHO (World Health Organization). (2015f). Regional strategy for patient safety in the WHO South-East Asia region (2016–2025). Retrieved from http://www.searo.who.int/entity/health_situation_trends/regional_strategy_for_patient_safety.pdf?ua=1.

WHO (World Health Organization). (2015g). United Arab Emirates: National expenditure on health. Retrieved from http://apps.who.int/nha/database.

WHO (World Health Organization). (2016a). European health for all database (HFA-DB): WHO/Europe. Retrieved from http://data.euro.who.int/hfadb/.

WHO (World Health Organization). (2016b). Life expectancy data by country. Retrieved from http://apps.who.int/gho/data/view.main.SDG2016LEXv?lang=en.

WHO (World Health Organization). (2016c). Nigeria develops new national health policy to accommodate emerging trends. Retrieved from http://www.afro.who.int/news/nigeria-develops-new-national-health-policy-accommodate-emerging-trends.

WHO (World Health Organization). (2016d). Resolution on strengthening people-centered health systems in the WHO European region framework on integrated, people-centered health services. Presented at 69th World Health Assembly, Geneva, Switzerland: 23–38 May, 2015.

WHO (World Health Organization). (2016e). Strengthening people-centered health systems in the WHO European region: Framework for action on integrated health services delivery. Retrieved from http://www.euro.who.int/__data/assets/pdf_file/0005/231692/e96929-replacement-CIHSD-Roadmap-171014b.pdf?ua=1.

WHO (World Health Organization). (2016f). Timor-Leste. Country cooperation strategy at a glance. Retrieved from http://www.searo.who.int/timorleste/publications/ccs-2015-19/en/.

WHO (World Health Organization). (2016g). Universal health coverage (UHC). Retrieved from http://www.who.int/mediacentre/factsheets/fs395/en/.

WHO (World Health Organization). (2017a). Country statistics, global health observatory. Retrieved from http://www.who.int/gho/countries/en/.

WHO (World Health Organization). (2017b). Universal health coverage. Retrieved from http://www.who.int/healthsystems/universal_health_coverage/en/.

WHO (World Health Organization). (2017c). Oman: Health promotion and community-based initiatives. Retrieved from http://www.emro.who.int/omn/programmes/health-promotion-and-community-based-initiatives.html.

WHO (World Health Organization). (2017d). Patient safety: Making healthcare safer. Retrieved from http://apps.who.int/iris/bitstream/10665/255507/1/WHO-HIS-SDS-2017.11-eng.pdf?ua=1.

WHO (World Health Organization). (2017e). Qualhat healthy village project, Oman. Retrieved from http://www.emro.who.int/ar/health-education/physical-activity-case-studies/qualhat-healthy-village-project-oman.html.

WHO (World Health Organization). (2017f). Strengthening national capacity for patient and community engagement and empowerment in healthcare in Oman. Retrieved from http://www.who.int/patientsafety/patients_for_patient/oman-report.pdf?ua=1.

WHO-AFRO (World Health Organization, Africa Regional Office). (2010). Integrated Disease Surveillance and Response (IDSR) documents—Technical guidelines. Retrieved from http://www.afro.who.int/en/integrated-disease-surveillance/idsr.

WHO–AFRO (World Health Organization, Africa Regional Office). (2015). Innovative approaches in education and training of health professionals. Retrieved from http://www.aho.afro.who.int/en/blog/2015/02/03/innovative-approaches-education-and-training-health-professionals.

WHO (World Health Organization), Global Health Data Observatory. (2017). Density per 1000: Data by country. Retrieved from http://apps.who.int/gho/data/node. main.A1444?lang=en.

Widen, S. & Haseltine, W. A. (2015). Case study: The Estonian eHealth and eGovernance system. ACCESS Health International. Retrieved from http://accessh. org/wp-content/uploads/2015/10/Estonian-eGovernance-Case-Study.compressed.pdf.

Wilkinson, J. E., Rushmer, R. K., & Davies, H. T. (2004). Clinical governance and the learning organization. *Journal of Nursing Management*, 12(2), 105–113.

Williams, D. I. (2000). RSM 1907: The acceptance of specialization. *Journal of the Royal Society of Medicine*, 93(12), 642–645.

Williams, G. (2015). Intersectorialidad en las Políticas de Recursos Humanos en Salud. Experiencia argentina y perspectivas. Regulación en el ejercicio profesional. *Reunión Regional de Recursos Humanos para la Salud*. Buenos Aires, Argentina: Ministerio de Salud.

Williams, H., Cooper, A., & Carson-Stevens, A. (2016). Opportunities for incident reporting. Response to: 'The problem with incident reporting' by Macrae et al. *BMJ Quality & Safety*, 25(2), 133–134.

Williams, H., Edwards, A., Hibbert, P., Rees, P., Evans, H. P., Panesar, S., et al. (2015). Harms from discharge to primary care: Mixed methods analysis of incident reports. *British Journal of General Practice*, 65(641), e829–e837.

Williams, I. & Dickinson, H. (2016). Going it alone or playing to the crowd? A critique of individual budgets and the personalisation of health care in the English National Health Service. *Australian Journal of Public Administration*, 75(2),149–158.

Witter, S., Jones, A., & Ensor, T. (2014). How to (or not to) … measure performance against the Abuja target for public health expenditure. *Health Policy and Planning*, 29(4), 450–455.

Wong, E. L., Coulter, A., Cheung, A. W. L., Yam, C. H., Yeoh, E. K., & Griffiths, S. M. (2012). Patient experiences with public hospital care: First benchmark survey in Hong Kong. *Hong Kong Medical Journal*, 18, 371–380.

Wong, E. L., Lui, S. F., Cheung, A. W. L., Yam, C. H. K., Huang, N. F., Tam, W. W. S., & Yeoh, E. K. (2017). Views and experience on patient engagement in healthcare professionals and patients—How are they different? *Open Journal of Nursing*, 7(6), 615–629.

Wong, E. L. Y., Coulter, A., Hewitson, P., Cheung, A. W., Yam, C. H., Lui, S. F., et al. (2015). Patient experience and satisfaction with inpatient service: Development of short form survey instrument measuring the core aspect of inpatient experience. *PLoS One*, 10(4), e0122299.

World Bank. (2014). GINI index (World Bank estimate). Retrieved from https://data. worldbank.org/indicator/SI.POV.GINI/.

World Bank. (2015a). Assessment of systems for paying health care providers in Mongolia: Implications for equity, efficiency and universal health coverage. Report No. 98790-MN. Retrieved from http://documents.worldbank.org/ curated/en/docsearch?query=98790-MN.

World Bank. (2015b). Seguro Popular: Health coverage for all in Mexico. Retrieved from http://www.worldbank.org/en/results/2015/02/26/ health-coverage-for-all-in-mexico.

World Bank. (2016a). Deepening health reform in China: Building high-quality and value- based service delivery. Retrieved from https://openknowledge. worldbank.org/bitstream/handle/10986/24720/HealthReformInChina.pdf.

World Bank. (2016b). GDP per capita (current US$). Retrieved from https://data. worldbank.org/indicator/NY.GDP.PCAP.CD.

World Bank. (2016c). Maternal mortality ratio (modeled estimate, per 100,000 live births). Retrieved from https://data.worldbank.org/indicator/SH.STA.MMRT.

World Bank. (2016d). World development indicators. Retrieved from http://data. worldbank.org/.

World Bank. (2017a). Data for Kazakhstan, Uzbekistan, Kyrgyz Republic, Turkmenistan, Tajikistan. Retrieved from https://data.worldbank. org/?locations=KZ-UZ-KG-TM-TJ.

World Bank. (2017b). Health expenditure, public (% of government expenditure). Retrieved from http://data.worldbank.org/indicator/SH.XPD.PUBL.GX.ZS.

World Bank. (n.d.a). Data: Out-of-pocket health expenditure. Retrieved from http:// data.worldbank.org/indicator/SH.XPD.OOPC.TO.ZS.

World Bank. (n.d.b). Data: Population. Retrieved from http://data.worldbank.org/ indicator/SP.POP.TOTL?name_desc=false.

World Economic Forum. (2014). Health systems leapfrogging in emerging economies. Retrieved from http://www3.weforum.org/docs/WEF_HealthSystem_ LeapfroggingEmergingEconomies_ProjectPaper_2014.pdf.

Worldometers. (2017). How many countries are there in the world? Retrieved from http://www.worldometers.info/geography/how-many-countries-are-there-in-the-world/.

Yang, J. S. (2011). Moving beyond traditional boundaries of health: Public health and multi-sectoral integration. *Californian Journal of Health Promotion*, 9(1), v–vi.

Yang, Y. (2016). Internet promotes health reforms, Minhang model solves difficulties in establishing stratified healthcare system. *Health News*. Retrieved from http:// www.jkb.com.cn/yzyd/2015/1106/379871.html.

Yarmohammadian, M. H., Shokri, A., Bahmanziari, N., & Kordi, A. (2013). The blind spots on Accreditation program. *Journal of Health Systems Research*, 9(11), 1158–1166.

Yen, J.-C., Chiu, W.-T., Chu, S.-F., & Hsu, M.-H. (2016). Secondary use of health data. *Journal of the Formosan Medical Association*, 115(3), 137–138.

Yende, S., Austin, S., Rhodes, A., Finfer, S., Opal, S., Thompson, T., et al. (2016). Long-term quality of life among survivors of severe sepsis: Analyses of two international trials. *Critical Care Medicine*, 44(8), 1461–1467.

Yeoh, P. L., Hornetz, K., Ahmad Shauki, N. I., & Dahlui, M. (2015). Assessing the extent of adherence to the recommended antenatal care content in Malaysia: Room for improvement. *PLoS ONE*, 10(8), e0135301.

Young, E. (2015). The importance of research in healthcare. *South Sudan Medical Journal*, 8(4), 90–91.

Younis, M. Z. (2013). Interview: Comparative effectiveness research and challenges to healthcare reform in the Middle East and USA. *Journal of Comparative Effectiveness Research*, 2(3), 223–225.

Younis, M. Z. (2017). Healthcare reform in the Middle East and the USA. *Journal of Comparative Effectiveness Research*, 6(1), 13–14.

Zachariah, R., Ford, N., Philips, M., Lynch, S., Massaquoi, M., Janssens, V., & Harries, A. (2009). Task shifting in HIV/AIDS: Opportunities, challenges and proposed actions for sub-Saharan Africa. *Transactions of the Royal Society of Tropical Medicine and Hygiene*, 103(6), 549–558.

Zhang, B. & Yu, H. (2016). Health insurance guides the establishment of stratified healthcare system. *China Social Security*. Retrieved from http://www.zgshbz.com.cn/Article9718.html.

Zhang, D. & Unschuld, P. U. (2008). China's barefoot doctor: Past, present, and future. *Lancet*, 372(9653), 1865–1867.

Zhao, L. & Huang, Y. (2010). China's blueprint for health care reform. *East Asian Policy*, 2, 51–59.

Zhu, H. (2015). How to make health insurance assisting the establishing of classified medical care system. *China Medical Insurance*, 6, 9–11.

Zoleikani, P. (2015). The impact of applying A&E accreditation standards on the performance of Sari hospitals' A&E departments. Unpublished thesis, Tehran University of Medical Sciences, Iran.

Index